STATISTICAL THERMODYNAMICS
AND STOCHASTIC THEORY OF
NONEQUILIBRIUM SYSTEMS

Werner Ebeling
Igor M. Sokolov

Humboldt-Universität, Germany

STATISTICAL THERMODYNAMICS AND STOCHASTIC THEORY OF NONEQUILIBRIUM SYSTEMS

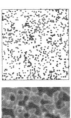

World Scientific

NEW JERSEY · LONDON · SINGAPORE · BEIJING · SHANGHAI · HONG KONG · TAIPEI · CHENNAI

Published by

World Scientific Publishing Co. Pte. Ltd.
5 Toh Tuck Link, Singapore 596224
USA office: 27 Warren Street, Suite 401-402, Hackensack, NJ 07601
UK office: 57 Shelton Street, Covent Garden, London WC2H 9HE

British Library Cataloguing-in-Publication Data
A catalogue record for this book is available from the British Library.

STATISTICAL THERMODYNAMICS AND STOCHASTIC THEORY OF NONEQUILIBRIUM SYSTEMS

Copyright © 2005 by World Scientific Publishing Co. Pte. Ltd.

All rights reserved. This book, or parts thereof, may not be reproduced in any form or by any means, electronic or mechanical, including photocopying, recording or any information storage and retrieval system now known or to be invented, without written permission from the Publisher.

For photocopying of material in this volume, please pay a copying fee through the Copyright Clearance Center, Inc., 222 Rosewood Drive, Danvers, MA 01923, USA. In this case permission to photocopy is not required from the publisher.

ISBN-13 978-981-02-1382-4
ISBN-10 981-02-1382-4

Printed in Singapore

Preface

This is a textbook about the relation between nonlinear dynamics, classical statistical physics and stochastic theory of systems out of thermodynamic equilibrium. The central concepts are instability, entropy, nonequilibrium and Brownian motion. Instability stands here as a central point of modern dynamics, it means the local (or global) divergence of trajectories, which are near one to another at the initial time. It will be shown that this property, which seems to be a characteristic feature of all many-particle (macroscopic) systems, is responsible for a statistical (stochastic) behaviour connected with the existence of probabilities and effects like ergodicity and mixing of microscopic trajectories. On the other hand entropy is the central statistical concept: entropy is a measure of uncertainity of the microscopic state. The idea that the instability of the microscopic trajectories generates uncertainty of the state, what means entropy and macroscopic properties, is the central point of this course. We believe, that this is the natural way to look at statistical physics now. In this respect we feel encouraged by other modern courses in statistical physics, say e.g. the books of R. Balecu (1975, 1997), L. Reichl (1980, 1992), P. Gaspard (1998), L. R. Dorfman (1999) etc. In explaining the main ideas of statistical physics, we prefer the inductive and sometimes the historical view. Telling about the history of the relations between thermodynamics, statistical physics and nonlinear science, we must apologize for some bias to historical events connected with Berlin, a town closely connected with the roots of thermodynamics and statistical physics, and with our own professional activities. However we do not claim for accuracy of the historical details; the historical research in this important field is still at the beginning. We are interested more in the origin and in the flow of ideas which became essential for the developement of our science. In connection with the important anniversaries of fundamental papers ob Brownian motion in the upcoming years 1905-1906, we give some more details on the development of this field, which is near to our own interests.

This work is based on lecture notes on general courses on "Thermodynamics" and "Statistical Physics" and special topics courses on "Stochastic Theory", "Brownian Motion", "Nonlinear Dynamics", "Self-Organization", "Entropy and

Information", "Physics of Evolutionary processes", "Physics of Macromolecules", "Kinetics" and related subjects given by the authors in close cooperation with Lutz Schimansky-Geier at the Humboldt University Berlin between 1980 and 2004, as well as at the University of Freiburg between 1995 and 2000. This way, basically this book it is to be considered as a textbook. However, the book contains also some original results of the research work of the authors (especially in Chapters 6–12). We thank the former and the present members of the Berlin group: Andreas Engel, Harald Engel, Udo Erdmann, Rainer Feistel, Ulrike Feudel, Jan Freund, Horst Malchow, Alexander Neiman, Thorsten Pöschel, Andrea Scharnhorst, Frank Schweitzer and others, who had a great impact on the ideas and results developed in this book. A great impact had also the discussions with friends and frequent visitors as Vadim Anishchenko, Michael Conrad, Hermann Haken, Peter Hänggi, Yuri Klimontovich, Frank Moss, Gregoire Nicolis, Ilya Prigogine, Yuri Romanovsky, Manuel Velarde, Michael Volkenstein, as well as Alex Blumen, Aleksei Chechkin, Katja Lindenberg, Ralf Metzler, Ramon Reigada, Francesc Sagues, Jose Maria Sancho, Len Sander, Manuel G. Velarde and others. Further we reflected the results of several conferences and workshops hold in or near Berlin, as e.g. "Selforganization by Nonlinear Irreversible Processes" (1979, 1984, 1986), "Stochastic Processes" (1990), "Models of Self-Organization in Complex Systems" (1991), "Statistical Physics and Thermodynamics of Nonlinear Systems" (1992), "Stochastic Processes and Granular Media" (1995), "Complexity and Self-Organization" (1995). Essentially, this is a textbook for physics students of the higher semesters and for graduate students in physics and related sciences which try to pave their way through the dschungel of modern research. The orientation is more on understanding than on details. In other words our aim is in the first line to contribute to the education of students and young scientists, to show them what holds physics and natural sciences together and not so much to fill their brains with thousands of equations.

The manuscript of this book was written in Berlin under conditions reminding of a *"far from equilibrium phase transition"*. The great political and social transitions which were taking place at the end of the 20th century especially in the eastern part of Europe were shaking also the scientific and universitary structures. In this way the present book about far from equilibrium (chaotic) states may profit from an intimate personal experience about instability and mixing processes in chaotic transitory states and the take-over by new order parameters. As we have learned by experience the most effective stabilizing factors which guarantee survival in turbulent human systems are good friends, enthusiastic students and last but not least the solidarity in the scientific community. We are very grateful that the solidarity between scientists, which was so self-understood for Helmholtz, Planck and Einstein is not completely forgotten in the hectic and competitive science business of our days.

This book is dedicated to our friends, coworkers and students, who helped to transform the chaos of ideas into the order of a book. In particular we have to thank our colleague Lutz Schimansky–Geier, who was involved in the early stage of this book, before he was forced to concentrate on important research projects. We are indebted to Yossi Klafter who carefully read the whole manuscript and made important suggestions which lead to a considerable improvement of the book.

Werner Ebeling, Igor Sokolov
Berlin, April 2005

Contents

Preface . . . v

Chapter 1 Introduction 1
1.1 The task of statistical physics . . . 1
1.2 On history of fundamentals of statistical thermodynamics . . . 3
1.3 On history of the concept of Brownian motion . . . 12

Chapter 2 Thermodynamic, Deterministic and Stochastic Levels of Description 19
2.1 Thermodynamic level: The fundamental laws . . . 19
2.2 Lyapunov functions: Entropy and thermodynamic potentials . . . 28
2.3 Energy, entropy, and work . . . 32
2.4 Deterministic models on the mesoscopic level . . . 39
2.5 Stochastic models on the mesoscopic level . . . 49

Chapter 3 Reversibility and Irreversibility, Liouville and Markov Equations 57
3.1 Boltzmann's kinetic theory . . . 57
3.2 Probability measures and ergodic theorems . . . 64
3.3 Dynamics and probability for one-dimensional maps . . . 69
3.4 Hamiltonian dynamics: The Liouville equation . . . 73
3.5 Markov models . . . 79

Chapter 4 Entropy and Equilibrium Distributions 83
4.1 The Boltzmann–Planck principle . . . 83
4.2 Isolated systems: The microcanonical distribution . . . 87
4.3 Gibbs distributions for closed and for open systems . . . 89
4.4 The Gibbs–Jaynes maximum entropy principle . . . 92

Chapter 5 Fluctuations and Linear Irreversible Processes — 97
5.1 Einstein's theory of fluctuations 97
5.2 Fluctuations of many variables 100
5.3 Onsager's theory of linear relaxation processes 102
5.4 Correlations and spectra of stationary processes near equilibrium 105
5.5 Symmetry relations and generalizations 109

Chapter 6 Nonequilibrium Distribution Functions — 111
6.1 A simple example — driven Brownian particles 111
6.2 Canonical-dissipative systems 115
6.3 Microcanonical non-equilibrium ensembles 118
6.4 Systems driven by energy depots 120
6.5 Systems of particles coupled to Nose–Hoover thermostats 126
6.6 Nonequilibrium distributions from information-theoretical methods 131

Chapter 7 Brownian Motion — 139
7.1 Einstein's concept of Brownian motion 139
7.2 The Langevin equation: Theme and variations 141
    7.2.1 The Langevin equation 141
    7.2.2 Thermalization of the velocity: The fluctuation-dissipation theorem 142
    7.2.3 The properties of the noise 144
7.3 Taylor–Kubo formula and velocity-velocity correlations. 146
7.4 The overdamped limit 148
7.5 Example: The Ornstein–Uhlenbeck process 150
7.6 The path integral representation 153
7.7 A non-Markovian Langevin equation 155

Chapter 8 Fokker–Planck and Master Equations — 161
8.1 Equations for the probability density 161
8.2 Special stochastic processes 165
    8.2.1 Example 1: The Ornstein–Uhlenbeck process revisited 165
    8.2.2 Example 2: The Klein–Kramers equation 167
8.3 The Fokker–Planck equation and the Liouville equation 169
8.4 Transition rates and master equations. 172
8.5 Energy diffusion and detailed balance 173
8.6 System in contact with several heat baths 176

Chapter 9 Escape and First Passage Problems — 183
9.1 General considerations 183
9.2 The renewal approach. 185
9.3 Example: Free diffusion in presence of boundaries 188

9.4	Mean life time in a potential well	189
	9.4.1 The flow-over-population approach	189
	9.4.2 The Arrhenius law	191
	9.4.3 Diffusion in a double-well	192
9.5	Moments of the first passage time	194
9.6	An underdamped situation	196

Chapter 10 Reaction Kinetics — 199

10.1	The mass action law	199
10.2	Classical kinetics	202
10.3	The Smoluchowski approximation	203
10.4	Fluctuation effects in chemical kinetics	207
	10.4.1 The Vaks–Balagurov kinetics in trapping	208
	10.4.1.1 The Vaks–Balagurov problem in one dimension	209
	10.4.1.2 General asymptotic behavior in trapping	212
	10.4.2 Fluctuation-dominated kinetics in $A + B \to 0$ Reaction	215
	10.4.3 Autocatalytic reactions and front propagation phenomena	221

Chapter 11 Random Walk Approaches — 225

11.1	The random walk approach to transport processes.	226
	11.1.1 Random walks on lattices	228
	11.1.2 The continuous-time random walks (CTRW)	232
	11.1.3 CTRW and the master equation	235
11.2	Power-law waiting-time distributions	236
11.3	Aging behavior of CTRW systems	241
11.4	CTRW and fractional Fokker–Planck equations.	245
11.5	Superdiffusion: Lévy flights and Lévy walks.	251

Chapter 12 Active Brownian Motion — 261

12.1	The general model	261
	12.1.1 Self-propelling of Brownian particles	261
	12.1.2 Equations of motions	263
12.2	Force-free motion of active particles and mean square displacement	268
	12.2.1 Distribution functions	268
	12.2.2 Mean-square displacement	270
12.3	Deterministic motion in external potentials	273
	12.3.1 Parabolic confinement	273
	12.3.2 Perturbed and transient limit cycles	276
12.4	Stochastic motion in symmetric external potentials	279
12.5	Collective dynamics of clusters and swarms	285
	12.5.1 Dynamics of active clusters and swarms	285
	12.5.2 Confined systems of active particles	287
	12.5.3 Dynamics of self-confined driven particles	292
	12.5.4 The influence of hydrodynamic Oseen-type interactions	300

References 305

Index 327

Chapter 1

Introduction

1.1 The task of statistical physics

Statistical Physics is that part of physics which derives emergent properties of macroscopic matter from the atomic structure and the microscopic dynamics. Emergent properties of macroscopic matter mean here those properties (temperature, pressure, mean flows, dielectric and magnetic constants etc.) which are essentially determined by the interaction of many particles (atoms or molecules). Emergent means that these properties are typical for many-body systems and that they do not exist (in general) for microscopic systems. The key point of statistical physics is the introduction of probabilities into physics and connecting them with the fundamental physical quantity entropy. The first task of this new scientific discipline was to connect atomistics with thermodynamics.

$$\text{atomistics} \rightarrow \text{statistical mechanics} \rightarrow \text{thermodynamics}$$

More generally, the task is to construct the bridge between microphysics, i.e. properties and dynamics of atoms and molecules with macrophysics, i.e. thermodynamics, hydrodynamics, electrodynamics of media

$$\text{microphysics} \rightarrow \text{statistical physics} \rightarrow \text{macrophysics}$$

In this connection macrophysical properties designate those properties which are determined by the interaction of very many particles (atoms, molecules), in contrast to properties which are characteristic for one or a few particles. From the dynamical point of view a macrosystem is a many-body system which is determined by the basic equations of classical or quantum mechanics. Further basic elements of the theory should be the laws of interaction of the particles as, e.g. the Coulomb law, the symmetry principle and the boundary conditions characterizing the macroscopic embedding. From this point of view, the problem seems to be practically insolvable, not only for the impossibility to solve more than 10^{24} coupled usual or partial differential equations but also due to the incomplete knowledge about the initial and boundary conditions. This is the point, where new concepts are needed and it has to be shown that probabilities and/or density operators and entropies may be introduced in a natural way if the dynamics is unstable. The point of view taken in this textbook is mainly classical but the quantum-statistical analysis goes in many aspects in a quite analogous way. By using probabilities instead of trajectories, which is indeed a basic change of concepts (Prigogine and Stengers, 1984; Prigogine, 1989; Prigogine et al., 1991), we come to a dynamics of probabilities. In the classical case we obtain in this way the Liouville equation and in the quantum case the von Neumann equation:

microdynamics + probabilities → Liouville–von Neumann dynamics

This is a great step in the right direction since now macroscopic properties may be described as mean values and a macroscopic dynamics seems to be in reach. The remaining problem is however, that the Liouville–von Neumann equation are formally completely equivalent to the original dynamical equations, they are just a different expression of the same dynamics. In so far, the Liouville–von Neumann equation have still the property of reversibility of the microscopic dynamics. All conclusions based on these equations would not be in accord with the second law of thermodynamics. So, still a second step has to be made to arrive at an appropriate macroscopic dynamics. The basic idea is, that macroscopic processes allow and require a coarse-grained description. This point was worked out first by Gibbs and the Ehrenfest's. It means at first, that it makes no sense to describe a macroscopic process in all microscopic detail, since it completely impossible to observe all the details and to follow the trajectories of all particles. Instead of the fine description, a coarse-grained description must be introduced which still keeps the relevant macroscopic informations but neglects the irrelevant microscopic details. This idea is rather old but it was only recently understood, that the whole concept of coarse-graining is intimately connected with the instabilities of the microscopic trajectories (Nicolis et al., 1991). Finally we arrive at equations for the coarse-grained probabilities which are irreversible and yield an appropriate basis for the macroscopic physics. These equations are called kinetic equations or master equations. Our scheme may now be completed in the following way:

coarse-graining + dynamic instability → kinetic/master eqs.

The following chapters are aimed to work out this program, with some special attention to the concept of Brownian motion. But, before going in the details we would like to have another more historical oriented look at the development of the basic ideas and at the historical facts — admittedly with some bias to the development in Berlin, largely following earlier work (Rompe et al., 1987; Ebeling & Hoffmann, 1990, 1991).

1.2 On history of fundamentals of statistical thermodynamics

Thermodynamics as a branch of science was established in the 19th century by pioneers as Sadi Carnot (1796–1832), Robert Mayer (1814–1878), Hermann Helmholtz (1821–1894), William Thomson (1824–1907) and Rudolf Clausius (1822–1888). Evidently Mayer was the first who formulated the law of energy conservation. His paper *"Bemerkungen über die Kräfte der unbelebten Natur"* published 1842 in Liebig's Annalen is clearly expressing the equivalence of work and heat. Joule came to similar conclusions which were based on direct measurements concerning the conversion of work into heat. A great role in the foundation of thermodynamics played a community of physicists which worked in the middle of the 19th century in Berlin. Without doubt it was the genius of Hermann Helmholtz who determined the direction and the common style of research (Ebeling & Hoffmann, 1991). On July 23, 1847 he reported to the *"Berliner Physikalische Gesellschaft"*, a new society founded by young physicists, about his research on the principle of conservation of energy. At 27 years of age Helmholtz was working as a military surgeon to a regiment of Hussars in Potsdam. He could follow his interest in physics only in his leisure time, since his family's financial situation did not allow him to enjoy full-time study. The experimental research which he carried out from the beginning of the 1840's in the laboratory of his adviser Professor Magnus was primarily devoted to the conversion of matter and heat in such biological processes as rotting, fermentation and muscular activity. Helmholtz's insight led him to infer a new law of nature from the complexities of his measurements on juices and extracts of meat and muscles. From experiments and brilliant generalization emerged the principle of conservation of energy or what is now called the first law of thermodynamics. Neither J.'R. Mayer nor J. P. Joule (not to speak of the other pioneers of the energy principle) recognized its fundamental and universal character as clearly as did Helmholtz, who must therefore be regarded as one of the discoverers of the principle, although his talk to the Berlin Physical Society was given later than the fundamental publications of Mayer and Joule. Both were unknown to Helmholtz at the time. Helmholtz had to fight hard for the recognition of his result — Professor Poggendorf, the influential editor of the *"Annalen der Physik und Chemie"*, had no wish to publish what seemed to him rather speculative and philosophical. Magnus also regarded it with disfavor, but at least recommend that it be printed as a separate brochure, as was very quickly managed with the help of the influential mathematician C. G. Jacobi. The new law of nature quickly demonstrated its fruitfulness and universal applicability. For instance Kirchhoff's second law for electrical circuits is essentially a particular case of the energy principle. Nowadays these laws are among the most frequently applied laws in the fields of electrical engineering and electronics. The discovery of the fundamental law of circuits was done early in Kirchhoff's life in Königsberg and Berlin.

Rudolf Clausius (1822–1888) also played an essential role in the history of the law of conservation of energy and its further elaboration (Ebeling & Orphal, 1990). After studying in Berlin, he taught for some years at the Friedrich-Werdersches Gymnasium in Berlin and was a member of the seminar of Professor Magnus at the Berlin University. His report on Helmholtz's fundamental work, given to Magnus' colloquium, was the beginning of a deep involvement with thermodynamical problems. Building on the work of Helmholtz and Carnot he had developed, and published 1850 in Poggendorff's Annalen his formulation of the second law of thermodynamics. Clausius was fully aware of the impact of his discovery. The title of his paper explicitly mentions "*laws*". His formulation of the second law, the first of several, that heat cannot pass spontaneously from a cooler to a hotter body, expresses its essence already. Unlike Carnot, and following Joule, Clausius interpreted the passage of heat as the transformation of different kinds of energy, in which the total energy is conserved. To generate work, heat must be transferred from a reservoir at a high temperature to one at a lower temperature, and Clausius here introduced the concept of an ideal cycle of a reversible heat engine. In 1851 William Thomson (Lord Kelvin) formulated independently of Clausius another version of the second law. Thomson stated that it is impossible to create work by cooling down a thermal reservoir. The central idea in the papers of Clausius and Thomson was an exclusion principle: "*Not all processes which are possible according to the law of the conservation of energy can be realized in nature. In other words, the second law of thermodynamics is a selection principle of nature*". Although it took some time before Clausius' and Thomson's work was fully acknowledged, it was fundamental not only for the further development of physics, but also for science in general. In later works Clausius arrived at more general formulations of the second law. The form valid today was reported by him at a meeting of the "*Züricher Naturforschende Versammlung*" in 1865. There for the first time, he introduced the quotient of the quantity of heat absorbed by a body and the temperature of the body $d'Q/T$ as the change of entropy. The idea to connect the new science with the atomistic ideas arose already in the fifties of the 19th century. August Karl Krönig (1822–1879), extended thermodynamics and started with statistical considerations. In this way Krönig must be considered a pioneer of statistical thermodynamics. In 1856, he published a paper in which he described a gas as system of elastic, chaotically moving balls. Krönig's model was inspired by Daniel Bernoulli's paper from 1738, where Bernoulli succeeded in deriving the equation of state of ideal gases from a billiard model. Krönig's early attempt to apply probability theory in connection with the laws of elastic collisions to the description of molecular motion, makes him one of the forerunners of the modern kinetic theory of gases. After the appearance of Krönigs paper, Clausius admitted, that he too had been thinking about related problems since the late 1840s. In a subsequent paper "*Über die Art der Bewegung, die wir Wärme nennen*", which appeared 1857 in Vol. 100 of the *Annalen der Physik*, Clausius published his ideas about the atomistic foundation of thermodynamics. In fact, his work from 1857 as well as a following paper published in 1858 are the first comprehensive survey of the kinetic theory of gases. As a result of his work Clausius developed new terms like the mean free path and cross section and introduced in 1865 the new fundamental quantity entropy. Further we mention the proof of the virial theorem for gases, which he discovered in 1870. Parallel to Clausius's work the statistical theory was developed in Great Britain by James Clerk Maxwell, who derived in 1860 in an article in Philosophical Magazine the probability distribution for the velocities of molecules in a gas. In 1866 Maxwell gave a new derivation of the velocity distribution based on a study of direct and reversed collisions and formulated a first version of a transport theory. In 1867 Maxwell considered first the statistical nature of the second law of thermodynamics and considered the connection between entropy and information. His *"Gedankenexperiment"* about a demon observing molecules we may consider as the first fundamental contribution to the development of an information theory. In 1878 Maxwell proposed the new term *"statistical mechanics"*.

This was, in brief, the situation when Ludwig Boltzmann (1844–1906) began his studies at the University of Vienna in 1863, the year when Josef Stefan (1835–1893) was appointed at the chair of physics of this University. He was deeply influenced by Stefan, who was a brilliant experimentalist and also by Johann Loschmidt (1821–1895) who was an expert in the kinetic theory of gases. In 1865 Loschmidt was the first who succeeded in estimating the diameter of molecules; based on this result, Planck calculated later the number of molecules in 1 cm^3 of a gas $N = 6.025 \cdot 10^{23}$ mol^{-1}. Influenced by the work of Stefan and Loschmidt as well as by studies of the papers of Clausius and Maxwell, Boltzmann started to work on the kinetic theory of gases. In 1866, he found the energy distribution for gases. In 1869, at the age of 25, Boltzmann was appointed as professor of mathematical physics in Graz. In 1871 he formulated the ergodic hypothesis, which is absolutely fundamental for the modern version of statistical physics and for the connection to nonlinear dynamics. The fruitful work in Graz culminated in 1872 with the formulation of Boltzmann's famous kinetic equation. In the same year 1872 he derived the H-theorem, which established a connection between mechanics and thermodynamics. The year 1872, which was so central for his work, was for Boltzmann also a year of traveling. He visited Bunsen and Königsberger in Heidelberg as well as Helmholtz and Kirchhoff in Berlin. The visit of Boltzmann to Berlin, which was rather influential for his further work, gives us the opportunity to consider, what happened in the mean time in Berlin. In 1871, after professorships in anatomy and physiology at several German universities, Helmholtz returned to Berlin to succeed Magnus as director of the physical institute of the university. Then began a very productive period in the history of physical research in Berlin. During the next two decades Helmholtz's direction of physical research in Berlin made it world famous. No burning questions of contemporary physics remained untouched by Helmholtz or his fellow workers, but thermodynamical problems remained central. Just as in the 1840s the social need for powerful and efficient power engines gave strong impetus to the discovery of the two laws of thermodynamics, so Helmholtz's late thermodynamical research was socially determined. The chemical industry, especially the large-scale production of fertilizers and dyes grew enormously during the second part of the nineteenth century, and necessitated a thermodynamic description of chemical processes. During Helmholtz's second period in Berlin his work revolved around pure and applied problems of thermodynamics. Using the general laws for thermodynamical and galvanic processes he advanced the theory and opened new fields of practical application. He developed the concept of free energy and investigated the relationship between the heat of reaction and the electromotive force of a galvanic cell. The thermodynamics of electrical double layers at boundaries also proved important when it became a keystone of modern physical chemistry and biophysics, as well as semiconductor electronics. We may also note that thermometric and calorimetric investigations dominated the activities of the Physikalische-Technische Reichsanstalt during Helmholtz's presidency. The studies of the properties and applications of light at the Reichsanstalt stimulated the scientific exploration of the physical basis of light generation, and led to the development of a thermodynamical theory of heat radiation. Helmholtz's pupil and coworker Wilhelm Wien (1864–1928) was a leader in this field and in collaboration with the Reichsanstalt he did very precise measurements of the spectral distribution of radiation, this way giving a sound basis for a theoretical interpretation. The critical analysis of Wien's formula and attempts to derive it from the basic laws of electrodynamics and thermodynamics formed the starting point for Max Planck's quantum theory. In 1889 Max Planck (1858–1947) was called to succeed Kirchhoff at the Berlin Chair of Theoretical Physics where he became one of the most famous of theoretical physicists at his time, in particular a world authority in the field of thermodynamics. He was a pioneer in understanding the fundamental role of entropy and its connection with the probability of microscopic states. Later he improved Helmholtz's chemical thermodynamics and his theory of double layers, as well as developed theories of solutions, including electrolytes, of chemical equilibrium and of the coexistence of phases. Planck was especially interested in the foundations of statistical thermodynamics. In fact he was the first who wrote down explicitly the famous formula

$$S = k \log W. \tag{1.1}$$

An independent and even more general approach to statistical thermodynamics and the role of entropy was developed by the great American physicists Josiah Willard Gibbs (1839–1903). Gibbs developed the ensemble approach, the entropy functional and was the first to understood the role of the maximum entropy method. His monograph on the principles of statistical physics is still the "bible" of statistical physics.

On the first glance the new field of physics developed by Boltzmann and Gibbs, called "statistical mechanics", "statistical thermodynamics" or "statistical physics" was beautiful, attractive to young scientists and very perspective. However soon it became clear that the new field is not free of contradictions and mathematical difficulties. This was criticized e.g. by one of the greatest mathematicians of that time Henri Poincare (1854–1912). Planck was also worried about these problems and asked his talented student, the mathematician Ernst Zermelo, to think about the mathematical foundation of Boltzmann's theory. In fact Boltzmann's theory was based on the idea that the equilibrium properties of a macroscopic system are the are the averages of phase functions of the microscopic state taken over an infinite time. Strictly speaking, the computation of time averages would require complete solution of the mechanical equation of an N-particle system. In practice, there is little to do with this concept for $N \gg 1$. Boltzmann sought to get around these difficulties by the introduction of the so-called "ergodic hypothesis". The latter says that the trajectory of a large system crosses every point of the energy surface. Zermelo found a serious mathematical objection against Boltzmann's theory which was based on the theorem of Poincare about the "quasi-periodicity of mechanical systems" published in 1890 in the paper "*Sur le probleme de trois corps les equations de la dynamique*". In this fundamental work Poincare was able to prove under certain conditions that a mechanical system will come back to its initial state in a finite time, the so-called recurrence time. Zermelo showed in 1896 in a paper in the "*Annalen der Physik*" that Boltzmann's H-theorem and Poincares recurrence theorem were contradictory. In spite of this serious objection, in the following decades statistical mechanics was dominated completely by ergodic theory. A deep analysis of the problems hidden in ergodic theory was given by Paul and Tatjana Ehrenfest in a survey article published 1911 in "*Enzyklopädie der Mathematischen Wissenschaften*". Much later it was recognized that the clue for the solution of the basic problem of statistical mechanics was the concept of instability of trajectories developed also by Poincare in 1890 in Paris. Before we study this new direction of research, we explain first the development of some other directions of statistical thermodynamics.

Planck reinforced the development of statistical thermodynamics through invitations to other leading specialists, such as Jacobus Henricus van't Hoff (1852–1911) whom, supported by Nernst and others, he nominated for a professorship at the Academy in Berlin. There van't Hoff had no teaching duties and could choose his topics of research absolutely freely. As the creator of a fruitful new branch of molecular physics he made important contributions to the thermodynamics of solutions and solution-salt equilibria. Furthermore he showed by thermodynamic methods the dependence of chemical equilibrium on temperature and pressure and discovered the fundamental laws of chemical kinetics, which remain the basis for the calculation of chemical processes. Finally he gave the first correct interpretation of osmotic pressure and introduced chemical affinity as the driving force for chemical reactions.

These last researches persuaded the Stockholm Nobel committee to award van't Hoff the first Nobel prize in chemistry (1901). Another important contribution to thermodynamics is connected with the work of Walther Nernst (1864–1941) who accepted in 1905 a call on a chair at the Berlin university. Nernst was one of the founders of physical chemistry and his work on the thermodynamical foundations of electrochemistry cemented his reputation as a leading contemporary scientist. In 1905 he detected the "missing stone in thermodynamics", the third law of thermodynamics. Nernst's seminal idea arose from the critical analysis of experimental data on chemical and electrochemical reactions in the liquid phase at low temperatures, where there appeared good correspondence between the free energy and the internal energy — M. P. Bertholet had already hypothesized the identity of these quantities, and Nernst found by his analysis that the correspondence improved at lower temperatures. This led him to suggest that the difference between the internal energy and free energy vanishes asymptotically at the zero temperature. Some years later Planck gave Nernst's new principles the following general and widely known formulation: *"The entropy of all bodies which are in internal equilibrium vanishes at the zero point of temperature"*. After postulating his new theorem Nernst and his collaborators took great pains to prove and develop further this new law of nature. The specific heat, being of special importance, was determined for several substances at low temperatures. This was a very difficult scientific problem which called for the construction of equipment and instruments from scratch.

At the same time (1905/06) that Nernst's group was working on the experimental verification of the heat theorem, Einstein and Smoluchowski worked out the theory of Brownian motion. The first theoretical work of Albert Einstein (1879–1955) was carried out during he was working at the Bern patent office. Einstein started his work on statistical physics in 1902/03 with two very interesting papers on *"The kinetic theory of thermal equilibrium and the second law of thermodynamics"*, published in leading physical journal of that time, the *"Annalen der Physik"*. Here independently of Gibbs, Einstein developed the basic ideas of ensemble theory and the statistics of interacting systems. In his dissertation, presented in 1905 to the Zürich University, he developed a first correct theoretical interpretation of Brownian motion. This work was published in volume 17 of the *"Annalen der Physik"*. Einstein was at that time only 26 years old, he published in the same volume of the "Annalen", also two other fundamental papers devoted to the theory of relativity and the theory of the photo effect. The first experimental check of Einstein's relations was published already in 1906 by Svedberg in the *"Zeitschrift für Elektrochemie"*. In the same year appeared the first paper of Marian von Smoluchowski (1872–1917) who worked at that time as a professor at the University of Lemberg (Lviv) and contributed many fundamental results to the theory of Brownian motion. Because of the fundamental importance of this direction of research to statistical physics and in connection with the upcoming anniversary of the papers of Einstein and Smoluchowski we will discuss more details on the history of the concept of Brownian motion in an extra section.

In 1907, Einstein proposed that quantum effects lead to the vanishing of the specific heat at zero temperature. His theory may be considered as the origin of quantum statistics. Einstein's work attracted the attention of Nernst and his collaborators and by 1910 they succeeded in confirming this prediction. In this way the third law of thermodynamics as well as the young and still controversial quantum theory found one of its first experimental verifications. Through these investigations Nernst became not only one of the earliest and most committed prophets of the quantum theory — he was the initiator of the first Solvay conference (1911) — but also a firm supporter of the young Einstein. In 1913, together with Planck, he was able to bring the *"new Copernicus"* into the exclusive circle of Berlin physicists and they could offer the unconventional genius excellent working and living conditions. As a *"paid genius"* in Berlin, Einstein could complete his general theory of relativity, and make further important contributions to thermodynamics and statistical physics. In 1924, he gave a correct explanation of gas degeneracy at low temperatures by means of a new quantum statistics, the so-called Bose–Einstein statistics. In addition to the Bose–Einstein condensation his ideas about the interaction between radiation and matter should be emphasized. In 1916 his discussion of spontaneous emission of light and induced emission and adsorption forms the theoretical basis of the nonlinear dynamics and stochastic theory of lasers. Concerning many other fundamental contributions to thermodynamics and statistical physics in the last century we must restrict ourselves to brief remarks. The German-Greek mathematician Constantin Caratheodory formulated thermodynamics on an axiomatic basis. His analysis of such fundamental concepts as temperature and entropy in terms of the mathematical theory of Pfaffian differential forms were not appreciated by most of his contemporaries, although Planck was an early supporter of what has become one of the important branches of modern thermodynamics.

Another important line of the development of thermodynamics is the foundation of irreversible thermodynamics. We mention only the early work of Thomson, Rayleigh, Duhem, Natanson, Jaumann and Lohr. The final formulation of the basic relations of irreversible thermodynamics we owe to the work of Onsager (1931), Eckart (1940), Meixner (1941), Casimir (1945), Prigogine (1947) and De Groot (1951). Irreversible thermodynamics is essentially a nonlinear science, which needs for its development the mathematics of nonlinear processes, the so-called nonlinear dynamics.

The great pioneers of nonlinear dynamics in the 19th century were Helmholtz, Rayleigh, Poincare and Lyapunov. John William Rayleigh (1842–1919) is the founder of the theory of nonlinear oscillations. Many applications in optics, acoustics, mechanics and hydrodynamics are connected with his name. Alexander M. Lyapunov was a Russian mathematician, who formulated in 1982 the mathematical conditions for the stability of motions. Henri Poincare (1854–1912) was a French mathematician, physicist and philosopher who studied in the 1890s problems of the mechanics of planets and arrived at a deep understanding of the stability of mechanical motion. His work *"Les methodes nouvelles de la mechanique celeste"* (Paris 1892/93) is a corner stone of the modern nonlinear dynamics. Important applications of the new concepts were given by the engineers Barkhausen and Duffing in Germany and van der Pol in Holland. Heinrich Barkhausen (1881–1956) studied physics and electrical engineering at the Technical University in Dresden, were he defended in 1907 the dissertation *"Das Problem der Schwingungserzeugung"* devoted to to problem of selfoscillations. He was the first who formulated in a correct way the necessary physical conditions for self-sustained oscillations. Later he found worldwide recognition for several technical applications as e.g. the creation of short electromagnetic waves. Georg Duffing worked at the Technical High School in Berlin-Charlottenburg. He worked mainly on forced oscillations; a special model, the Duffing oscillator was named after him. In 1918 he published the monograph *"Erzwungene Schwingungen bei veränderlicher Eigenfrequenz und ihre technische Bedeutung"*. Reading this book, one can convince himself that Duffing had a deep knowledge about the sensitivity of initial conditions and chaotic oscillations. A new epoch in the nonlinear theory was opened when A.A. Andronov connected the theory of nonlinear oscillations with the early work of Poincare. In 1929 he published the paper *"Les cycles limites de Poincare et la theorie des oscillations autoentretenues"* in the *Comptes Rendus Acad. Sci. Paris*. The main center of the development of the foundations of the new theory theory evolved in the 1930s in Russia connected with the work of Mandelstam, Andronov, Witt and Chaikin as well as in the Ukraina were N. M. Krylov, N. N. Bogoliubov and Yu. A. Mitropolsky founded a school of nonlinear dynamics.

That there existed a close relation between statistical thermodynamics and nonlinear science was not clear in the 19th century when these important branches of science were born. Quite the opposite, Henri Poincare, the father of nonlinear science, was the strongest opponent of Ludwig Boltzmann, the founder of statistical thermodynamics. In recent times we have the pleasure to see that Poincare's work contains the keys for the foundation of Boltzmann's ergodic hypothesis. The development of this new science had important implications for statistical thermodynamics. We have mentioned already the new concept of instability of trajectories developed by Poincare in Paris in 1890. This concept was introduced into statistical thermodynamics by Fermi, Birkhoff, von Neumann, Hopf and Krylov. The first significant progress in ergodic theory was made through the investigations of G. Birkhoff and J. von Neumann in two subsequent contributions to the Proceedings of the national Academy of Science U.S. in 1931/32. The Hungarian Johann von Neumann (1903–1957) came in the 1920s to Berlin attracted by the sphere of action of Planck and Einstein in physics and von Mises in mathematics. Von Neumann, who is one of the most influential thinkers of the 20th century made also important contributions to the statistical and quantum-theoretical foundations of thermodynamics. Von Neumann belonged to the group of "surprisingly intelligent Hungarians" (D. Gabor, L. Szilard, E. Wigner), who studied and worked in Berlin around this time. The important investigations of von Neumann on the connection between microscopic and macroscopic physics were summarized in his fundamental book "*Mathematische Grundlagen der Quantenmechanik*" (published in 1932). It is here that he presented the well known von Neumann equation and other ideas which have since formed the basis of quantum statistical thermodynamics. Von Neumann formulated also a general quantum-statistical theory of the measurement process, including the interaction between observer, measuring apparatus an the object of observation. This brings us back to Maxwell and in this way to another line of the historical development.

Information-theoretical considerations in statistical physics start with Maxwell's speculations about a demon observing the molecules in a gas. Maxwell was interested in the flow of information between the observer, the measuring apparatus and the gas. In fact this was the first investigation about the relation between observer and object, information and entropy. This line of investigation was continued by Leo Szilard, prominent assistant and lecturer at the University of Berlin and a personal friend of von Neumann. His thesis (1927) "*Über die Entropieverminderung in einem thermodynamischen System bei Eingriffen intelligenter Wesen*" investigated the connection between entropy and information. This now classic work is probably the first comprehensive thermodynamical approach to a theory of information processes and, as the work of von Neumann, deals with thermodynamical aspects of the measuring process. The first consequent approach to connect the foundations of statistical physics with information theory is due to Jaynes (Jaynes, 1957; 1985). Jaynes method was further developed and applied to nonequilibrium situations by Zubarev (Subarev, 1976; Zubarev et al., 1996, 1997)). The information-theoretical method is of phenomenological character and connected with the maximum entropy approach.

We have to mention also of the important contribution of Erwin Schrödinger to the foundation of statistical and biological thermodynamics. In 1927, Schrödinger succeeded Planck in the chair of theoretical physics. In the fall of 1933 he resigned from this post and after some years of travelling (England, Belgium, Austria) in 1939 he found his final refuge in Dublin. Here in 1944 he published the two little but very influential books "*Statistical Thermodynamics*" and "*What is Life?*", which considerably influenced the development of science and especially statistical thermodynamics and its applications to life sciences. The merit of the first comprehensive representation of the ideas of statistical thermodynamics in the framework of the ensemble theory belongs indisputably to Richard C. Tolman (1881–1948), who wrote the "*bible*" of modern statistical physics, the book "*The Principles of Statistical Mechanics*" (Oxford 1938). Another fundamental book which summarized the development of nonlinear mechanics was written at nearly the same time by the Russian scientists Andronov, Witt and Chaikin. The history of this book is a tragedy. After the book was in print, one of the authors, Witt, was arrested in Moscow for political reasons. He was placed in prison, where he eventually died. The first edition of the book appeared in 1941 under the Russian title *"Nelineinye Kolebaniya"* (i.e. *"Nonlinear Oscillations"*) and the two named authors were only Andronov and Chaikin. Only the first German translation which appeared in 1959 corrected this crime (Andronov, Witt, Chaikin, 1959). The comprehensive books of Tolman at one hand and Andronov, Witt and Chaikin on the other stand at the end of the first century of development of our science. There is no room to discuss here the "explosion" of work after these two pioneering books, so we must restrict ourselves to occasional remarks during the text of the next chapters. Let us mention however that the first book on the connection between statistical physics and nonlinear science was published in 1950 by the Russian scientist N. M. Krylov. The title of his fundamental monograph was *"Works on the Foundation of Statistical Physics"*.

1.3 On history of the concept of Brownian motion

Having the aim to derive macroscopic properties we have to look first at microscopic dynamics. As observed first by Ingenhousz and Brown, the microscopic motion of particles is essentially erratic. These observations led to the concept of Brownian motion which is basic to Statistical Physics. Moreover, the discussion of Brownian motion introduced quite new concepts of microscopic description, pertinent to stochastic approaches. The description put forward by Einstein in 1905/1906, Smoluchowski in 1906 and Langevin in 1908 is so much different from the one of Boltzmann: it dispenses from the description of the system's evolution in phase space and relies on probabilistic concepts. Marc Kac put is as follows: *"...while directed towards the same goal how different the Smoluchowski approach is from Boltzmann's. There is no dynamics, no phase space, no Liouville theorem — in short none of the usual underpinnings of Statistical Mechanics. Smoluchowski may not have been aware of it but he begun writing a new chapter of Statistical physics which in our time goes by the name of Stochastic processes"* (Fulinski, 1998). The synthesis of the approaches leading to the understanding of how the properties of stochastic motions are connected to deterministic dynamics of the system and its heat bath were understood much later in works by Mark Kac, Robert Zwanzig and others. A big part of this book is devoted to Brownian motion. For this reason and also having in mind the anniversary of the fundamentals of stochastic theory to be noticed in the years 2005–2008, we will discuss now the history of this important concept in some more detail.

The perpetual erratic motion of small particles immersed in a fluid was first observed as early as in 1785 by a Dutch physician Jan Ingenhousz; however, the phenomenon stayed unknown to the non-Dutch speaking community until it was rediscovered and studied in some detail by the Scottish botanist Robert Brown in

1827. Working on mechanisms of fertilization in plants, he turned his attention to the structure of pollen and concentrated at their microscopical characterization. Having observed the unceasing motion of the pollen in water he thought at first that the movement must be due to the living nature of the particles under observation. However, being a cautious scientist, he repeated his experiments with pollen kept in alcohol for several months (presumably dead) and with non-organic particles, which all showed similar behavior. (See one of the the Brown's original reports *"Additional Remarks on Active Molecules"* (Brown, 1829) and the discussion given by Brian J. Ford (1992).

The qualitative explanation of the Brownian motion as a kinetic phenomenon was put forward by several authors. In 1877 Desaulx wrote: *"In my way of thinking the phenomenon is a result of thermal molecular motion in the liquid environment (of the particles)."* In 1889 G. Gouy gave an account of detailed qualitative studies of the phenomenon (Gouy, 1889). He found that the Brownian motion is really a phenomenon which is not due to to random external influences (vibration, electric or magnetic fields) and that the magnitude of the motion depends essentially only on two factors: on the particles' size and on the temperature. In 1900 F. M. Exner undertook the first quantitative studies, measuring how the motion depends on these two parameters.

The ingenious microscopic derivation of the diffusion equation by Albert Einstein (which contained, in a nutshell, several different approaches) and the discussion by Paul Langevin put forward the two different approaches, one based on the discussion of the deterministic equations for the probability densities, another one based on the discussion of single, stochastic realizations of the process. These approaches, refined both from physical and from the mathematically point of view build now the main instrument of description of both equilibrium and nonequilibrium processes on a mesoscopic scale. Albert Einstein's first theoretical work was was carried out during he was working at the Bern patent office. Einstein started his work on statistical physics in 1902–1903 with two very interesting papers published in his favorite journal the "Annalen der Physik". Here independently of Gibbs, Einstein developed the basic ideas of ensemble theory and the statistics of interacting systems. In 1905 he presented a dissertation to the Zürich University which contains a theory of Brownian motion. Einsteins work which appeared in 1906 in the Annalen der Physik is the true origin of stochastic theory in physics, one of the corner stones of modern statistical thermodynamics.

In fact Einstein work was a theoretical discussion of one of possible consequences of the molecular-kinetic theory of heat: *"Über die von der molekularkinetischen Theorie der Wärme geforderte Bewegung von in ruhenden Flüssigkeiten suspendierten Teilchen"*. In this work Einstein discusses that the kinetic theory of heat predicts the unceasing motion of small suspended particles (Einstein, 1905). He was not sure that the phenomenon discussed is exactly the Brownian motion, but considered this as a reasonable hypothesis. After publication of the work, Siedentopf and

Gouy pointed out that the effect he discussed was really the Brownian motion, since not only the qualitative properties, but also the predicted orders of magnitude of the effect were correct, as discussed in the Einstein's second work (Einstein, 1906).

Let us briefly discuss now Einstein's approach. More technical details will be explained in Chapter 7. Einstein, in his analysis of the situation has connected the motion of suspended particles with diffusion and showed that this diffusive behavior follows from the three postulates. First, the particles considered are the assumed not to interact with each other: their trajectories are independent. Second, one assumes that the motion of the particles lacks the long-time memory: one can chose such a time interval τ, that the displacements of the particle during two subsequent intervals are independent. Third, the distribution of the particle's displacements s during the subsequent time intervals $\phi(s)$ is symmetric and possesses at least two moments. The displacement of the particle can thus be considered as a result of many small, independent, equally distributed steps. The further line of his reasoning is very close to what is called now a Kramers–Moyal expansion. This line of reasoning was also adopted by Smoluchowski, who however, assumed a more radical approach based on combinatorics. Next step was made by Adriaan Daniel Fokker (1887–1972) in his dissertation. However, the short publication (Fokker, 1914) did not contain the details of his derivation and concentrated essentially only on the treatment of the stationary case. A more general approch was developed by Max Planck, who formulated a general transport equation for diffusive motion, now known as the Fokker–Planck equation (Planck, 1918).

Essential contributions, independently of Einstein, to this line of research we owe to the work of the great Polish physicist Marian von Smoluchowski (1872–1917) published in subsequent contributions to the Annalen der Physik beginning with 1906. In the first published work (Smoluchowski, 1906) he claimed that his method, consequently based on probabilistic ideas, is *"more direct, simpler and thus more convincing than one of Einstein"*. The theory of Brownian motion stayed the main topic of the scientific activity of M. Smoluchowski until his sudden death in 1917. One of his last published works *"Versuch einer mathematischen Theorie der Kogulationskinetik kolloidaler Lösungen"* published in 1917 (Smoluchowski, 1917) has put foundation to the theory of diffusion-controlled reactions, and this line of argumentation was pursued by Peter Debye, Hendrik Anthony Kramers, and many others, see Chapters 10–11.

The macroscopically measurable fluctuations gave new experimental possibilities and several famous experimentalists were fascinated by the behavior of mesoscopic particles (Perrin measured Loschmidt's number and determined Planck's constant, Millikan and Ehrenhaft made experiments to define the elementary charge, Houdijk and Zeeman observed the motion of platin wires). The French physicist Paul Langevin (1872–1946) has to be mentioned who added random forces to the dynamical laws initiating thereby a new mathematical fruitful field, that of stochastic differential equation. Further we mention the dissertations of two young scientists,

L. Ornstein and G. Uhlenbeck, presented in 1908 and 1927 respectively to the University of Leiden as well as their subsequent publications.

The formulation of the quantitative theory by Einstein and by Smoluchowski (1906) motivated new, quantitative experiments performed by J.-B. Perrin (starting from 1908), A. Westgren, E. Kappler and many others. Perrin's work was crowned by the Nobel prize in 1926. These works put the firm foundation to the modern understanding of both equilibrium and non-equilibrium phenomena.

We will say now a few words on the completely different alternative approach to Brownian motion which is based on the concept of the Langevin equation. In his article published in *Comptes Rendus* in 1908 Paul Langevin proposed another approach to description of Brownian motion, than the one of Einstein and Smoluchowski. This one was assumed to be "infinitely simpler" than the Einstein's one and seemed to be only based on the equipartition theorem (Langevin, 1908).

The fruitful idea of Langevin was that the motion of the Brownian particle in fluid is governed by the Newtonian dynamics under friction, which for a for a macroscopic spherical particle follows the Stokes law. However, this behavior, leading to the continuous decay of the particle's velocity, holds only on the average. In order to describe the erratic motion of the particle, resulting from random, uncompensated impacts of the molecules of surrounding fluid, one has to introduce an additional, fluctuating force ("noise"). Adding fluctuating noise to deterministic equation is now a starting point of many fruitful theoretical approaches; the mathematics of such approaches, however, had first to be clarified by works of Kiyosi Ito (born 1915) and of Ruslan Stratonovich (1930–1997).

So far we talked about standard Brownian motion. This is the passive stochastic motion of microscopic objects due to random collisions from the side of the surrounding medium. In recent times the concept of Brownian motion is actively developing. At first we mention the development of the theory of active Brownian motion which considers self-propelling particles (Schweitzer et al., 1998; Ebeling et al., 1999; Erdmann et al., 2000; Mikhailov and Cahlenbuhr, 2002; Schweitzer, 2003). This new theory and several applications will be explained in Chapter 12. A funny note about the difference between standard and active Brownian motion is on place. Microbiologists claim, they can visually distinguish the Brownian motion of the essentially passive particles from the motion of the self-propelling ones. The students are trained to see the difference. However, the differences in the motion pattern is not so evident, so that many well-trained and famous scientists described the essentially immobile bacteria as mobile ones: This mistake happened to the great Japanese microbiologist Kitasato with the pest bacterium, as well as to his pupil Shiga and to the American Flexner with two different types of dysentery bacteria: All these scientists described the essentially immobile bacteria as self-propelling (Winkle, 1999).

The interrelation between theory of stochastic processes and biological problems is older and deeper than it might seem. In 1905 Karl Pearson, engaged in biostatistical problems, wrote a short letter to the journal Nature (Pearson 1905): "*Can any of you readers refer me to a work wherein I should find a solution of the following problem, or failing the knowledge of any existing solution provide me with an original one? I should be extremely grateful for the aid in the matter. A man starts from the point O and walks l yards in a straight line; he then turns through any angle whatever and walks another l yards in a second straight line. He repeats this process n times. Inquire the probability that after n stretches he is at a distance between r and r + δr from his starting point O*". This was the birth of the term "random walk", of a discrete stochastic model which got to be one of the most successful and important models of the whole statistical physics. The asymptotic solution of the problem for large number of steps was immediately given by Rayleigh, who referred to his old publication on summation or random oscillations (Rayleigh, 1880). The whole nature of the random walk approaches (taking into account discrete events on lowest scales, either as jumps, or as a continuous, homogeneous motion interrupted by scattering events (in the sense of the Paul Drude's approach to metals), as well as mathematical approaches to the problem, look very much different from typical continuous approaches used in theoretical physics. The model, however, is more adequate than, say, the Langevin approach fully neglecting local dynamics, for the description of such processes as hopping conductivity in strongly disordered semiconductors, transport processes in hydrology, and many others, and leads to a powerful and beautiful mathematics of fractional diffusion and Fokker–Planck equations (see Hilfer, 2000, Sokolov et al., 2002, West, 2003). The corresponding results will be discussed in Chapter 11 of our book.

The first random walk model was, however, put forward five years before Einstein's and Smoluchowski's work in the doctoral thesis of Louis Bachelier (1870–1946), defended and published in Paris in 1900. The thesis worked out mathematically the idea that the stock market prices with their unceasing ups and downs are essentially sums of independent, bounded random changes (Bachelier, 1900). The report on this thesis written by H. Poincaré can be found in the work of Taqqu (2002). It is sometimes claimed that the thesis was written under the supervision of H. Poicaré; this statement doesn't seem to be true: Bachelier worked and studied at the same time, he took courses occasionally, and probably presented his theses as the "external" candidate. Poincare, who disliked probabilistic approaches, made a positive note that the author "does not exaggerate the range of his results, and I do not think that he is deceived by his formulas."

Bernard Bru puts the situation as follows: "It was a thesis on mathematical physics, but since it was not physics, it was about the Stock Exchange, it was not a recognized subject" (see Taqqu, 2002). Note that application of physical methods to analysis of economical systems is now a rather well established branch of statistical physics; and a word "Econophysics" was coined to describe this field, see Stanley, 2003. Thus, the issues 1 and 2 of Volume 324 of journal **Physica A: Statistical Mechanics and its Applications** published the proceedings of the International

Econophysics Conference. We, however, will not discuss econophysical applications in this book, and we shall confine ourselves to classical application fields pertinent to physics, and partly to chemistry and biology.

The results put forward by Bachelier were of highest importance, and (partly used, partly rediscovered by later workers) lead to a flash of interest to stochastic processes and corresponding probabilistic approaches. The mathematical approaches (culminating in work by Norbert Wiener, Paul Lévy and Andrei Nikolaevich Kolmogorov) in the direction of formulating and refining stochastic approaches lead to a large body of knowledge giving a solid basis for modern applications in statistical physics. The purely mathematical discussion of limit theorems for the sums of independent, identically distributed random variables, connected mostly with the name of Lévy, lead to the results pertinent to the distributions lacking the dispersion, or even the mean. Such distributions are not as unphysical as they may seem, some of the modern applications in physics are discussed in the book of Shlesinger (1996).

Chapter 2

Thermodynamic, Deterministic and Stochastic Levels of Description

2.1 Thermodynamic level: The fundamental laws

On the thermodynamic level, matter is described by concepts like thermodynamic system, thermodynamic state, components, phases, etc. and by thermodynamic variables like energy, entropy, temperature, and quantities (masses) of the different kinds of matter. Following Schottky, a thermodynamic system is a finite macroscopic body with a well defined border (surface) through with the system is exchanging matter, heat and work with the surrounding. A classification of thermodynamic systems may be based on the types of admitted exchanges with the surrounding:

- *Open systems*: Matter, heat and work may be exchanged as well.
- *Closed systems*: The system may exchange heat and work, but not matter.
- *Adiabatic systems*: Work is exchanged, but heat and matter cannot penetrate the border of the system.
- *Isolated systems*: No macroscopic exchange with the surrounding is admitted.

In order to have at least one example for each of the classes let as think on a glass of water as an open system, a gas enclosed in a vessel embedded into a heat bath with a movable piston as an example for a closed system. Further we consider a gas with fixed volume as an example for an adiabatic system and a gas enclosed into a Dewar vessel as an example for an isolated system. Speaking about the state of such a system we mean so far just a set of measurable physical quantities which somehow characterize the macroscopic properties of the system. A thermodynamic system is called homogeneous if measurements at different positions inside the system yield always the same quantities, otherwise it is called inhomogeneous. Components of a thermodynamic system are the chemical units (elements, compounds) we find in it. These units are denoted by a Latin index i which is running from 1 to s ; i.e. s is the total number of components. The composition is the set of the chemical constituents as e.g. N_2, O_2, CO_2, H_2O in the case of air. The phases of a thermodynamic system are the relatively uniform parts of a thermodynamic system. In a vessel containing water we may find e.g. vapor and liquid. The phases will be denoted by a Greek index α and will be numbered from 1 to σ. A quantity x is called a thermodynamic variable if it has the following properties:

- The quantity x is expressed by a real number or by a set of real numbers.
- There exists a prescription for the measurement of the quantity x.
- The quantity x characterizes certain macroscopic properties of the system.

There are three groups of conjugated pairs of variables: *dependent and independent variables, extensive and intensive variables, internal and external variables*. A thermodynamic variable is called dependent, if we succeed to find a relation which reduces it to the values of other variables, otherwise it is independent. For example the pressure of a gas may be calculated from the volume and the temperature, if an equation of state is available. A variable is called extensive if its value for a system consisting of the parts 1 and 2 is the sum of the values of the parts

$$X = X_1 + X_2. \tag{2.1}$$

Examples are the volume V and the mass M. Intensive variables on the other hand have the property to have the same value in each part of the system

$$x = x_1 = x_2. \tag{2.2}$$

Examples are the pressure p and the chemical potential μ. In general extensive variables will be denoted by capital letters and intensive variables by small letters. An exclusion is the temperature, which traditionally is denoted by T in spite of its intensive character. Internal variables are variables depending on the internal state as the positions and velocities of the molecules only, examples are the temperature and the pressure. External variables depend on the state of the bodies in the surrounding, an example is the gravitational field. A variable is called a state variable if it depends only on the actual state but not on the history of the system. Let us now be more concrete. The quantity of matter contained in a system is given by the mass M and the quantities of the components by the individual masses $M_1, M_2, \ldots, M_s$. The measurement of masses is no problem, the units are Kg or g. Instead of these units we may use the molar masses m_i known from chemistry. According to *Avogadro* (1811), *Loschmidt* (1865) and *Planck* (1899) the number of particles corresponding to the molar mass is

$$N_L = 6.0230 \cdot 10^{23} \text{ mol}^{-1}.$$

We introduce further the mol numbers of a sort denoted by the index i in a system by

$$N_i = M_i/m_i$$

and the particle numbers by

$$N_i = N_i * N_L.$$

The total particle number is denoted by N. Beside the extensive variables discussed so far we introduce also the following intensive variables,

$n_i = N/V$ – particle densities,
$c_i = N/V$ – molar concentrations,
$\rho_i = M_i/V$ – mass densities,
$x_i = N_i/N$ – mol fractions.

In the case that our system consists of several phases we add an upper index denoting the phase, e.g. x_i^α would denote the mole fraction of the component i in the phase α. Now we shall discuss briefly the mechanical variables V and p. The volume V of a system is measured either by using geometrical formulae, or the Archimedes principle. In order to measure the pressure we introduce a small cylinder into the system having a movable piston of area S at one end. If F is the force exerted by the body on the piston, $p = F/S$ is the pressure. The unit of the pressure is 1 N/m^2 = 1 Pa (Pascal). In most cases the pressure is independent of the position and the direction of the piston used for the measurement, i.e. p is a scalar. However there exist also more complicated cases where the pressure is a tensor and where the properties depend not only on volume but also on the shape of the system. The central thermal variable of the system is the temperature T. The temperature of systems in thermodynamic equilibrium is a well defined quantity, which has the same value for two systems being in thermal equilibrium one with another. A scientific method for measuring temperatures was first introduced in 1592 by *Galilei* during a lecture where he presented a "*thermoskop*". By means of this measuring device he reduced temperature measurements to length measurements. Using a similar principle we introduce here absolute temperatures by means of a gas thermometer. It consists of a vessel with a movable piston containing a gas with a very low density n and a possibility to measure the position of the piston and in this way the volume. Then the relative temperature is defined by

$$T = T_t \cdot \lim_{n \to 0} \left(\frac{V}{V_t}\right). \qquad (2.3)$$

Here T_t is a reference temperature and V_t the volume of the gas at that temperature. In this way the measurement of the temperature is reduced to a comparison of volumes for a low density gas. As the reference temperature we take the temperature of water at the critical point which is (now by definition)

$$T_t = 273.16 \text{ K}. \qquad (2.4)$$

Our temperature definition is based on the empirical observation that dilute gases show a kind of universality: At low density all gases behave in the same way, independent of the chemical composition. This property guarantees the universality of our temperature definition. Coming back to the definition of the state of a system we may say that the thermodynamic state is a set of independent thermodynamic variables which completely characterize the thermodynamic system. There exists a special state of thermodynamic systems which has unique properties, the equilibrium state: A system is in the state of *thermodynamic equilibrium* if the following conditions are fulfilled:

- The system is characterized by a unique set of extensive and intensive variables which do not change in time.
- After isolation of the system from its environment, all the variables remain unchanged.

All states which do not satisfy these conditions are called non-equilibrium states. For example a gas kept for a long time under fixed environmental conditions in a vessel will be in equilibrium. Isolation from the environment will not change its state. Summarizing the findings given above we formulate now a first working principle centered around the equilibrium state and the temperature variable. Following *Fowler* this principle is called the

Zeroth Law of thermodynamics:

Thermodynamic systems possess a special state, called thermodynamic equilibrium. For systems in thermodynamic equilibrium there exists a scalar intensive variable, called the temperature, which is uniquely defined. The temperature has equal values for any two systems in thermal equilibrium. When two systems (1) and (2) are each in equilibrium with a third one (3), then system (1) is also in equilibrium with system (2).

The property described by the last part of our principle is called transitivity of the temperature variable. A universal relative temperature scale may be defined by means of low density gas thermometers. The temperature measurement is fundamentally based on the property of transitivity. Systems not in equilibrium change their state in time. Sequences of states in time are called *processes*. We introduce the notations:

- *Quasistatic processes* are sequences of equilibrium states. Processes which are sufficiently slow may be considered approximately as quasistatic.
- *Reversible processes* may proceed in both directions. A film of a reversible process shown in opposite time sequence will not contradict any law of nature. Especially we have to require that the reversed process does not violate the second law of thermodynamics.
- All processes not belonging to the classes defined above are called *irreversible processes*.

Evidently all *real thermodynamic processes* are belonging to the third class. However the concepts of quasistatic and reversible processes are very useful as abstractions modeling limiting situations. Real thermodynamic processes show a fundamental asymmetry between future and past; they have the property of irreversibility. Let us discuss now our definition of irreversibility in more detail. We consider an arbitrary real process. In order to decide, whether this process is reversible or irreversible we make a film of the process. If the macroscopic state at time t is denoted by $x(t)$, the film corresponds to a time series

$$x(t_1), x(t_2), \ldots, x(t_n).$$

In the next step we show our film in the opposite direction. This defines an inverse process or backward process, corresponding to the time series.

$$x(t_n), x(t_{n-1}), \ldots, x(t_1).$$

If this inverse process is in no contradiction to any known law of nature, the original process is called a reversible process. In all other cases we speak about irreversible processes. In order to illustrate this idea let us consider two examples. First we consider the orbit of an electron around an atomic nucleus. In this case there is no natural law, which would forbid the backward motion. Forward and backward motion are both possible, both are allowed motions according to the laws of classical or quantum mechanics. Classical or quantum-mechanical motions are of reversible type. Our second example is a stone falling down to earth, where it comes to rest converting kinetic energy into heat and producing in this way entropy. The backward motion would correspond to a stone collecting heat from the surrounding and converting it to kinetic energy. This process would destroy entropy, what is impossible according to the second law. The second law postulates that macroscopic processes may produce entropy, but in no case they may destroy entropy. This proves that all macroscopic processes are irreversible. The backward process which is associated to a macroscopic process would show a process which destroys entropy. In this way the backward process conjugated to any macropscopic process is forbidden by the second law. This is the proof, that all macroscopic processes are irreversible. Reversible processes are a limiting case of irreversible processes; the opposite however is not true.

How may we describe the difference between reversibility and irreversibility in a more technical (mathematical) way? We will show in the next section that the entropy and other thermodynamic functions provide us with the tools for finding characteristic functions, the so-called *Lyapunov functions*, which are the appropriate tools for the quantitative description of irreversibility. Now we shall introduce the central variable of thermodynamics, the *energy*. The energy, however, is not only the central variable of thermodynamics, it is also the central variable of all branches of physics and natural sciences, a quantity which links all real processes we know. Energy is subject to a conservation law, called the

First Law of thermodynamics:
Energy can neither be created nor be destroyed. It is conserved in isolated systems. Energy can be converted to other forms, or moved to another system, without changing its total amount.

It is extremely difficult or even impossible to give a strict definition of the term *energy*. From a principal point of view energy is given "*a priori*", i.e. it must be considered as an element of the axiomatics of thermodynamic systems. Due to the fundamental character of the energy it is very difficult to avoid tautologies in its definition. However, the same difficulty was already discussed by *Newton* with respect to mass and force. *Poincare* discussed the difficulties to define the term *energy* in his lectures on thermodynamics (1893). Poincare said that in every special instance it is clear to us what energy is, and we can give at least a provisional definition of it. However it seems to be impossible to give a general definition of energy. Being aware of the difficulties with any definition of energy we restrict ourselves to a description of the properties of energy. The first property of energy is extensivity. Further we find that any finite system of physical or non-physical origin contains certain amount of energy. In this respect energy has similar properties as mass. The deeper reason for this close relation may be Einstein's formula

$$E = mc^2$$

which relates the total energy E contained in a body to its total mass m, including the rest mass. Therefore, strictly speaking, only one of these quantities is independent. Another property of energy is, that it accompanies any process in our world we know. Any process is connected with a transfer or with a transformation of energy. Energy transfer may have different forms as heat, work and chemical energy. The unit of energy is 1 J = 1 Nm, corresponding to the work needed to move a body 1 meter against a force of 1 Newton. An infinitesimal heat transfer we denote by $d'Q$ and the infinitesimal work transfer by $d'A$. If there is no other form of transfer, i.e. the system is closed, we find for the energy change

$$dE = d'Q + d'A. \tag{2.5}$$

In other words, the infinitesimal change of energy of a system equals to the sum of of the infinitesimal transfers of heat and work. This is a mathematical expression of the principle given above: A change of energy must be due to a transfer, since creation or destruction of energy is excluded. If the system is open, i.e. the exchange of matter in the amount dN_i per sort i is admitted, we assume

$$dE = d'Q + d'A + \sum_i \mu_i dN_i. \tag{2.6}$$

Here the so-called chemical potential μ_i denotes the amount of energy transported by a transfer of a unit of the particles of the chemical sort i. Here μ_i has the dimension of energy per particle or per mole. The infinitesimal work has in the simplest case the form

$$d'A = -pdV. \tag{2.7}$$

In the case that there are also other forms of work we find more contributions having all a bilinear form

$$d'A = \sum_k l_k dL_k \tag{2.8}$$

where l_k is intensive and L_k is extensive and in particular for the case of Eq. (2.7) we have $l_1 = -p$ and $L_1 = V$. Strictly speaking this expression for the infinitesimal work is valid only for reversible forms of work. Later we shall come back to the irreversible case. In this way the balance equation for the energy changes (2.6) assumes the form

$$dE = d'Q + \sum_k l_k dL_k + \sum_i \mu_i dN_i. \tag{2.9}$$

In this equation there remains only one quantity which is not of the bilinear structure, namely the infinitesimal heat exchange $d'Q$. The hypothesis, that bilinearity holds also for the infinitesimal heat, leads us to the next fundamental quantity, the entropy. We shall assume that $d'Q$ may be written as the product of an intensive quantity and an extensive quantity. The only intensive quantity which is related to the heat is T and the conjugated extensive quantity will be denoted by S. In this way we introduce entropy as the extensive quantity which is conjugated to the temperature:

$$d'Q = TdS. \tag{2.10}$$

This equation may be interpreted also in a different way by writing

$$dS = \frac{d'Q}{T}. \tag{2.11}$$

The differential of the state variable entropy is given by the infinitesimal heat $d'Q$ divided by the temperature T. In more mathematical terms, the temperature T is an integrating factor of the infinitesimal heat.

The variable entropy was introduced in 1865 by Clausius. The unit of entropy is 1 J/K. One can easily show that this quantity is not conserved. Let as consider for example two bodies of different temperatures T_1 and T_2 being in contact. Empirically we know that there will be a heat flow from the hotter body 1 to the cooler one denoted by 2. We find

$$d'Q_1 = T_1 dS_1 = d'Q = T_2 dS_2. \tag{2.12}$$

Due to our assumption $T_1 > T_2$ we get $dS_1 < dS_2$, i.e. heat which flows down a gradient of temperature produces entropy. The opposite flow against a gradient of temperature is never observed. A generalization of this observations leads us to the

Second Law of thermodynamics:
Thermodynamic systems possess the extensive state variable entropy. Entropy can be created but never be destroyed. The change of entropy in reversible processes is given by the exchanged heat divided by the temperature. During irreversible processes entropy is produced in the interior of the system. In isolated systems entropy can never decrease.

Let us come back now to our relation (2.9) which reads after introducing the entropy by Eq. (2.10)

$$dE = TdS + \sum_k l_k dL_k + \sum_i \mu_i dN_i. \qquad (2.13)$$

This equation is called *Gibbs fundamental relation*. Since the Gibbs relation contains only state variables it may be extended (with some restrictions) also to irreversible processes. In the form (2.13) it may be interpreted as a relation between the differentials dE, dS, dL_k and dN_i. Due to the Gibbs relation one of those quantities is a dependent variable. In other words we may write e.g.

$$E = E(S, L_k, N_i)$$

or

$$S = S(E, L_k, N_i).$$

In order to avoid a misunderstanding, we state explicitly:

"*The Gibbs fundamental relation (2.13) should not be interpreted as a balance, but as a relation between the extensive variables of a system*".

Since this point is rather important, let as repeat it again: Eq. (2.6) expresses a balance between the energy change in the interior dE (the l.h.s of the equation) and the transfer of energy forms through the border (the r.h.s.). On the other hand Eq. (2.13) expresses the dependence between state variables, i.e. a completely different physical aspect. For irreversible processes the Gibbs relation (2.13) remains unchanged, at least in cases where the energy can still be expressed by the variables S, L_k, N_i. On the other hand the balance (2.6) has to be modified for irreversible processes. This is due to the fact that there exists a transfer of energy which passes the border of the system as work and changes inside the system into heat. Examples are Ohms heat and the heat due to friction. In the following we shall denote these terms by $d'A$. Taking into account those contributions, the balance assumes the form

$$dE = d'Q + \sum_k l_k dL_k + d'A_{dis} + \sum_i \mu_i dN_i \,. \tag{2.14}$$

For later applications we formulate now the first and the second law in a form due to Prigogine (1947). The balance of an arbitrary extensive variable X may be always written in the form

$$dX = d_e X + d_i X \tag{2.15}$$

where the index "e" denotes the exchange with the surrounding and the index "i" the internal change. Then the balances for the energy and the entropy read

$$dE = d_e E + d_i E \,, \tag{2.16}$$
$$dS = d_e S + d_i S \,. \tag{2.17}$$

Further the first and the second laws respectively assume the mathematical forms

$$dE = d_e E; \qquad d_i E = 0$$

$$d_e E = d'Q + d'A + \sum_i \mu_i d_e N_i$$

$$d_e S = \frac{d'Q}{T}$$

$$dS \geq d_e S; \qquad d_i S \geq 0 \,. \tag{2.18}$$

This way of writing is especially useful for our further considerations. We mention that the relation for $d_e S$ is to be considered as a definition of exchanged heat. The investigations of De Groot, Mazur and Haase have shown that for open systems other definitions of heat may be more useful. As to be seen so far, the "most natural" definition of heat exchange in open systems is

$$d_e S = \frac{d^*Q}{T} + \sum_i s_i d_e N_i \,. \tag{2.19}$$

The new quantity d^*Q is called the "reduced heat"; s_i is the specific entropy carried by the particles of kind i. The idea which led to the definition (2.19) is, that the entropy contribution which is due to a simple transfer of molecules should not be considered as a proper heat. Another advantage of the reduced heat is, that it possesses several useful invariance properties (Haase, 1963; Keller, 1977). The first and the second laws of thermodynamics formulated above are a summary of several hundred years of physical research. They constitute the most general rules of prohibition in physics.

The Third Law of thermodynamics is more technical but it constitutes also a summary of a long experience. We will formulate this law in the following form:

Third Law of thermodynamics:
Energy and entropy are extensive scalar quantities, being finite for finite systems and bounded from below. For real macroscopic systems $E > 0$ and $S > 0$ holds. In the limit of zero temperature $T \to 0$, the entropy as well as its derivatives with respect to extensive variables disappear asymptotically.

The four principles of thermodynamics are general laws of nature. For the formulation of the zeroth and of the third principle respectively we required the concept of temperature. This means that these principles are applicable to thermal systems only. Much more general are the first and the second principle. The community of physicists claims that *the fundamental laws are valid for any macroscopic process in nature and society*. The idea is surprising, that the laws of physics provide hard bounds (like the positive cone in relativity theory) for any possible process in nature and society (Ebeling & Feistel, 1982, 1986).

2.2 Lyapunov functions: Entropy and thermodynamic potentials

As stated already by *Planck*, the most characteristic property of irreversible processes is the existence of so-called *Lyapunov functions*. This type of function was defined first by the Russian mathematician Lyapunov more than a century ago. A Lyapunov function is a non-negative function with the following properties

$$L(t) \geq 0, \qquad \frac{dL(t)}{dt} \leq 0. \tag{2.20}$$

As a consequence of these two relations, Lyapunov functions are per definition never increasing in time. Our problem is now to find a Lyapunov function for an arbitrary macroscopic system. Let us assume that the system is initially ($t = 0$) in a nonequilibrium state, and that we are able to isolate the given system for $t > 0$ from the surrounding. From the the definition of equilibrium follows that, after isolation, changes will occur. Under conditions of isolation, the energy E will remain fixed, but the entropy will monotoneously increase according to the second law. Irreversible processes connected with a positive entropy production $P \geq 0$ will drive the system finally to an equilibrium state located at the same energy surface. In thermodynamic equilibrium, the entropy assumes the maximal value

$$S_{eq}(E, X)$$

which is a function of the energy and certain other extensive variables. The total production of entropy during the process of equilibration of the isolated system may be obtained by integration of the entropy production over time.

$$\Delta S(t) = \int_0^t P(t)dt. \qquad (2.21)$$

Due to the condition of isolation, there is no exchange of entropy during the whole process. Due to the non-negativity of the entropy production

$$\frac{d_i S}{dt} = P(t) \geq 0 \qquad (2.22)$$

the total production of entropy $\Delta S(t)$ is a monotonously non-decreasing function of time. The concrete value of $\Delta S(t)$ depends on the path γ from the initial to the final state and on the rate of the transition processes. However the maximal value of this quantity $\Delta S(\infty)$ should obey some special conditions.

Just for the case that the transition occurs without any entropy exchange, this quantity should be identical with the total entropy difference between the initial state and the equilibrium state at time $t \to \infty$

$$\delta S = S_{eq}(E, X) - S(E, t = 0). \qquad (2.23)$$

This is the so-called entropy lowering which is simply the difference between the two entropy values. By changing parameter values infinitely slow along some path γ we may find a reversible transition and calculate the entropy change in a standard way e.g. by using the Gibbs fundamental relation (2.13). An important property of the quantity δS is, that it is independent on the path γ from the initial state to the equilibrium state. On the other hand the entropy change on an irreversible path may depend on details of the microscopic trajectory. In average over many realizations (measurements) should hold

$$\langle \Delta S(\infty) \rangle = \left\langle \int_0^\infty P(t)dt \right\rangle = \delta S. \qquad (2.24)$$

This equality follows from the fact that S is a state function, its value should be independent on the path on which the state has been reached. Assuming for a moment that the equality (2.24) is violated, we could construct a cyclic process which contradicts the second law. The macroscopic quantity

$$\Delta S = \langle \Delta S(\infty) \rangle \qquad (2.25)$$

may be estimated by averaging the entropy production for many realizations of the irreversible approach from the initial state to equilibrium under conditions of strict isolation from the outside world. The result

$$\Delta S = \delta S \qquad (2.26)$$

is surprising: It says that we can extract equilibrium information δS from nonequilibrium (finite-time) measurements of entropy production. We will come back to this point in the next section. Here let us proceed on the way of deriving Lyapunov functions. By using the relations given above we find for macroscopic systems the following Lyapunov function

$$L(t) = \Delta S - \Delta S(t) \tag{2.27}$$

which yields for isolated systems (Klimontovich, 1992, 1995)

$$L(t) = S_{eq} - S(t).$$

Due to

$$\frac{dL(t)}{dt} = -P(t) \leq 0 \tag{2.28}$$

and

$$\Delta S \geq \Delta S(t) \tag{2.29}$$

the function $L(t)$ has indeed the necessary properties (2.20) of a Lyapunov function. Let us consider now a system which is in contact with a heat bath of temperature T. Following Helmholtz we define the thermodynamic function

$$F = E - TS \tag{2.30}$$

which is called the *free energy*. According to Gibbs' fundamental relation the differential of the free energy is given by

$$\begin{aligned} dF &= dE - TdS - SdT \\ &= \sum_k l_k dL_k + \sum_i \mu_i dN_i - SdT \,. \end{aligned} \tag{2.31}$$

In this way we see that the proper variables for the free energy are the temperature and the extensive variables L_k (note $L_1 = V$) and N_i with $(i = 1, \ldots, s)$. In other words we have

$$F = F(T, L_k, N_i) \,.$$

The total differential (2.31) may also be considered as a balance relation for the free energy change for a quasistatic transition between two neighboring states. Let us now consider the transition under more general situations admitting also dissipative elements. Then we find

$$\begin{aligned} dF &= dE - SdT - TdS \\ &= d'A - d'Q - SdT - Td_eS - Td_iS \,, \end{aligned}$$

$$dF = d'A - SdT - Td_iS. \qquad (2.32)$$

At conditions where the temperature is fixed and where the exchange of work is excluded we get

$$dF = -Td_iS \leq 0; \qquad \frac{dF}{dt} = -TP \leq 0. \qquad (2.33)$$

As a consequence from Eq. (2.33), the free energy is a nonincreasing function for systems contained in a heat bath which excludes exchange of work. At these conditions the free energy assumes its minimum F at the thermal equilibrium. Consequently the Lyapunov function of the system is given by

$$L(t) = F(t) - F_{eq} \qquad (2.34)$$

which possesses the necessary Lyapunov properties (2.20). Another important situation is a surrounding with given temperature T and pressure p. The characteristic function is then the free enthalpy (Gibbs potential)

$$G = E + pV - TS \qquad (2.35)$$

with the total differential

$$dG = dE - pdV - Vdp - SdT - TdS \qquad (2.36)$$

and the balance relation

$$dG = d'A - d'Q - Vdp - pdV - SdT - Td_eS - Td_iS$$
$$= Vdp - SdT - Td_iS. \qquad (2.37)$$

For given temperature and pressure we get

$$dG = -Td_iS \leq 0$$

and

$$\frac{dG}{dt} = -TP \leq 0. \qquad (2.38)$$

Consequently the free enthalpy G is a non-increasing function for systems embedded in an isobare and isothermal reservoir. In thermal equilibrium the minimum G_{eq} is assumed. Therefore

$$L(t) = G(t) - G_{eq} \qquad (2.39)$$

is a Lyapunov function possessing the necessary properties (2.20). Let us consider now the most general situation, where our system is neither isolated nor in a reservoir with fixed conditions during its course to equilibrium. In the general case the

Lyapunov function may be defined by

$$L(t) = S_{eq}(E(t=0), X) - S(E, t=0) - \int_0^t P(t')dt'. \qquad (2.40)$$

This function has again the necessary properties of a Lyapunov function (2.20), i.e. it is non-negative and non-increasing. However it will tend to zero only under the condition of total isolation during the time evolution.

The definition of the entropy production is in general a non-trivial problem. In the special case however that the only irreversible process is a production of heat by destruction of mechanical work the definition of $P(t)$ is quite easy. Since then $P(t)$ is given as the quotient of heat production and temperature, a calculation of $L(t)$ requires only the knowledge of the total mechanical energy which is dissipated.

2.3 Energy, entropy, and work

Energy, entropy and work are the central categories of thermodynamics and statistical physics. The fundamental character of these phenomenological quantities requires our full attention. The entropy concept closes the gap between the phenomenological theory and the statistical physics. In spite of the central position of energy, entropy and work in physics, there exist many different definitions and interpretations (Zurek, 1990). Below we will be concerned with several these interpretations. We introduce mechanical energy, heat and work. Furtheron we talk about the Clausius entropy, the Boltzmann entropy, the Gibbs entropy, the Shannon entropy and Kolmogorov entropy. Extending the categories of energy and entropy to other sciences some confusion may arise. In order to avoid any misinterpretation one has to be very careful when talking about these categories. However there should be no doubt, that energy and entropy are central quantities. But due to their fundamentality a specific difficulty of philosophical character arises: It is extremely difficult or even impossible to avoid tautologies in their definition. In conclusion we may say that energy and entropy should be elements of an axiomatics of science. As we mentioned above, the difficulty in defining fundamental quantities was already discussed by Poincare with respect to energy. In his lectures on thermodynamics (1893) Poincare' says:

"*In every special instance it is clear what energy is and we can give at least a provisional definition of it; it is impossible however, to give a general definition of it. If one wants to express the (first) law in full generality, ... one sees it dissolve before one's eyes, so to speak leaving only the words: There is something, that remains constant (in isolated systems).*"

We may translate this sentence to the definition of entropy in the following way:

In every special instance it is clear what entropy is and we can give at least a provisional definition of it; it is impossible however, to give a general definition of it. If one wants to express the second law in full generality, ... one sees it dissolve before one's eyes, so to speak leaving only the words: *There is something, that is non-decreasing in isolated systems.*

In this way our definition of entropy is finally:
Entropy is that fundamental and universal quantity characterizing a real dynamical system, that is non-decreasing in isolated systems.

Energy and entropy are not independent, but are connected in a rather deep way. Our point of view is based on a valoric interpretation (Ebeling, 1993). This very clear interpretation will be taken as the basis for a reinterpretation of the various entropy concepts developed by Clausius, Boltzmann, Gibbs, Shannon and Kolmogorov. As the key points we consider the value of energy with respect to work. The discussion about this relation started already in the 19th century and is continuing till now. The valoric interpretation was given first by *Clausius* and was worked out by *Helmholtz and Ostwald*. But then, due to a strong opposition from the side of *Kirchhoff, Hertz, Planck* and others, it was nearly forgotten except by a few authors (Schöpf, 1984; Ebeling and Volkenstein, 1990). As a matter of fact however, the valoric interpretation of the entropy, was itself for Clausius the key point for the introduction of this new concept in 1864–65. What many physicists do not know is that the entropy concept taught in universities as the Clausius concept is much nearer to the reinterpretation given by Kirchhoff than to the original Clausius' one (Schöpf, 1984). Here we try to develop the original interpretation in terms of a value concept in connection with some more recent developments. We concentrate on processes with fixed energy. Let us start with a comparison of the entropy concepts of *Clausius, Boltzmann and Gibbs*. In classical thermodynamics the entropy difference between two states is defined by Clausius in terms of the exchanged heat

$$dS = \frac{d'Q}{T}; \qquad \delta S = S_2 - S_1 = \int_1^2 \frac{d'Q}{T}. \tag{2.41}$$

Here the transition $1 \to 2$ should be be carried out on a reversible path and $d'Q$ is the heat exchange along this path. In order to define the entropy of a nonequilibrium state we may construct a reversible "Ersatzprozess" connecting the nonequilibrium state with an equilibrium state of known entropy. Let us assume in the following that the target state 2 is an equilibrium state. By standard definition an equilibrium state is a special state of a system with the properties that the variables are uniquely defined, constant in time and remain the same after isolation from the surrounding (compare Section 2.1). The state 1 is by assumption a nonequilibrium state, i.e. a state which will not remain constant after isolation. Due to internal irreversible processes, the process starting from state 1 will eventually reach the equilibrium state 2 which is located (macroscopically) on the same energy level. This is due to the condition of isolation which is central in our picture. Now we may apply Eq. (2.41) finding in this way the nonequlibrium entropy.

$$S_1(\boldsymbol{y}; E, \boldsymbol{X}, t = 0) = S_{eq}(E, \boldsymbol{X}) - \delta S(\boldsymbol{y}; E, \boldsymbol{X}). \tag{2.42}$$

The quantity δS is the so-called entropy lowering in comparison to the equilibrium state with the same energy. It was introduced by *Klimontovich* as a measure of organization contained in a nonequilibrium system (Klimontovich, 1982, 1984, 1990; Ebeling and Klimontovich, 1984; Ebeling, Engel and Herzel, 1990). Several examples were given as e.g. the entropy lowering of oscillator systems, of turbulent flows in a tube and of nonequlibrium phonons in a crystal generated by a piezoelectric device.

We shall assume in the following that the entropy lowering depends on a set of order parameters $\boldsymbol{y} = y_1, y_2, \ldots, y_n$ as well as on the energy E and on other extensive macroscopic quantities $\boldsymbol{X}$. The equilibrium state is characterized by $y_1 = y_2 = \cdots = y_n = 0$.

There are some intrinsic difficulties connected with the construction of an "Ersatzprozess". Therefore Muschik (1990) has developed the related concept of an accompanying process. By definition this is a projection of the real path on a trajectory in an equilibrium subspace. Since the entropy is as a state function independent on the path, the concepts "Ersatzprozess" or accompanying process give at least a principal possibility of calculating the nonequilibrium entropy. In practice these concepts work well for nonequilibrium states which are characterized by local equilibrium, which is valid e.g. for many hydrodynamic flows and chemical reactions (Glansdorff and Prigogine, 1971). In more general situations the exact definition of the thermodynamic entropy remains an open question, which is the subject of intensive discussions (Ebeling and Muschik, 1992). Let us consider now another approach which is based on the concept of entropy production. Assuming again that the initial state 1 is a nonequilibrium state, we know from the definition of equilibrium that, after isolation, changes will occur. Under conditions of isolation the energy E will remain fixed, within the natural uncertainties, but the entropy will monotonously increase due to the second law. Irreversible processes connected with a positive entropy production $P > 0$ will drive the system finally to an equilibrium state located at the same energy surface. In thermodynamic equilibrium, the entropy assumes the maximal value $S_{eq}(E, \boldsymbol{X})$ which is a function of the energy and certain other extensive variables. According to Eq. (2.28) the entropy change is a Lyapunov functions and may be obtained by integration of the entropy production over time:

$$\delta S = S_{eq}(E, \boldsymbol{X}) - S(E, t = 0) = \int_1^2 P(t) dt. \qquad (2.43)$$

In the special case that production of heat by destruction of mechanical work is the only irreversible process application of Eq. (2.43) is quite easy. Since then $P(t)$ is given as the quotient of heat production and temperature a calculation of δS requires the knowledge of the total mechanical energy which is dissipated. Equation (2.43) is another way to obtain the entropy lowering and in this way the entropy of any nonequilibrium state. As above we may consider this difference as a measure of order contained in the body in comparison with maximal disorder in equilibrium. Equation (2.43) suggests also the interpretation as a measure of distance from equilibrium. So far the thermodynamic meaning of entropy was discussed, but entropy is like the face of Janus, it allows other interpretations. The most important of them with respect to statistical physics is the interpretation of entropy as measure of uncertainty or disorder. In the pioneering work of Boltzmann, Planck and Gibbs it was shown that in statistical mechanics the entropy of a macrostate is defined as the logarithm of the thermodynamic probability W

$$S = k_B \log W \qquad (2.44)$$

which is defined as the total number of equally probable microstates corresponding to the given macrostate. Further k_B is the *Boltzmann constant*. In the simplest case of classical systems, the number of states with equal probability corresponds to the volume of the available phase space $\Omega(A)$ divided by the smallest accessible phase volume h^3 (h — Planck's constant). Therefore the entropy is given by

$$S_{BP} = k_B \log \Omega^*(A), \qquad (2.45)$$

$$\Omega^*(A) = \Omega(A)/h^3. \qquad (2.46)$$

Here A is the set of all macroscopic conditions. In isolated systems in thermal equilibrium, the available part of the phase space is the volume of the energy shell enclosing the energy surface

$$H(\boldsymbol{q}, \boldsymbol{p}) = E. \qquad (2.47)$$

If the system is isolated but not in equilibrium only certain part of the energy shell will be available. In the course of relaxation to equilibrium the probability is spreading over the whole energy shell filling it finally with constant density. Equilibrium means equal probability, and as we shall see, least information about the state on the shell. In the nonequilibrium states the energy shell shows regions with increased probability (attractor regions). We may define an effective volume of the occupied part of the energy shell by

$$S(E, t) = k_B \log \Omega^*_{\text{eff}}(E, t), \qquad (2.48)$$

$$\Omega^*_{\text{eff}}(E, t) = \exp(S(E, t)/k_B). \qquad (2.49)$$

In this way, the relaxation on the energy shell may be interpreted as a monotonous increase of the effective occupied phase volume. This is connected with a devaluation of the energy.

Let us discuss now in more detail the relation between free energy and work. The energetic basis of all human activities is *work*, a term which is also difficult to define. The first law of thermodynamics expresses the conservation of the energy of systems. Energy may assume various forms. Such forms of energy as heat or work appear in processes of energy transfer between systems. They may be of different value with respect to their ability to perform work. The (work) value of a specific form of energy is measured by the entropy of the system. As shown first by Helmholtz, the free energy

$$F = E - TS \tag{2.50}$$

represents the amount of energy in a body at fixed volume and temperature which is available for work. Before going to explain this in more detail we go back for a moment to a system with fixed energy and with fixed other external extensive parameters. Then the capacity to do work (the work value) takes it minimum zero in thermodynamic equilibrium, where the entropy assumes the maximal value $S_{eq}(E,X)$. Based on this property Helmholtz and Ostwald developed a special entropy concept based on the term "value". In the framework of this concept we consider the difference

$$\delta S = S_{eq}(E,X) - S(E,X) \tag{2.51}$$

as a measure of the "value" of the energy contained in the system. In dimensionless units we may define a "lowering of entropy" by

$$Le = [S_{eq}(E,X) - S(E,X)]/Nk_B \tag{2.52}$$

where N is the particle number. We consider Le as a quantity which measures the distance from equilibrium or as shown above the (work) value of the energy contained in a system. Further the lowering of entropy Le should also be connected with the non-occupied part of the phase space. As shown above, any nonequilibrium distribution is concentrated on certain part of the energy surface only. Therefore the relaxation to equilibrium is connected with a spreading of the distribution and a decrease of our knowledge on the microstate. An equivalent dimensionless quantity which is bounded between zero and one is

$$w(t) = 1 - \exp(-Le), \tag{2.53}$$

$$w(t) = 1 - \exp[(S(E,X) - S_{eq}(E,X))/Nk_B]. \tag{2.54}$$

In terms of the phase space volume of statistical mechanics this measure has the following meaning: It gives the relative part of the phase space in the energy shell which is occupied by the system. The second law of thermodynamics tells us that entropy can be produced in irreversible processes but never be destroyed. Since entropy is a measure of value of the energy this leads to the formulation that the distance from equilibrium and the work value of energy in isolated systems cannot increase spontaneously. In other words Le and $w(t)$ are Lyapunov functions expressing a tendency of devaluation of energy.

In order to increase the value of energy in a system one has to export entropy. In this way we have shown, that the meaning of the thermodynamic concept of entropy may be well expressed in terms of distance from equilibrium, of value of energy or of relative phase space occupation instead of the usual concept of entropy as a measure of disorder.

Now let us come back to the free energy, the term introduced into thermodynamics by Helmholtz. As we will shown in detail, the concept of Helmholtz may be interpreted in the way that the total energy consists of a free part which is available for work and a bound part which is not available. A related concept is the exergy, which is of much interest for technical applications.

Due to the relation

$$E = F + TS = E_f + E_b \tag{2.55}$$

the energy in a body consists of two parts.

$$E = E_f + E_b; \qquad E_f = F; \qquad E_b = TS. \tag{2.56}$$

Correspondingly, the first part $E_f = F$ may be interpreted as that part of the energy, the "free energy", that is available for work. The product of entropy with the temperature $E_b = TS$ may be interpreted as the bound part of the energy. On the other hand due to

$$H = G + TS = H_f + H_b \tag{2.57}$$

it gives also the bound part of the enthalpy (G being the free enthalpy). From the second law follows, as shown in the previous section, that under isothermal conditions the free energy F is a non-increasing function of time

$$\frac{dF}{dt} \leq 0 \tag{2.58}$$

and that under isobaric-isothermal condition the free enthalpy G is non-increasing

$$\frac{dG}{dt} \leq 0. \tag{2.59}$$

The tendency of F and G to decrease is in fact determined by the general tendency expressed by the second law, to devaluate the energy (or the enthalpy) with respect to their ability to do work.

Let us study now the work W performed on a system during a finite-time transition from an initial nonequilibrium state to a final equilibrium state. Then, as we have shown

$$W = \delta F = F_{ne} - F_{eq} \tag{2.60}$$

is the work corresponding to a process when the parameters are changed infinitely slowly along the path γ from the starting nonequilibrium point to the final equilibrium state. This relation is not true, if the parameters are switched along the path γ at a finite rate. At that conditions the process is irreversible and the work W will depend on the microscopic initial conditions of the system and the reservoir, and will, on average exceed the free energy difference (Jarzynski, 1996)

$$\langle W \rangle \geq \delta F = F_{ne} - F_{eq}. \tag{2.61}$$

The averaging is to be carried out over an ensemble of transitions (measurements). The difference

$$[\langle W \rangle - W] \geq 0 \tag{2.62}$$

is just the dissipated work W_{dis} associated with the increase of entropy during the irreversible transition.

In recent work Jarzynski (1996) discussed the above relations between free energy and work from a new perspective. The new relation derived by Jarzynski (1996) instead of the inequality (2.61) is an equality

$$\langle \exp[-\beta W] \rangle = \exp[-\beta \langle W \rangle] \tag{2.63}$$

where $\beta = 1/k_B T$. This nonequilibrium identity, proven by Jarzynski (1997) using different methods, is indeed surprising: It says that we can extract equilibrium information

$$\delta F = W = -k_B T [\ln \langle \exp(-\beta W) \rangle] \tag{2.64}$$

from an ensemble of nonequilibrium (finite-time) measurements. In this respect Eq. (2.64) is an equivalent of Eqs. (2.24)–(2.26).

Finally let us come back to the splitting of the energy into two parts proposed by Helmholtz. Recently the idea was developed that entropy may also be splitted in a similar way as energy into a free part and into a bound part (Ebeling, 1993)

$$S = S_f + S_b. \tag{2.65}$$

The bound part reflects the entropy bound in the microscopic motion, the free part is that part of the entropy available for information processing. The splitting of entropy into two parts reminds the decomposition

$$S = Z + Y \tag{2.66}$$

given originally by *Clausius* in 1864 in his book *"Mechanische Wärmetheorie I"*. Here Clausius defines the part Y as the change value of the heat and Z as the disgregation. However since Clausius argumentation is notoriously difficult, we do not know exactly, what the great pioneer had in mind with his decomposition. However Schöpf's deep analysis makes plausible, that Clausius considered the valoric interpretation as the central point of his entropy concept (Schöpf, 1978, 1984). The relation of entropy and information, i.e. the splitting of entropy into different contributions was discussed already in the papers of Szilard and Brillouin and later by many other workers (Weber et al., 1988; Zurek, 1990). A similar relation as we have described for the entropy itself should hold for the entropy transfer. There exist many forms of entropy transfer, such as heat conduction, and entropy transferred by matter; information transfer appears to be a special form of entropy transfer (Ebeling, 1993). We shall come back to the problem of the relation between entropy and information in Chapters 4–5.

2.4 Deterministic models on the mesoscopic level

We will introduce here only several basic concepts of the deterministic dynamics. Some details will be discussed in the next Chapter. Let us first introduce the concepts of state space, dynamical models and trajectories. The geometrical interpretation of the trajectories of a dynamical system as orbits in an appropriate state space (phase space) appears to be one of the most important instruments for the investigation of dynamical processes. In fact, these concepts go back to the classical mechanics. Generalizing the concepts of mechanics, Poincare, Lyapunov, Barkhausen, Duffing, van der Pol, Andronov, Witt, Chaikin, Birkhoff, Hopf and others laid the basis for the modern theory of dynamical systems. Parallel research was started in the biosciences by Lotka, Volterra, Rashevsky and others. In the second half of this century dynamical systems theory was devoloped quite independently of physics and biosciences and found applications in most branches of science. In particular, the theory of dynamical systems is also the heart of the science of synergetics founded by Haken starting from laser physics (Haken, 1970; 1977, 1983, 1988) and the theory of self-organization developed by Prigogine, Glansdorff and Nicolis backing on the concepts of irreversible thermodynamics (Glansdorff and Prigogine, 1971; Nicolis and Prigogine, 1977). We will take here a rather general view following Neimark (Neimark, 1972; Butenin, Neimark & Fufaev, 1976; Neimark & Landa, 1987). Let us consider a dynamical system assuming that its state at time t is given by a set of time-dependent parameters (coordinates). We consider this set as a vector and write

$$\boldsymbol{x}(t) = [x_1(t), x_2(t), \ldots, x_n(t)]. \tag{2.67}$$

The set of state vectors $\boldsymbol{x}(t)$ spans a vector space $\mathbf{X}$ which will be called the state space or phase space of the system. Let us assume now that the state at time t which is $\boldsymbol{x}(t)$ and the state at a later time $t+\delta t$ which is $\boldsymbol{x}(t+\delta t)$ are connected in a causal way. In particular we assume that the state $\boldsymbol{x}(t+\delta t)$ is given as a function or a more general mapping of the state $\boldsymbol{x}(t)$ and certain parameters which we denote by the vector

$$\boldsymbol{u}(t) = [u_1, u_2, \ldots, u_m]. \tag{2.68}$$

The set of all possible parameter values $\boldsymbol{u}$ forms the control space of the system. The parameters $\boldsymbol{u}$ take into account the influence of the environment and possibly also actions of external control. We assume that the connection between $\boldsymbol{x}(t)$ and $\boldsymbol{x}(t+\delta t)$ is given by a dynamical map $\mathbf{T}$

$$\boldsymbol{x}(t+\delta t) = \mathbf{T}(\boldsymbol{x}(t), \boldsymbol{u}, \delta t). \tag{2.69}$$

The set $(\mathbf{X}, \mathbf{T})$ is called the dynamical model of the system (Neimark, 1972; Neimark & Landa, 1987; Ebeling, Engel & Herzel, 1990). The choice of the phase space is not unique and is in general a rather difficult problem. Even if we have a set of observations and measurements, there are still many possibilities to make a choice for the phase space. For example one may think about the movements of the planets, which first were modeled by so-called epicycles and only much later by *Newtons laws*. The time increments δt may be continuous or discrete. In the case of continuous time the states form a continuous orbit. The representation of processes by orbits (trajectories) in a phase space requires a continuous model of the process. Sometimes the trajectories are given only by a smoothed sequence of (discrete) observations. In the case that the observations are made at fixed time distances δt the trajectory is a sequence of events,

$$x(t_1), x(t_2), \ldots, x(t_n). \tag{2.70}$$

Such time series may be given by recording the outputs of some measuring instruments in certain time intervals, but also by a stroboscopic observation of continuous trajectories or by periodic reports. There are processes as e.g. the annual reports about the production of a state, which are intrinsically discrete. Examples of models with discrete time will be considered in Chapter 3. In the rest of this Chapter we restrict ourselves to the simpler case of continuous time. As we have mentioned already, the existence of a dynamical map $\boldsymbol{T}$, which defines the state at the time $t+\delta t$, i.e. $\boldsymbol{x}(t+\delta t)$ by means of the state at an earlier time t, i.e. $\boldsymbol{x}(t)$, expresses the causality of the dynamical process under consideration. We observe in the real word two different cases for the causal relation between $x(t)$ and $x(t+\delta t)$. If the relation between $x(t)$ and $x(t+\delta t)$ is unique, i.e. the future state is given by the map $\mathbf{T}$ in a unique way from the initial one, we speak about deterministic models. If the map is not unique, i.e. if there are several possibilities for the state $x(t+\delta t)$ depending on chance, we speak about stochastic models. In this paragraph we restrict ourselves to the case of deterministic models.

One simple example of deterministic systems are iterated maps, corresponding to discrete time (Collet & Eckmann, 1980; Schuster, 1984). We will discuss this case in more detail in Section 3.3. From the point of view of physics the simplest example of a deterministic dynamics is given by the overdamped motion of a particle with the coordinates x in a potential field $U(x)$ which has the equations of motion (ρ — friction coefficient):

$$\frac{dx(t)}{dt} = -\frac{1}{\rho}\frac{dU(x)}{dx}. \tag{2.71}$$

In this case the orbits are given by the gradient lines (the lines of steepest descent of the potential). The orbits are attracted by the minima of the potential. In general the minima correspond to point which will be called then point attractors. Another (less simple) example are mechanical systems including inertia. Most of thsi class of mechanical systems are described by a *Hamilton dynamics* which is defined by a scalar function H on a space of f coordinates $\boldsymbol{q} = (q_1, \ldots, q_f)$ and f momenta $\boldsymbol{p} = (p_1, \ldots, p_f)$ which is called a Hamiltonian

$$\frac{dq_i}{dt} = \frac{\partial H}{\partial p_i}, \qquad \frac{dp_i}{dt} = -\frac{\partial H}{\partial q_i} \tag{2.72}$$

with $i = 1, 2, \ldots, f$. By integration of the Hamiltonian equations at given initial state $q_i(0), p_i(0), (i = 1, \ldots, f)$ we may calculate the state at $t + \delta t$ in a unique way. More about Hamiltonian systems we shall learn in Section 3.4. One of the most important results which will be discussed there is that in spite of the deterministic connection between initial and future states, the predictability of future states is quite limited (Hoover, 2001; Landa, 2001). The dynamics of Hamilton type is quite special. In many cases, we may think for example about chemical and ecological processes, the dynamics is given by more general types of autonomous differential equations which cannot be derived from just one generating function (as e.g. a potential or a Hamilton function). For example the time evolution of the state vector $\boldsymbol{x}(t)$ may be generated by n functions F_i defining a system of autonomous differential equations

$$\dot{x}_i = F_i(x_1, \ldots, x_n; \boldsymbol{u}), \tag{2.73}$$

with $i = 1, 2, \ldots, n$ or in more compact vector form

$$\frac{d\boldsymbol{x}}{dt} = \boldsymbol{F}(\boldsymbol{x}, \boldsymbol{u}).$$

However even a dynamics defined by such a quite general differential equation is still a rather special form of the map (2.69). One may obtain a differential equation if the map $\boldsymbol{T}$ has for small δt a Taylor expansion

$$\boldsymbol{x}(t + \delta t) = \boldsymbol{x}(t) + \boldsymbol{F}(\boldsymbol{x}(t); \boldsymbol{u}(t))\delta t + 0((\delta t)^2). \tag{2.74}$$

The class of systems described by a system of differential equations is rather large. We have to remember that all the differential equations of higher order as well as the non-autonomous differential equations may be reduced to systems of type (2.73). Even partial differential equations may be reduced at least in some approximation to usual d.e. The standard method which allows the reduction of an (infinite-dimensional) partial d.e. to a finite-dimensional system of differential equations is the procedure proposed by the Russian mathematician Galerkin. For the given reasons we consider the systems of type (2.73) as a sufficiently general basis for our further considerations of dynamical systems. We will assume further, that the conditions of the theorem of Cauchy, which guarantee the existence and the uniqueness of the solutions of (2.73) are fulfilled, then at given initial conditions x_0 at time $t = t_0$ their exists exactly one solution

$$\boldsymbol{x}(t) = \boldsymbol{x}(t; u, t, x), \qquad \boldsymbol{x}(t_0) = \boldsymbol{x}_0. \qquad (2.75)$$

Since $\boldsymbol{x}(t)$ is time-dependent, we get a trajectory in the state space. The theorem about uniqueness has an important topological consequence: Any crossing of trajectories is forbidden, since the crossing point would correspond to two solutions, what is excluded by the theorem. The manifold of motions described by (2.73) defines a field of trajectories. In general the functions F are nonlinear and may be quite complicated. Thus the solution cannot be given analytically. However several general conclusions about the properties of the dynamical system may be obtained even without knowledge of the explicit solution. This is possible on the basis of the so-called qualitative theory of dynamical systems (Andronov et al., 1965, 1967; Neimark et al. 1987). One may answer questions such as:

What are the stationary states of the dynamical system, are they stable?
Do periodical motions exist, are they stable?
Can we expect chaotic motions?
Is it possible that the system reaches certain boundaries in the state space?
What is the dependence of the motion on the control parameters?

Answers on these questions may be achieved by a purely qualitative analysis of the geometry of the field of trajectories. Of special interest are the stationary points, where the system is at rest and especially the stable stationary points which may be the targets of the motion at long time. We call these states point attractors. There may exist other attracting manifolds of dimension $d \leq n$. Attractors are closed bounded sets which are attracting and invariant with respect to the dynamics (Anishchenko et al., 2002). In other words, if a point belongs at time t to an attractor, it will stay on it for any time. Attractors have an attracting basin consisting of all trajectories which approach the attractor in the limit of long time ($t \to \infty$). To the class of attractors belong the stable stationary points discussed above, the stable limit cycles, stable tori and the so-called strange or chaotic attractors. What kind of attractors are possible in a dynamical system depends on the dimension n of the state space and on the functions F. The investigation of the attractors of a dynamical system and their properties is the main aim of the qualitative theory (Anishchenko, 1989, 1994; Anishchenko et al., 2002). Of special interest is also the localization of separatrices and saddle points. The saddles are the crossing points of the separatrices. They, on the other hand, partition the state space into the regions of attraction of different stable manifolds. To acquire knowledge about the dynamical system, the information about saddles and separatrices is therefore crucially. Saddles and separatrices can be reached only if the initial condition is located at a saddle or at the separatrix. The situation may change when stochastic influences are included. In the frame of stochastic theory deterministically unstable manifolds may be reached and may be crossed due to fluctuations. Therefore stochastic perturbations may lead to transitions between several regions of (deterministic) attraction. A very useful method of qualitative dynamics is the investigation of the vector field $\boldsymbol{F}(\boldsymbol{x}; \boldsymbol{u})$. The system of differential equations (2.73) defines a flux in the state space. The trajectories are tangential to this vector field. Investigating the divergence of this field we find 3 principal cases for the local behavior of the vector field:

(1) Locally expanding dynamics with

$$\operatorname{div} \boldsymbol{F}(\boldsymbol{x}; \boldsymbol{u}) > 0, \tag{2.76}$$

(2) Locally conservative dynamics with

$$\operatorname{div} \boldsymbol{F}(\boldsymbol{x}; \boldsymbol{u}) = 0, \tag{2.77}$$

(3) Locally contracting dynamics with

$$\operatorname{div} \boldsymbol{F}(\boldsymbol{x}; \boldsymbol{u}) < 0. \tag{2.78}$$

We underline, that this is a local property, the character may change in other regions of the space, in particular it may change along trajectories. The sign of $\operatorname{div} \boldsymbol{F}$ determines the relative change of small volume elements of the state space, which is given by

$$\Delta \Omega(x; u; t + \delta t) = \Delta \Omega(x; u; t) + \operatorname{div} \boldsymbol{F}(x; u) \delta t. \tag{2.79}$$

If the vector field is everywhere free of sources (the divergence is everywhere zero), the dynamics is called conservative. Reversible motions, as the Hamiltonian dynamics (2.72) are necessarily conservative. For all members of this class the relation

$$\operatorname{div} \boldsymbol{F}(\boldsymbol{x}; u) = 0 \tag{2.80}$$

holds for all points of the state space. A typical example of a contracting dynamics is a dynamical system with an attractor. In this case a small element of the state space is always decreasing. For the attractor holds (Anishchenko, 1987, 1989, 1995; Anishchenko et al., 2002)

$$\langle \operatorname{div} \boldsymbol{F}(\boldsymbol{x}; u) \rangle < 0.$$

Dynamical systems with this property are called dissipative (Lichtenberg et al., 1983). Irreversible motions correspond to dissipative systems, their dynamics corresponds to a movement towards the attractor. The most simple example is an oscillator with linear friction

$$\dot{q} = v, \qquad \dot{p} = -\gamma_0 p - kq \tag{2.81}$$

where $p = mv$ and $\gamma_0 = \rho/m$ is the collision frequency connected with the friction constant ρ. Accordingly we get

$$\mathrm{div}\,\boldsymbol{F}(\boldsymbol{x};\boldsymbol{u}) = -\gamma_0 \leq 0. \tag{2.82}$$

Two other important classes of dynamical systems are the gradient systems

$$\frac{d\boldsymbol{x}}{dt} = -\mathrm{grad}V(\boldsymbol{x};\boldsymbol{u}) \tag{2.83}$$

and the canonical-dissipative systems (H — Hamiltonian)

$$\frac{d\boldsymbol{q}}{dt} = \frac{\partial H}{\partial \boldsymbol{p}}$$

$$\frac{d\boldsymbol{p}}{dt} = -\frac{\partial H}{\partial \boldsymbol{q}} + f(H)\frac{\partial H}{\partial \boldsymbol{p}}. \tag{2.84}$$

Here $f(H)$ is certain function characterizing the dissipative properties of the system. A simple case with

$$H = \frac{p^2}{2m} + \frac{m}{2}\omega_0^2 q^2; \qquad f(H) = -\gamma_0 = const.$$

leads us to a damped Hamiltonian dynamics of the type of Eq. (2.81). Another interesting case, closely related to Rayleigh's theory of self-sustained sound oscillations, is $f(H) = a - bH$. These dynamical systems form a limit cycle located on the ellipse

$$H = H_0 = \frac{a}{b}$$

with the exact stable periodic solutions

$$p(t) = \sqrt{\frac{2ma}{b}} \cos(\omega_0 t + \delta); \qquad q(t) = \sqrt{\frac{2a}{\omega_0^2}} \sin(\omega_0 t + \delta).$$

Sections 5.2 and 12.3 will be devoted to a detailed study of this quite interesting class of dissipative systems. Gradient systems may have only a very special type of attractors, the stationary points, corresponding to the minima of the potential. The potential $V(\boldsymbol{x};\boldsymbol{u})$ is a Lyapunov-function of the motion, since from Eq. (2.83) follows

$$\frac{dV}{dt} \leq 0. \tag{2.85}$$

Any motion is therefore accompagnied by a monotoneous decrease of V. This process continues until a stable point (a minimum of V) is reached. In the case of canonical-dissipative systems the dynamics consists of a conservative part and a dissipative part. For two-dimensional dynamical systems ($n = 2$) this corresponds to a representation of the vector field $\boldsymbol{F}$ by a divergence-free and a rotation-free component. For the (modified) Rayleigh system of stable oscillations discussed above, the attractor is a limit cycle and $(H - H_0)^2$ is a Lyapunov function.

In the following part we are concerned with the topics stability of motion and Lyapunov exponents. Since the states and the trajectories of dynamical systems are never exactly known and are subject to stochastic perturbations, the stability of motion with respect to small changes is of large interest. The first mathematical investigation of dynamical stability was given by the Russian mathematician Lyapunov in 1892 and nearly at the same time by the French theoretician Poincare. A remarkable contribution to the stability theory was given in the thirties of the 20th century by the russian school founded by Mandelstam, Andronov, Witt and Chaikin in Moscow and Gorki and by another school founded by Krylov, Bogoliubov and Mitropolsky in Kiev. In order to give the main ideas of stability theory let us consider two trajectories $\boldsymbol{x}(t; \boldsymbol{x}, t)$ and $\boldsymbol{x}(t; \boldsymbol{x} + \boldsymbol{q}, t)$, which at the initial time t differ by a small vector $\boldsymbol{q}$. In order to investigate the stability we calculate according to Lyapunov the time developement of the distance vector

$$\boldsymbol{q}(t) = \boldsymbol{x}(t; \boldsymbol{x} + \boldsymbol{q}, t) - \boldsymbol{x}(t; \boldsymbol{x}, t), \qquad \boldsymbol{q}(t_0) = \boldsymbol{q}_0. \tag{2.86}$$

The motion is called globally stable in the sense of Lyapunov if for all t and any $\epsilon > 0$ there exists an $\eta(\epsilon, t_0)$ such that from

$$|\boldsymbol{q}(t_0)| < \eta$$

follows

$$|\boldsymbol{q}(t)| < \eta$$

for any $t > t_0$. If such an η does not exist, the motion is called unstable in the sense of Lyapunov. In the special case that

$$\lim_{t \to \infty} |\boldsymbol{q}(t)| = 0, \tag{2.87}$$

the motion is called asymptotically stable. From the equations of motion (2.73) we get in linear approximation

$$\dot{q}_i(t) = F_i(\boldsymbol{x} + \boldsymbol{q}) - F_i(\boldsymbol{x}) = \sum_j J_{ij}(\boldsymbol{x}) q_j, \tag{2.88}$$

where the elements of the Jacobi matrix are given by

$$J_{ij}(\boldsymbol{x}) = \frac{\partial F_i(\boldsymbol{x})}{\partial x}. \tag{2.89}$$

In the special case that we are interested in the stability of a singular point

$$\boldsymbol{F}(\boldsymbol{x}_0) = 0$$

which may be considered as a special (constant) trajectory, the Jacobi matrix has constant matrix elements. In this case Eq. (2.88) is a system of linear homogeneous differential equations with constant cofficients. As well known, the stability is then determined by the eigenvalues of the Jacobi matrix. With the standard ansatz

$$q(t) \simeq \exp(\lambda t)$$

we get the characteristic equation

$$\det(J_{ij} - \lambda \delta_{ij}) = 0. \qquad (2.90)$$

The roots of this equation are in general complex. A necessary and sufficient condition for the asymptotic stability of the stationary point (singular point) is that all roots have negative real parts

$$\operatorname{Re} \lambda_i < 0$$

for all $i = 1, \ldots, n$. Already one eigenvalue with positive real part is sufficient to make the stationary point unstable, since any small deviation in the corresponding direction will be amplified. If the eigenvalue has an imaginary part the increase ($\operatorname{Re} \lambda > 0$) or decrease ($\operatorname{Re} \lambda < 0$) will be oscillatory. In the case of $\operatorname{Re} \lambda = 0$ no conclusion about stability is possible in the framework of the linear theory (Butenin et al., 1976; Arnold et al., 1983). Since for gradient systems the Jacobi matrix is symmetric

$$J_{ji}(x) = J_{ij}(x)$$

the corresponding eigenvalues are always real. Correspondingly the stationary points are always nodes or saddles. For Hamiltonian systems the trace of the Jacobian is zero

$$\operatorname{Tr} \mathbf{J}(\mathbf{x}) = 0.$$

The eigenvalues come in pairs located on the imaginary axis. This means that Hamiltonian systems do not possess stable singular points, all unstable points are of saddle type. If one is interested in the stability of periodic orbits

$$\mathbf{x}(t + T) = \mathbf{x}(t)$$

we may start again from (2.88). However the Jacobi matrix is now a periodic function of time with the period T. In this way the perturbations satisfy a system of linear diferential equations with periodic coefficients. As known, the stability is in that case depending on the so-called Floquet coefficients (Berge et al., 1984). A stable closed orbit is called a limit cycle. The real parts of the eigenvalues of the Jacobi matrix and the Floquet exponents are special cases of a more genaral concept, the so-called Lyapunov exponents. In order to explain this we go back to Eq. (2.88), which symbollically reads

$$\frac{d\mathbf{q}}{dt} = \mathbf{J}(x(t))\mathbf{q}. \qquad (2.91)$$

Here $x(t)$ is the trajectory which we like to investigate. This trajectory is the solution of

$$\frac{d\boldsymbol{x}(t)}{dt} = \boldsymbol{F}(\boldsymbol{x}; \boldsymbol{u}). \tag{2.92}$$

In general the simultaneous solution of Eqs. (2.91) and (2.92) will be impossible, one has to refer to numerical calculations. The norm of the solution of Eq. (2.91) will in general behave exponentially in time.

$$\boldsymbol{q} \simeq \exp(\lambda t).$$

Therefore we define the characteristic quantities

$$\lambda = \lim_{t \to \infty} \cdot \frac{1}{t} \log \frac{|\boldsymbol{q}(t)|}{|\boldsymbol{q}(0)|} \tag{2.93}$$

as the *Lyapunov exponents*. They characterize the long time behavior of the deviations in linear approximation. In dependence on the initial conditions for $\boldsymbol{q}(t)$ the exponents λ may assume a number of discrete values λ_i ($i = 1, \ldots, n$), which are called the spectrum of the Lyapunov exponents. The method described here is based on the original investigations of Lyapunov and on a statement proved in 1968 by Oseledec. The sum over all exponents is connected with the divergence of the vector field averaged along the trajectories

$$\langle \mathrm{div} \boldsymbol{F} \rangle = \sum_i \lambda_i. \tag{2.94}$$

Therefore we have in the case of conservative systems

$$\sum_i \lambda_i = 0$$

and for dissipative systems

$$\sum_i \lambda_i < 0.$$

The fundamental work of Lorenz (1963), Ruelle and Takens (1971) has shown that for dimensions of the state space $n \geq 3$ there exist systems having a positive largest Lyapunov exponent. The corresponding attractors were called strange attractors (Eckmann & Ruelle, 1985; Anishchenko, 1987, 1990; Ott, 1993; Anishchenko, 1987, 1989, 1995; Anishchenko et al., 2002).

In the following we shall assume that the Lyapunov exponents are ordered with respect to their index $1, \ldots, n$ in the form of a non-decreasing series. In symbolic notation the following possibilities exist for the signs of the Lyapunov exponents:

(i) Stable singular points:

$$(-, -, -, \ldots, -).$$

(ii) Stable limit cycles:
$$(0, -, -, \ldots, -).$$

(iii) Stable m-torus:
$$(0, 0, \ldots, 0, -, - \ldots, -), \qquad \text{(m-zeros)}.$$

(iv) Chaotic attractor:
$$(+, 0, -, -, \ldots, -).$$

(v) Chaos of higher order
$$(+, \ldots, +, 0, -, \ldots, -).$$

If at least one (the largest) Lyapunov exponent is positive, we will say that the motion is chaotic. The Hausdorff dimension D_H of chaotic attractors is in general a non-integer number (Schuster, 1995; Ott, 1993). If at least one exponent is positive we know that for the dimension $D_H > 1$ holds. If j exponents are positive, i.e. the the ordered exponents satisfy the inequality

$$\lambda_1 > \cdots > \lambda_j > 0 > \lambda_{j+1} > \cdots > \lambda_n$$

then the dimension of the attractor will be, in general, between j and $j+1$ i.e. the inequality $j < D_H < j+1$ holds. This is due to the following rule: If the sum of the j largest Lyapunov exponents is positive, then a small volume of dimension j around a phase point of the (chaotic) trajectory is expanding. The quantity

$$D_L = j + \frac{\sum_{i=1}^{j} \lambda_i}{|\lambda_{j+1}|} \qquad (2.95)$$

with the j defined by the inequality given above is called the Lyapunov dimension of the attractor (Eckmann and Ruelle, 1985). For stable singular points holds $D_L = 0$, for stable limit cycles holds $D_L = 1$ and for stable m-tori we find $D_L = m$. For chaotic attractors of systems defined by n d.e. the dimension D is a non-integer number with $2 < D < n$. The sum of the positive Lyapunov exponents

$$H_P = \sum_{i}^{+} \lambda_i \qquad (2.96)$$

we call the *Pesin entropy* (Steuer et al., 2001). In many cases the Pesin entropy is equal to the *Kolmogorov entropy*. The Kolmogorov entropy was originally defined in terms of information theoretical methods and is closely connected with the problem of the predictability of motions (Schuster, 1995; Kantz & Schreiber, 1997).

2.5 Stochastic models on the mesoscopic level

The description of processes on the deterministic level is in some sense incomplete since under real condition stochastic influences are always present (Stratonovich, 1961, 1963, 1967; van Kampen, 1992; Ebeling & Feistel, 1982). We will introduce here only several principal aspects of stochastic descriptions. Details and examples of stochastic descriptions will be found in Chapters 7–12. There are several reasons for the importance of stochastic models:

- Mesoscopic or macroscopic variables are always representative for a large number of microscopic degrees of freedom, which are subject to thermal fluctuations.
- All variables depending on the number of particles are necessarily discrete. This gives rise to a special kind of noise, the so-called shot noise.
- The intrinsic quantum character of the microscopic dynamics leads to stochastic effects.
- A mesoscopic or macroscopic system is as a rule imbedded in a very complex surrounding leading to stochastic external forces
- Any part of our universe is filled with thermal photons forming the sea of background radiation (about 500 photons with a temperature of 2.7 K in 1 cubic centimeter). These photons interact stochastically with any system under invesitigation.

Due to these stochastic influences the future state of a dynamical system is in general not uniquely defined by its previous state. In other words the dynamic map defined by (2.58) is non-unique. A given initial point $x(0)$ may be the source of several different trajectories. The choice between the different possible trajectories is a random event. In this way the term trajectory looses its precise meaning and should be supplemented in terms of probability theory. Stochastic models will be described in detail in Chapters 7–12 of this book. Here we will introduce only the basic ideas. We describe in the framework of stochastic descriptions the state of the system at time t by a probability density $P(\boldsymbol{x}, t; \boldsymbol{u})$. Per definition $P(\boldsymbol{x}, t; \boldsymbol{u})d\boldsymbol{x}$ is the probability of finding the trajectory at time t in the interval $(\boldsymbol{x}, \boldsymbol{x} + d\boldsymbol{x})$. Instead of the deterministic equation for the state we get now a differential equation for the probability density $P(\boldsymbol{x}, t; \boldsymbol{u})$. In order to find this equation we introduce the idea of stochastic forces due to Paul Langevin (1908). Instead of Eq. (2.73) we write

$$\dot{x}_i(t) = F_i(\boldsymbol{x}, \boldsymbol{u}) + \xi_i(t) \tag{2.97}$$

where $\xi_i(t)$ is the i-th component of a stochastic force with the mean value zero

$$\langle \xi_i(t) \rangle = 0\,.$$

The latter condition guarantees, that the averaged trajectories satisfy

$$\langle \dot{x}_i(t) \rangle = \langle F_i(\boldsymbol{x}(t); \boldsymbol{u}) \rangle \simeq F_i(\langle \boldsymbol{x}(t) \rangle; \boldsymbol{u})\,. \tag{2.98}$$

In order to derive the desired equation for the probabilities we consider an ensemble of N points in the state space, each corresponding to one special system, distributed at $t = 0$ corresponding to the initial probability $P(x, t = 0; u)$. The number of points in a given volume V at time t will be

$$N_V(t) = \int_V dx P(x, t; u)$$

and the time derivative will be

$$\frac{d}{dt} N_V(t) = -N \int_S dO \cdot G(x, t; u)$$

where G is the probability flow vector. Due to the Gauss theorem we get

$$\partial_t P(x, t; u) = -\operatorname{div} G(x, t; u). \tag{2.99}$$

By multiplying with x and integrating we find

$$\frac{d}{dt} \langle x \rangle = \int_S dx \cdot G.$$

In the special case that there are no stochastic forces the flow is proportional to the deterministic field i.e.

$$G_i(x, t, ; u) = F_i(x, t; u) P(x, t; u).$$

Including now the influence of the stochastic forces we assume here *ad hoc* an additional diffusive contribution to the probability flow which is directed downwards the gradient of the probability

$$G_i(x, t; u) = F_i(x, t; u) P(x, t; u) - D \frac{\partial}{\partial x_i} P(x, t; u). \tag{2.100}$$

This is the simplest "Ansatz" which is consistent with Eq. (2.73) for the mean values. The connection of the "diffusion coefficient" D with the properties of the stochastic force will be discussed later. Introducing Eq. (100) into Eq. (2.99) we get a partial differential equation which is called the *Smoluchowski–Fokker–Planck equation*. In this way we have found a closed equation for the probabilities. The Smoluchowski–Fokker–Planck equation is consistent with the deterministic equation and will, at least approximately, take into account stochastic influences. More rigorous statistical-mechanical derivations of diffusion-type equations based on the Liouville equation and the Zwanzig projection technique were first given in (Falkenhagen & Ebeling, 1965; Ebeling, 1965).

On the basis of a given probability distribution $P(x, t; u)$ we may define mean values of any function $f(x)$ by

$$\langle g(x) \rangle = \int dx g(x) P(x, t; u).$$

Further we may define the standard statistical expressions as e.g. the dispersion and in particular the mean uncertainty (entropy) which is defined as

$$H = -\langle \log P(\boldsymbol{x}, t; u) \rangle. \tag{2.101}$$

The approach based on the Fokker–Planck equation is the simplest but not the only one. There exists a different approach which originally goes back to Markov who studies first discrete time stochastic processes, nowadays called Markov chains. Markovs model was generalized to continuous time processes by Chapman and Kolmogorov. The basic idea of Markov, Chapman and Kolmogorov are the transition probabilities. In order to explain the idea we study in the simplest case a discrete state space consisting of λ states: $x_n = 1, 2, \ldots, \lambda$ and a stochastic hopping between the states at the discrete times

$$t = t_1, t_2, \ldots, t_i, \ldots, t_n. \tag{2.102}$$

Full information we would obtain from the knowledge of the joint probability distribution density

$$P_N(\boldsymbol{x}_1, t_1; \boldsymbol{x}_2, t_2; \ldots; \boldsymbol{x}_n, t_n) = \langle \delta(\boldsymbol{x}_1 - \boldsymbol{x}_1(t)) \cdots \delta(\boldsymbol{x}_n - \boldsymbol{x}_n(t)) \rangle \tag{2.103}$$

there $\boldsymbol{x}_1, \boldsymbol{x}_2, \ldots, \boldsymbol{x}_n$ are n coordinates describing the system at subsequent times $t_1 < t_2 < \cdots < t_n$.

$$P_N(\boldsymbol{x}_1, t_1; \boldsymbol{x}_2, t_2; \ldots; \boldsymbol{x}_n, t_n) d\boldsymbol{x}_1 d\boldsymbol{x}_2 \cdots d\boldsymbol{x}_n \tag{2.104}$$

gives the probability to find the system at time t_1 in the infinitesimal neighbourhood of $\boldsymbol{x}_1$, at time t_2 in the infinitesimal neighbourhood of $\boldsymbol{x}_2$ etc. In Eq. (2.103) the brackets on the r.h.s. indicates averaging over an ensemble. $\boldsymbol{x}_1(t_1), \boldsymbol{x}_2(t_2), \ldots$ stands for the stochastic realizations of the number belonging to the ensemble at different times. The delta-functions map that part of realizations with values $\boldsymbol{x}_i$ to the coordinate $\boldsymbol{x}_i(t_i)$. The joint probability (2.103) defines an infinite hierarchy of probability densities $N = 1, 2, \ldots$. Full knowledge means that for example arbitrary time dependent moments of the stochastic process may be calculated.

$$\langle A \rangle = \int A P_k(x_1, t_1; x_2, t_2; \ldots; x_k, t_k) dx_1 \ldots dx_k.$$

Note, for simplicity we use from here a scalar notation. Thereby, how many times are necessary to include, i.e. the value of k depends on the variable standing inside the brackets. For the most physical applications the values $k \leq 2$ are realized. To give a precise definition of a Markov-Process the conditional probability density is introduced. Following Risken (1984) it reads

$$P(x_n, t | x_{n-1}, t_{n-1} \cdots x_1, t_1) = \langle \delta(x_n - x_n(t)) | x_{n-1} = x_{n-1}(t), \ldots, x_1 = x_1(t) \rangle. \tag{2.105}$$

In difference to (2.103) it gives the probability to be at time t in the neighbourhood of x, but under the condition that at former times $t_1 < t_2, \ldots; t_{n-1} < t_n$ exactly $x_1, x_2, \ldots, x_n$ is realized. Then we find the following identity

$$P_n(x_1, t_1; \ldots; x_n, t_n) = P(x_n, t_n | x_{n-1}, t_{n-1}; \ldots; x_1, t_1) P_{n-1}(x_1, t_1; \ldots, x_{n-1}, t_{n-1}). \tag{2.106}$$

On the basis of the relation (2.105) a Markovian process can be defined. If the conditional probability depends on the value x_{n-1} of the nearest time in the past only

$$P(x_n, t_n | x_{n-1}, t_{n-1}; \ldots x_1, t_1) = P(x_n, t_n | x_{n-1}, t_{n-1}) \tag{2.107}$$

but does not depend on the states at former times, we call the process Markovian (of 1st order). In that case the joint-probability density can be traced back to a simple product of conditional probabilities at different times and to an initial probability density at time t_1

$$P_N(x_1, t_1; \ldots x_N, t_N) = P(x_N, t_N | x_{N-1}, t_{N-1}) \ldots P(x_2, t_2 | x_1, t_1) P_1(x_1, t_1). \tag{2.108}$$

Markovian (first order), therefore, means the described process possesses the memory of one single transition. The step from x_{N-1} at time t_{N-1} to x_N at t_N is independent of the path by which the state at time t_{N-1} was approached. In other words, the transition probability is not affected by any knowledge about the states at earlier times. $P(x_2, t_2 | x_1, t_1)$ is often called the transition probability and we also will make use of this notation. Let us still underline that the concept of the Markov process is rather a property of the model we apply for the description than a property of the physical system under consideration (Van Kampen, 1981). If a certain physically given process cannot be described in a given state space by a Markov relation, often it may be possible, by introducing additional components, to embed it into a Markov model. This way a non-Markovian model can be converted, by enlargement of the number of variables, into another model with Markovian properties. We note already here, that the basic equation of statistical physics, the Liouville equation which will be introduced in the next Chapter, is of markovian character.

In order to derive a kinetic equation for the time evolution of the probability density we consider the following identy

$$P_3(x_3, t_3; x_1, t_1) = \int P_3(x_3, t_3; x_2, t_2; x_1, t_1) dx_2. \tag{2.109}$$

Assuming Markovian character of the considered process we obtain

$$P(x_3, t_3 | x_1, t_1) P_1(x_1, t_1) = \int dx_2 P(x_3, t_3 | x_2, t_2) P(x_2, t_2 | x_1, t_1) P_1(x_1, t_1). \tag{2.110}$$

For arbitrary $P(x_1, t_1)$, therefore, the Chapman–Kolmogorov equation for the transition probabiltities follows

$$P(x_3, t_3 | x_1, t_1) = \int P(x_3, t_3 | x_2, t_2) P(x_2, t_2 | x_1, t_1) dx_2 \,. \tag{2.111}$$

Similar equations hold true for the P_1 and P_2

$$P_2(x_3, t_3; x_1, t_1) = \int dx_2 P(x_3, t_3 | x_2, t_2) P_2(x_2, t_2; x_2, t_2) \,. \tag{2.112}$$

Additional integration over x_1 leads us to back to

$$P(x, t) = \int dx_2 P(x_3, t_3 | x_2, t_2) P_1(x_2, t_2) \,. \tag{2.113}$$

For Markovian processes Eqs. (2.111)–(2.113) are the starting points to find dynamical laws for the time evolution of the special probability densities. Now a second simplification will be made. Most of the stochastic problems are uniform in time, i.e. they are invariant under a time shift $t \to t + \Delta t$. Events at subsequent times do not depend on the absolute time but in most cases on the time difference. Such stochastic processes are called stationary ones. In conclusion the P_1 — density will be independent of time. Some situations when this is not true will be considered in Chapter 11. The two time-joint density as well as the transition probability will be a function of the time difference only

$$P_2(x_2, t_2; x_1, t_1) = P_2(x_2, t_2 - t_1; x_1) P(x_2, t_2 | x_1, t_1) = P(x_2, t_2 - t_1 | x_1) \,. \tag{2.114}$$

Since both functions are equivalent in their dynamical behviour we omit furtheron the subscript if posible and the initial condition to avoid many functional arguments. In the limit that the time difference between the two subsequent events is small $\delta t = t_2 - t_1 \to 0$ the transition probability in the kernel of the integrals (2.111) and (2.112) can be expanded in a series of small δt

$$P(x_3, \delta t | x_2) = \left(1 - \delta t \int dx W(x | x_2) \delta(x_3 - x_2)\right) + W(x_3 | x_2) \delta t + o(\delta t^2) \,. \tag{2.115}$$

$W(x'|x)$ occurs here as the probability for a transition $x \to x'$ per unit time which existence we assume. The first item in front of the delta-function is the probability for remaining in the state x_2 during δt. The second item stands for the transition from x_2 to x_3. Inserting expansion Eq. (2.115) into the Chapman–Kolmogorov equation brings the dynamical equation for the probability density.

$$\frac{\partial P(x, t)}{\partial t} = \int dx' [W(x|x') P(x', t) - W(x'|x) P(x, t)] \,. \tag{2.116}$$

This equation is called *Pauli equation* or *master equation* since it plays a fundamental role in the theory of stochastic processes. The integration is performed over all possible states x' which are attainable from the state x by a single jump. It is a linear equation with respect to P and determines uniquely the evolution of the probability density. The r.h.s. consists of two parts, the first stands for the gain of probability due to transitions $x' \to x$ whereas the second describes the loss due to reversed events. Equation (2.116) needs still further explanation by the determination of the transition probabilities per unit time corresponding to the special physical situation. It will be the subject of Chapters 7 and 8. The transition probability is in many cases a quickly decreasing function of the jump $\Delta x = x - x'$. By using a Taylor expansion with respect to Δx and moments of the transition probability one can transform Eq. (2.116) to an infinite Taylor series. This is the so-called *Kramers–Moyal expansion*.

$$\frac{\partial P(x,t)}{\partial t} = \sum_1^\infty \frac{(-1)^m}{m!} \sum \frac{\partial^m M(x) P(x,t)}{\partial x_{i_1} \cdots \partial x_{i_m}} \qquad (2.117)$$

with

$$M_{i_1 \ldots i_m}(x) = \int d\Delta x \, d\Delta x_{i_1} \, d\Delta x_{i_m} W(x + \Delta x | x) \qquad (2.118)$$

being the moments of the transition probabilities per unit time. According to the *Pawula theorem* there are just two possibilities considering homogeneous Markov processes:

(1) All coefficients of the Kramers–Moyal expansion are different from zero.
(2) Only two coefficients in the expansion are different from zero.

In the first case we have to deal with the full master equation. In the latter one the Markovian process is called difusive, which is of special interest to us, and leads to the following second-order partial differential equation:

$$\frac{\partial}{\partial t} P(x,t) = \frac{\partial}{\partial x_i}[M_i(x) P] + \sum_j \frac{\partial^2}{\partial x_i \partial x_j}[M_{ij}(x) P]. \qquad (2.119)$$

This is a generalization of the Smoluchowski–Fokker–Planck equation given above. For its solution we need of course initial conditions $P(x, t = 0)$ and boundary conditions which take into account the underlying physics. Writing Eq. (2.119) again in the form of a continuity equation (2.99) we find for the vector of the probability flow the components

$$G_i(x,t) = M_i(x) P(x,t) + \sum_j \frac{\partial}{\partial x_i}[M_{ij}(x) P(x,t)]. \qquad (2.120)$$

The strong mathematical theory of Eq. (2.119) was developed by Kolmogorov and Feller; therefore one speaks often about the *Kolmogorov–Feller equation*. In physics however, this equation was used much earlier by Einstein, Smoluchowski, Fokker and Planck for the description of diffusion processes and Brownian motion respectively (Chandrasekhar, 1943). Due to this original physical relation the coefficients $M_i(x)$ and $M_{ij}(x)$ are often called drift coefficients and diffusion coefficients respectively.

Let us still mention that an alternative mathematical foundation of the theory of stochastic processes may be based on the theory of stochastic differential equations.

Another large class of Markovian processes contains systems with a discrete state space. This concerns the atomic processes or extensive thermodynamic variables, like e.g. particle numbers in chemical reacting systems. The Pauli equation than transforms to

$$\frac{\partial}{\partial t} P(\boldsymbol{N}, t) = \sum \left[W(\boldsymbol{N}|\boldsymbol{N}') P(\boldsymbol{N}', t) - W(\boldsymbol{N}'|\boldsymbol{N}) P(\boldsymbol{N}, t) \right] \qquad (2.121)$$

where $\boldsymbol{N}$ is the vector of possible discrete events (particle numbers). As example we refer to the large class of birth and death processes where $\boldsymbol{N}$ are natural numbers which change during one transition by $\Delta N = \pm 1$. Let us still summarize the new tools in comparison with the deterministic models. Obviously the stochastic approach contains more information about the considered systems due to the inclusion of fluctuations into the description. Besides moments of the macroscopic variables it enables us to determine correlation functions and spectra which will give knowledge between the functional dependence of the fluctuational behaviour at different times.

Summarizing we may state: Some physical phenomena can be explained only by taking into account fluctuations. The stochastic approach on a mesoscopic level delivers often more elegant solutions than the microscopic statistical approach. Inclusion of fluctuations of the macroscopic variables does not necessarily enlarge the number of relevant variables but changes only their character by transforming them into stochastic variables. The main difference compared with the deterministic models is the permeability of separatrices. This statement concerns especially nonchaotic dynamics, for instance if dealing with one or two order parameter. With certain probability stochastic realizations reach (or cross) unstable points, saddle points, and separatrices what is impossible in the deterministic description. Stochastic effects make possible to escape regions of attraction around stable manifolds. Physical situations which make use of that circumstamce are e.g. nucleation or chemical reactions where energetically unfavourable states has to be overwhelmed. In Chapters 7–12 we will present a more detailled analysis of stochastic phenomena and many examples of processes induced by noise.

Chapter 3

Reversibility and Irreversibility, Liouville and Markov Equations

3.1 Boltzmann's kinetic theory

As we stated in the introduction, Boltzmann is the father of statistical physics. He formulated the basic tasks of this scientific discipline: How to derive the macroscopic properties of matter and especially thermodynamic potentials from atomistics and the laws of mechanics. He introduced the new natural constant k_B which connects the basic macroscopic quantity, the entropy S, with the probabilities of microscopic states. Boltzmann's approach was in contradiction to most contemporary views. His arguments and the controversy with Loschmidt, Zermelo and Poincare played a great role for the formation of modern statistical physics.

In classical thermodynamics the entropy difference between two states was defined by Clausius in terms of the exchanged heat $d'Q$ and the temperature T:

$$dS = \frac{d'Q}{T} \tag{3.1}$$

$$\delta S = S_2 - S_1 = \int_1^2 \frac{d'Q}{T}. \tag{3.2}$$

Here the transition $1 \to 2$ should be carried out on a reversible path and $d'Q$ is the heat exchange along this path. Boltzmann first formulated the basic link between Clausius' entropy and probability. In his first work on kinetic theory he introduced the concept of the phase space X of a macroscopic system consisting of N molecules, each of them described by a set of generalized coordinates and momenta:

$$[\bm{q}, \bm{p}] = [q_1, q_2, \ldots, q_f, p_1, p_2, \ldots, p_f]. \tag{3.3}$$

The phase space X is the $2f$-dimensional space of the f coordinates and f momenta which describe the state of one molecule. Here f has in the simplest case of Cartesian coordinates of the molecule the value 3, including further internal degrees of freedom it may be of the order 5–6. Often this space is denoted as the γ-space of statistical mechanics. The state of one molecule in this space corresponds to a point and the state of the ensemble of all molecules of the body, which is under consideration, is a cloud of points (see Fig. 3.1).

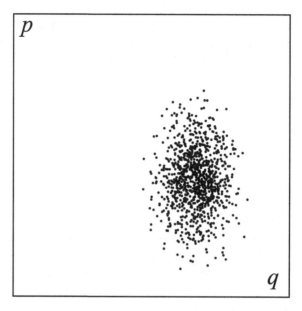

Fig. 3.1 A cloud of points in the phase space of molecules corresponding to the state of a macroscopic molecular system.

Let us define, as did already Boltzmann, the function $f(\boldsymbol{q},\boldsymbol{p},t)$ as the density of the points in the γ-space. Boltzmann concentrated on dilute gases and in this case the molecules and their corresponding points are independent. Due to this we may interpret the density as a probability to find at time t, some molecule represented by a point at $\boldsymbol{q}$ and $\boldsymbol{p}$; more precise we have to consider not phase points but volume elements $d\boldsymbol{q}d\boldsymbol{p}$ in the phase space. We consider $f(\boldsymbol{q},\boldsymbol{p},t)$ as dimensionless and take into account that the volume element in the phase space $d\boldsymbol{q}d\boldsymbol{p}$ has the dimension of an action to the power of the space dimension. Therefore we have chosen Plancks constant h^3 as a normalization factor. Note, that we we have chosen as the most natural unit of action Planck's constant, which was of course not yet known to Boltzmann. With our choice $f(\boldsymbol{q},\boldsymbol{p},t)$ is dimensionless and

$$f(\boldsymbol{q},\boldsymbol{p},t)\frac{d\boldsymbol{q}d\boldsymbol{p}}{h^3} \tag{3.4}$$

can be interpreted as the density (probability) of finding at time t a point in the interval $d\boldsymbol{q}d\boldsymbol{p}$. The normalization is assumed to be

$$N = \int \frac{d\boldsymbol{q}d\boldsymbol{p}}{h^3} f(\boldsymbol{q},\boldsymbol{p},t) \tag{3.5}$$

where N is the total number of molecules in the gas. In 1866, Boltzmann was able to derive an expression for the distribution function for the special case of thermal equilibrium $f^{eq}(\boldsymbol{q},\boldsymbol{p})$. Instead of repeating the derivation let us simply state the central ideas. We consider two particles with the states $\boldsymbol{q},\boldsymbol{p}$ and $\boldsymbol{q}_1,\boldsymbol{p}_1$ before a collision. Assuming that $\boldsymbol{q}',\boldsymbol{p}'$ and $\boldsymbol{q}_1',\boldsymbol{p}_1'$ are the corresponding states after the collision, we expect that the products of probabilities before and after the collision are equal

$$f^{eq}(\boldsymbol{q},\boldsymbol{p})f^{eq}(\boldsymbol{q}_1,\boldsymbol{p}_1) = f^{eq}(\boldsymbol{q}',\boldsymbol{p}')f^{eq}(\boldsymbol{q}_1',\boldsymbol{p}_1') \tag{3.6}$$

where both sets of states are connected by the equations of motion for the collisions. The products in Eq. (3.6) express the independence of the molecules far before and after the collision. Equation (3.6) may be rewritten as

$$\ln f^{eq}(\boldsymbol{q},\boldsymbol{p}) + \ln f^{eq}(\boldsymbol{q}_1,\boldsymbol{p}_1') = \ln f^{eq}(\boldsymbol{q}',\boldsymbol{p}') + \ln f^{eq}(q_1',p_1'). \tag{3.7}$$

Equation (3.7) suggests that the function $\log f^{eq}(\boldsymbol{q},\boldsymbol{p})$ depends only on invariants of motion. Assuming that the relevant invariant is the Hamiltonian and that the dependence is linear, we arrive at

$$\ln f^{eq}(\boldsymbol{q},\boldsymbol{p}) = -\beta H(\boldsymbol{q},\boldsymbol{p}) \tag{3.8}$$

where β is a constant. For the identification of β we may use known relations from the thermodynamics of ideal gases. From Eq. (3.8) we obtain for the mean kinetic energy

$$\langle E_{kin}\rangle = \left\langle \frac{p^2}{2m}\right\rangle = \frac{3}{2\beta}. \tag{3.9}$$

Comparing this with the energy of ideal gases the constant β can be identified with the reciprocal temperature multiplied with a constant.

$$\beta = \frac{1}{k_B T},$$

$$k_B = 1.381 \cdot 10^{-23}\,J/K. \tag{3.10}$$

Boltzmann's constant k_B is a universal constant which characterizes the connection between microphysics and macrophysics. The fact that in statistical physics a new universal constant appears, makes clear that statistical physics is indeed a new physics in comparison with microphysics. This situation corresponds to the philosophical idea that the whole is more than the sum of its parts. Boltzmann's constant stands for emergent properties of macrosystems.

This way we get finally for point particles in a potential field the famous *Maxwell–Boltzmann distribution*

$$f^{eq}(\boldsymbol{q},\boldsymbol{p}) = const \times \exp\left(-\frac{p^2/2m + U(q)}{k_B T}\right). \tag{3.11}$$

By using the probabilistic concepts discussed above, Boltzmann introduced also a new function of the probabilities which possesses very interesting properties:

$$H_B = -\int \frac{d\boldsymbol{q}d\boldsymbol{p}}{h^3} f(\boldsymbol{q},\boldsymbol{p},t) \cdot \ln f(\boldsymbol{q},\boldsymbol{p},t). \qquad (3.12)$$

In fact, he used the opposite sign for H_B in his original definition; we introduced the "minus" to be consistent with the standard notation in mathematics and informatics. Indeed a similar function (with the minus sign) was suggested already in the eighteenth century by the mathematician DeMoivre for the characterization of the mean uncertainty in the outcome of games. In modern times Claude Shannon founded the information theory on an H-function. In any case, Boltzmanns H_B measures the mean uncertainty of the location of the molecules in the phase space. Later we shall come back to this point several times. Following the basic postulate of Boltzmann the H_B-function is connected with the thermodynamical entropy by the relation

$$S = k_B H_B. \qquad (3.13)$$

At least in equilibrium this assumption proves to be correct since introducing Eq. (3.9) into Eqs. (3.12)–(3.13) leads to

$$S = -k_B N[\ln(n\Lambda^3) + const]. \qquad (3.14)$$

Here Λ is the thermal De Broglie wave length defined by

$$\Lambda = \frac{h}{\sqrt{2\pi m k_B T}}.$$

The entropy obtained this way, corresponds up to a constant to the standard expression from equilibrium thermodynamics. In the phenomenological thermodynamics the expression for the entropy contains the so-called Sacur-Tetrode constant, which is estimated from experiments. In the statistical theory the new constant is given explicitly in a natural way by the normalization procedure; remarkably this constant depends on h. Originally, Boltzmann's hypothesis was essentially based on a theorem on the time evolution of H_B. Indeed, he succeeded in deriving first an equation for $f(\boldsymbol{q},\boldsymbol{p},t)$ which has the form of an integro-differential equation. For simplicity we restrict ourselves from now to the case that the distribution does not depend on $\boldsymbol{q}$, which is true for spatially homogeneous systems. Then we get according to Boltzmann

$$\frac{\partial f(\boldsymbol{p},t)}{\partial t} = I[f(\boldsymbol{p},t)] \qquad (3.15)$$

with a certain nonlinear functional I of the distribution function:

$$I[f(\boldsymbol{p},t)] = \int \sigma \left[f(\boldsymbol{p}')f(\boldsymbol{p}'_1) - f(\boldsymbol{p})f(\boldsymbol{p}_1)\right] d\boldsymbol{p}' d\boldsymbol{p}'_1 d\boldsymbol{p}_1 \qquad (3.16)$$

where σ is the so-called cross section. We see immediately that the equilibrium distribution (3.9) is a stationary solution of Eq. (3.15). The concrete form of this functional is not essential for our consideration, for smaller deviations from equilibrium we may approximate it in the form

$$I[f(\mathbf{p},t)] \simeq -\nu[f(\mathbf{p},t) - f^{eq}(\mathbf{p},t)]. \tag{3.17}$$

The physics behind this expression is the plausible assumption that the effect of collision occuring with the frequency ν is proportional to the deviation from equilibrium. By using this so-called relaxation time approximation we immediately find a solution of the form

$$f(p,t) = f^{eq}(p) + \exp(-\nu t)[f(p,0) - f^{eq}(p)]. \tag{3.18}$$

This result means physically that there is an exponential relaxation of all deviations from the Maxwell–Boltzmann distribution. By introducing (3.18) into the formula for the entropy

$$S = -k_B \int (d\mathbf{q}d\mathbf{p}/h^3) f(\mathbf{q},\mathbf{p},t) \ln f(\mathbf{q},\mathbf{p},t), \tag{3.19}$$

we may show that

$$\delta S(t) = S_{eq} - S(t) \tag{3.20}$$

is non-negative and is a monotonously decreasing function, giving

$$\delta S(t) \geq 0 \tag{3.21}$$

and

$$\frac{d}{dt}\delta S(t) \leq 0. \tag{3.22}$$

As shown by Boltzmann, one may prove this directly from Eqs. (3.15)–(3.16) by using several tricks. Therefore, $\delta S(t)$ is a Lyapunov function; it follows that $S(t)$ always increases. This is in full agreement with the second law of thermodynamics. It was exactly this point, which Boltzmann considered as the main success of his theory, that was later the target of the heavy attacks from other experts as Poincare, Loschmidt and Zermelo. Before we explain this point in more detail, let us first discuss one essential generalization of Boltzmann's approach, which is due to Gibbs. The most essential restriction of Boltzmann's theory was the assumption of weak interactions between the particles. This assumption could be removed by the great American theoretician Josiah Willard Gibbs (1839–1903) who published in 1902 a fundamental book "*Elementary Principles in Statistical Mechanics*". Gibbs considered a more general class of macroscopic systems. He introduced a high-dimensional phase space, the so-called Γ-space, which is given by all the $3N$ (or fN respectively) coordinates $q_1, \ldots, q_{3N}$ and the $3N$ momenta $p_1, \ldots, p_{3N}$ of the macroscopic system. Gibbs' generalization of the entropy to interacting systems of point-like molecules reads:

$$S_G = -k_B \int (d\boldsymbol{q}d\boldsymbol{p}/h^{3N}) \rho(\boldsymbol{q},\boldsymbol{p}) \cdot \ln \rho(\boldsymbol{q},\boldsymbol{p}) \qquad (3.23)$$

where $\rho(\boldsymbol{q},\boldsymbol{p})$ is the normalized probability density in the $6N$-dimensional phase space. The Gibbs expression includes all interaction effects which in general lead to a decrease of the value of the entropy in comparison to the ideal gas. For the special case of equilibrium systems with fixed energy E the probability density is assumed to be constant in a shell around the surface

$$H(q_1, \ldots, q_{3N}, p_1, \ldots, p_{3N}) \simeq E. \qquad (3.24)$$

Gibbs calls this the microcanonical distribution or the microcanonical ensemble. In principle Gibbs assumption goes back to Boltzmann's hypothesis, that the trajectory fills the whole energy shell in a uniform way. We will come back to this idea of ergodicity in the next section.

Boltzmann assumed that in the case of equal probabilities of the microstates the entropy of the corresponding macrostate is the logarithm of the thermodynamic probability

$$S_{BP} = k_B \ln W. \qquad (3.25)$$

where W is defined as the total number of equally probable microstates corresponding to the given macrostate. Strictly speaking, the first explicit writing of formula Eq. (3.25) is due to Planck, therefore we will speak sometimes about the Boltzmann–Planck entropy. As we see, the Gibbs' expression for the entropy is for a constant probability density (microcanonical distribution) in full agreement with the Boltzmann–Planck formula

$$S_G(E,V) = k_B \ln \Omega^*(E,V): \qquad (3.26)$$

$$\Omega^*(E,V) = W = \Omega(E,V)/h^{3N}.$$

Here $\Omega(E)$ is the volume of the energy shell. We assume as earlier, that Planck's constant defines the appropriate unit cell, in order to make the argument of the log dimensionless. As above W is the number of equally probable microstates in the energy shell. All these arguments will be explained in much more detail in Chapter 4. In principle, with expressions for the entropy either in the Boltzmann–Planck or in the Gibbs form, we reached already at our aim, to derive thermodynamics from microphysics. However, the solution is not as simple, there remain open problems.

Boltzmann's first paper on the connection between mechanics and thermodynamics appeared in 1871; it had the remarkable title "*Analytical Proof of the Second law . . .*". In a later (main) paper which appeared in 1872, he worked out his arguments in more detail and presented further results. However in 1876 Boltzmann's teacher and colleague Loschmidt published a serious objection against Boltzmann's theory, which became known as the Loschmidt's paradox. Loschmidt considered a gas in a box with completely plane elastic surfaces. During the time evolution of this system Boltzmann's H-function at subsequent times should form a nondecreasing time series

$$H_B(t_1) \leq H_B(t_2) \leq \cdots \leq H_B(t_n). \quad (3.27)$$

Loschmidt then proposed the following "Gedankenexperiment". Consider at certain time t_n an inversion of all the velocities of the molecules. Corresponding to the reversibility of the laws of mechanics we would observe a backward trajectory leading to a decreasing H-function.

$$H_B(t_n) \geq H_B(t_{n-1}) \geq \cdots \geq H_B(t_1). \quad (3.28)$$

However this is in clear contradiction to Boltzmann's H-Theorem and to the second law. The next critical objection against Boltzmann's theory was based on the theorem of Poincare about the *"quasi-periodicity of mechanical systems"* published in 1890 in the famous paper *"Sur le probleme de trois corps les equations de la dynamique"*. Poincare was able to prove under certain conditions, that a mechanical system will come back to its initial state in a finite time, the so-called recurrence time. Zermelo, a student of Planck, showed in 1896 in a paper in the *"Annalen der Physik"* that Boltzmann's H-theorem and Poincares recurrence theorem are contradictory. On the basis of these arguments Chandrasekhar (1943) concluded that *"a process would appear to be irreversible (or reversible) according as whether the initial state is characterized by a long (short) average time of recurrence compared to the time during which the system is under observation"*. Poincare himself was very critical about Boltzmann's work, which he believed to be completely wrong. At that time, Poincare could not know that he had already created the tools for the solution to that deep controversy. The clue was the concept of the instability of trajectories, developed by Poincare in 1890. Recent results on chaotic dynamics lead us to revise Poincare's conception (Prigogine, 1980, 1989, Petrosky & Prigogine, 1988; Gaspard, 1998; Dorfman, 1999, Hoover, 2001). Most systems of statistical mechanics such as systems of hard spheres are characterized by positive Lyapunov exponents, which implies the existence of a finite time horizon. As a result, the concept of classical trajectories is lost for long times, and the existence of a Poincare's recurrence time becomes irrelevant for times much longer than the Lyapunov time (Prigogine, 1989).

Other quite different, but indeed convincing arguments in favour of Boltzmann's approach are based on computer simulations. First in 1957 Alder and Wainwright started to simulate the dynamics of molecules (beginning with hard core models) on a computer. Now this method is getting more and more a central part of statistical physics with very fruitful implications (Hoover, 2001). This can be said also on the closely related Monte Carlo method (Binder, 1987). Computer simulations based on molecular dynamics are most useful to clarify the relations between the irreversibility and molecular dynamics, as well as between probability and fluctuations (Marechal and Kestemont, 1987; Hoover, 1988, 2001; Morosov et al., 2001; Norman & Stegailov, 2002).

3.2 Probability measures and ergodic theorems

Boltzmann's approach to introduce probabilities into physics has proven to be one of the most fruitful ideas of science and yet, in his day, Boltzmann was heavily attacked by mathematicians and physicists. The reasons for these attacks were, that Boltzmann was forced to introduce some probabilistic assumptions which were in contradiction to the principles of mechanics. In an effort to place his theory on on firm ground, Boltzmann founded the subject of ergodic theory. The aim of ergodic theory is to derive probabilities from the study of the flow of trajectories (Balescu, 1963; Arnold & Avez, 1968; Sinai, 1977).

In order to explain the key point of Boltzmann's idea, let us first remind several notations and results obtained in the framework of the classical mechanics of Hamiltonian systems. We consider a Hamilton dynamics which is defined by a scalar function H, called the Hamiltonian, which is defined on a space of f coordinates $q_1, \ldots, q_f$ and f momenta $p_1, \ldots, p_f$:

$$H(q_1, \ldots, q_f, p_1, \ldots, p_f). \tag{3.29}$$

The equations of motions of our Hamiltonian dynamics are

$$\frac{dq_i}{dt} = \frac{\partial H}{\partial p_i}, \quad \frac{dp_i}{dt} = -\frac{\partial H}{\partial q_i}. \tag{3.30}$$

By integration of the Hamiltonian equations at given initial state $q(t), p(t), i = 1, \ldots, f$ we may calculate the future state at $t + \delta t$ in a unique way. A Hamiltonian system given by (3.29)–(3.30) is either "integrable" or "non-integrable" dependent on the behavior of the integrals of motion

$$I_k(q_1, \ldots, q_f, p_1, \ldots, p_f) = C_k \tag{3.31}$$

where the C_k are certain constants. The Hamiltonian system is called "integrable" if there exist f constants of motion which are single valued differentiable (analytic) functions. These f functions must be independent of each other and exist globally, i.e. for all allowed values of the coordinates and momenta. As well known, a mechanical systems with f degrees of freedom has in total $2f - 1$ integrals of motion (3.31). This expresses just the uniqueness of the trajectory. Namely, if $q_1(t), \ldots, q_f(t), p_1(t), \ldots, p_f(t)$ are given explicitly as functions of time and the initial values, one may (in principle) exclude the time and find this way $f - 1$ relations of type of Eq. (3.31). For integrable systems exactly f of these integrals are well defined smooth functions and each of them defines a smooth surface in the phase space. The f single-valued constants of motion restrict the $2f$-dimensional phase space to an f-dimensional surface which one can prove to be an f-dimensional torus. Therefore the solution of (3.30) can be expressed in terms of f cyclic variables (angle variables) and f action variables. The Hamilton–Jacobi equation corresponding to Eq. (3.30) possesses a global solution. As examples of "integrable" systems we may consider for $f = 1$ the linear oscillator and for $f = 2$ the Kepler problem. For the linear oscillator the constant of motion is the energy $H(q,p) = E$, correspondingly, the motion is restricted to an ellipse. For the Kepler problem the constants of motion are the Hamiltonian itself and the angular momentum. Other examples of integrable high-dimensional systems are chains of coupled harmonic oscillators. In connection with the great importance of coupled oscillators to many branches of physics as e.g. solid state theory, these systems were carefully studied and we arrived nearly at a full understanding. However linear coupling is just a theoretical model and cannot be considered as a realistic model for the actual interactions in many-body systems. Therefore the main interest of statistical physics is devoted to systems with nonlinear interactions as e.g. hard-core and Lennard–Jones interactions. For such complicated systems, however, theoretical results for $f \gg 1$ are very rare.

A well-studied nonlinear problem of high dimension is the linear chain of Toda oscillators (Toda, 1981, 1983), a system of $N = f$ equal masses moving in a 1-d phase space. In equilibrium all of the masses are situated at their rest positions with mutual distances fixed at certain equilibrium values. The interactions are given by the strong anharmonic potentials:

$$V(r) = \frac{a}{b}\left[\exp(-br) - 1 + ar\right]. \tag{3.32}$$

Here r is the deviation from the equilibrium distance between two of the masses. The forces derived from this potential tend to a constant for expansions much larger than the equilibrium distance, and are exponentially increasing for strong compressions with respect to the equilibrium position. Closely related is the exponential potential which is purely repulsive exponential potential

$$V(r) = A\exp(-br).$$

In the limit $b \to \infty$, $(ab = const)$ we get the well known hard core forces which are zero for expansions and infinitely strong for compressions. Let us mention another potential, the Morse potential, which is also closely related to the Toda potential. The Morse potential, which is defined as the difference of two exponential potentials (with different sign) shows an attracting region similar as the Lennard–Jones potential. The Morse potential which possesses very interesting properties was treated in several recent papers (Dunkel et al., 2001, 2002; Chetverikov & Dunkel, 2003; Chetverikov et al., 2004, 2005). Here we consider only the simpler Toda potential. In the static equilibrium state of the chain, all the molecules are at rest in equidistant positions and the total energy is zero. By a collision we may accelerate a mass at the border of the system and introduce in this way kinetic energy which will run in form of an excitation through the system. In a thermal regime we may excite even a whole spectrum of excitations (Bolterauer & Opper, 1981; Jenssen, 1991; Jenssen & Ebeling, 1991, 2000). In the case of a purely linear coupling we know all about these excitations: We will observe sinusoidal oscillations and waves, acoustical and optical phonons etc. Eventually local excitations i.e. wave packets will be observed which however show strong dispersion. In other words local excitations are not stable in linear systems. In the case of Toda interactions however these excitations are stable, there exist soliton solutions which are based on the integrability of the Toda system. The strong interest in local excitations of soliton type is especially inspired by the theory of reaction rates (Ebeling and Jenssen, 1988, 1991, Hänggi et al., 1990). In the context considered here, the Toda systems serve as an example of integrable many-particle systems. It is just the special type of the interactions which allows an analytical treatment of the equations of motion. Now we are going to consider the dynamics and the integrals of motion of this system. We study a uniform chain of masses at the positions y_n which are connected to their nearest neighbors by Toda springs with the nonlinear spring constant b. The Hamiltonian reads

$$H = \sum_n \frac{p_n^2}{2m} + \frac{a}{b} \left[\exp\left(-b(y_{n+1} - y_n)\right) - 1 + a(y_{n+1} - y_n)\right]. \quad (3.33)$$

For an infinite uniform chain ($-\infty < n < +\infty$) Toda (1981, 1983) found the integrals of motion

$$\exp\left[-b(y_{n+1} - y_n)\right] - 1 = \sinh^2 \chi \operatorname{sech}^2 \cosh\left(\chi n - \sqrt{ab/m} \sinh \chi t\right), \quad (3.34)$$

corresponding to the soliton energy

$$E^s = \frac{2a}{b} \left(\sinh \chi \cosh \chi - \chi\right). \quad (3.35)$$

The soliton corresponds to a wandering local compression of the lattice with spatial "width" χ^{-1}. The quantity

$$\tau = \left((ab/m)^{1/2} \sinh \chi\right)^{-1} \quad (3.36)$$

defines a characteristic excitation time of a spring during soliton passage (Toda, 1983, Ebeling & Jenssen, 1988, 1991). The energy of a strongly localized soliton satisfying the condition

$$\frac{\sinh^2 \chi}{\chi} \gg 1 \quad (3.37)$$

reads according to Eq. (3.35)

$$E^s \simeq \frac{2a}{b} \sinh^2 \chi. \quad (3.38)$$

In this way we have demonstrated that there exists indeed a class of many-particle systems which are integrable. However, integrability is connected always with rather special interactions. In our example most variations of the Toda law (as e.g. the Morse interaction law) destroy the integrability. An interesting property of the Toda system is, that all statistical functions according to the Gibbs theory may be exactly calculated (Toda & Saitoh, 1983). The question however, whether this completely integrable system will assume a thermodynamic equilibrium in the limit of long time, remains completely open. The situation changes drastically if we add a small coupling to a thermal heat bath (Jenssen & Ebeling, 2000). Then we have no problems with irreversibility since the heat bath drives our system to an equilibrium state with exactly known properties. We see that the coupling to a heat bath is at least one possible solution of the irreversibility problem.

We consider now more general many-particle systems in physical space obeying classical mechanics, assuming nonlinear interactions of Lennard–Jones or Morse type. Then the integrals of motion can be divided into two kinds, isolating and nonisolating ones. Isolating integrals define a connected smooth surface in the phase space, while nonisolating integrals are not defining a smooth surface. The phase trajectory is a cross-section of isolating integrals defining a surface. In this way the cross section of isolating integrals defines that part of the phase space which is filled by trajectories. Boltzmann's hypothesis stated that for statistical-mechanical systems the energy surface

$$H(q_1, \ldots, q_f, p_1, \ldots, p_f) = E \tag{3.39}$$

is the only isolating integral of motion. Further Boltzmann stated that in the course of time the trajectory will fill the whole energy surface and will come close to any point on it. Further Boltzmann stated that in thermodynamic equilibrium the time average of given phase functions $F(q_1, \ldots, q_f, p_1, \ldots, p_f)$ exists which is defined by

$$\langle F(q_1, \ldots, q_f, p_1, \ldots, p_f) \rangle = \lim_{T \to \infty} \frac{1}{T} \int_t^{t+T} dt' F(q_1(t'), \ldots, q_f(t'), p_1(t'), \ldots, p_f(t')). \tag{3.40}$$

For stationary processes, the time average will not depend on the initial time t. However it may possibly depend on the initial state. The state space is called connected (nondecomposable) if it cannot be decomposed into two parts having different time averages. This property guarantees the independence on the initial state of the averaging. The dynamics is called mixing if the average of a product of two phase function equals the product of the averages

$$\langle F(q_1, \ldots, q_f, p_1, \ldots, p_f) G(q_1, \ldots, q_f, p_1, \ldots, p_f) \rangle \tag{3.41}$$
$$= \langle F(q_1, \ldots, q_f, p_1, \ldots, p_f) \rangle \langle G(q_1, \ldots, q_f, p_1, \ldots, p_f) \rangle. \tag{3.42}$$

A system is expected to have this property if the trajectories are well mixed. Boltzmann considered the time average as a theoretical model for the result of a measurement of the physical quantity F. Further Boltzmann introduced an ensemble average as the integral over the energy surface

$$\langle F(q_1,\ldots,q_f,p_1,\ldots,p_f)\rangle$$
$$= \int_{E-\Delta/2 \leq H \leq E+\Delta/2} dq_1 \ldots dq_f dp_1 \ldots dp_f F(q_1,\ldots,q_f,p_1,\ldots,p_f). \quad (3.43)$$

Here the integral is to be extended over a thin sheet around the energy surface. The finite width Δ of the surface was introduced for mathematical convenience, physically it may be considered as an uncertainty of the energy measurement. The final part of Boltzmann's so-called ergodic hypothesis (which is formulated here in a more recent notation) states, that time and ensemble averages may be identified, i.e.

$$\langle F(q_1,\ldots,q_f,p_1,\ldots,p_f)\rangle_t = \langle F(q_1,\ldots,q_f,p_1,\ldots,p_f)\rangle_s. \quad (3.44)$$

In other words, the result of measurements of F (the time average) may be predicted on the basis of an ensemble averaging. So far there is no general proof of this statement for the case of arbitrary interactions. The modern theory has shown, however, that there are indeed many-body systems, as Sinai's billiard which possess the properties stated hypothetically by Boltzmann. We will see later that ergodicity is related to the chaotic character of the motion of complex Hamiltonian systems. This means, that the practical predictability is limited, in spite of the fact that the initial conditions (if there are known exactly) fully determine the future states. Before we go to a discussion of this fundamental relation, let us first generalize the notation of ergodicity following the work of Birkhoff and others (Ruelle, 1987; Steeb, 1991). A dynamical system is called ergodic if

- for phase functions $F(q_1,\ldots,q_f,p_1,\ldots,p_f)$ the time average is well defined and,
- a probability measure $\rho(q_1,\ldots,q_f,p_1,\ldots,p_f)$ exists (called also invariant density) such that

$$\langle F(q_1,\ldots,q_f,p_1,\ldots,p_f)\rangle_t = \langle F(q_1,\ldots,q_f,p_1,\ldots,p_f)\rangle_s. \quad (3.45)$$

The ensemble average used in this condition is defined as

$$\langle F(q_1,\ldots,q_f,p_1,\ldots,p_f)\rangle_s = \int dq_1 \ldots dq_f dp_1 \ldots dp_f \quad (3.46)$$
$$F(q_1,\ldots,q_f,p_1,\ldots,p_f)\rho(q_1,\ldots,q_f,p_1,\ldots,p_f). \quad (3.47)$$

Let us conclude this section with a few general remarks about the ergodicity problem: The works of Boltzmann, Gibbs and the Ehrenfests raised the ergodicity problem: to find conditions under which the result of measurements on many-body systems may be expressed by probability measures. Since Boltzmann, Gibbs and Ehrenfest, the subject of ergodic theory was primarily the domain of mathematicians. In 1931 Birkhoff proved an ergodic theorem showing the necessary and sufficient condition for an ergodic behavior of dynamic systems. Nevertheless Birkhoff's result did not close the problem, since for the complex time evolutions which occur in many-body Hamiltonian systems, ergodicity remains a property which is difficult to prove. However on the positive side of the ledger is, that ergodic systems exist indeed, as Sinai's billiard. Sinai has shown in a remarkable paper, that systems consisting of two or more hard spheres enclosed in a hard box are ergodic. From this example as well as from other investigations we know that ergodicity is closely connected with the instability of complex mechanical systems, i.e. with the chaotic character of their dynamics. Very important contributions to this field of research we owe to the early work of Krylov and Born. The main idea of the investigation pioneered by their work is the following: Due to the instability of the motion in phase space the trajectories are becoming very complex. This leads to the mixing character of the trajectories and to integrals of motion which are nonisolating. We mention also that this view is supported by the simulations of the N-particle dynamics by means of powerful computers. So far a complete solution of the problem of ergodicity is still missing, but anyhow we may state that our understanding of ergodicity has much increased since the times of Boltzmann, Gibbs, Ehrenfest and Birkhoff.

3.3 Dynamics and probability for one-dimensional maps

The most simple dynamical systems, which already show a whole universe of beautiful phenomena including statistical and thermodynamical aspects are 1-dimensional maps (Schuster, 1984, Lasota & Mackey, 1985; Anishchenko, 1989; Ebeling, Steuer & Titchener, 2001). Let us first consider a 1-dimensional map T defined by the iteration

$$T: \quad x(t+1) = f(x(t)). \tag{3.48}$$

The state $x(t)$ is a point on the one-dimensional x-axis or of certain interval on it. The time t is an integer

$$t = 0, 1, 2, 3, \ldots .$$

The trajectory $x(t)$ forms a set of points, one point for each integer time. Let us consider for example the famous logistic map (Fig. 3.2)

$$x(t+1) = rx(t)[1 - x(t)] \tag{3.49}$$

and the tent map (Fig. 3.2)

$$x(t+1) = rx(t) \quad \text{if} \quad x \leq 1/2$$

$$x(t+1) = r(1 - x(t)) \quad \text{if} \quad x \geq 1/2. \tag{3.50}$$

Both these examples which map the interval $[0, 1]$ into itself depend on *one*

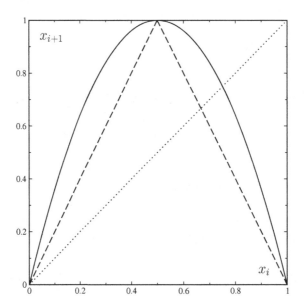

Fig. 3.2 Representation of the logistic map for $r = 4$ (a solid line) and the tent-map for $r = 2$ (a dashed line).

parameter only. The careful study of the dependence of the trajectories $x(t)$ on the value of these parameters, which was pioneered by Feigenbaum, Grossmann and Thomae in the 1970's led us to deep insights. One can get a nice overview about the map by plotting the points generated by 100 interactions in dependence on the r - parameter. Let us give now an elementary consideration of the bifurcation scenario of our nonlinear map (3.49) (see e.g. Holden, 1986). In the region $0 < r < 1$ the state $x = 0$ is stable and the formal solution $x = 1 - (1/r)$ makes no sense for $r < 1$. At $r = 1$ we observe an exchange of stability. The state $x = 0$ is getting unstable and a stable stationary state $x = 1 - (1/r)$ appears. At $r = r_1 = 3$ the stable state $x = 1 - (1/r)$ loses stability and a stable 2-cycle flipping between the states x_2 and x_3 with

$$x_3 = rx_2(1 - x_2), \qquad x_2 = rx_3(1 - x_3) \tag{3.51}$$

is born. At $r = r_2 = 1 + \sqrt{6} = 3.449...$ the two branches of the 2-cycle lose stability and a stable 4-cycle is created. This procedure is continuing in an infinite sequence of bifurcations occurring at the parameter values

$$r_3 = 3.544090, \qquad r_4 = 3.564407 \tag{3.52}$$
$$r_5 = 3.568759, \qquad r_6 = 3.569692, \ldots \tag{3.53}$$

The sequence of period-doubling bifurcations appears to converge to a limit

$$r_\infty = 3.569946\ldots$$

in geometric progression

$$r_k = r_\infty - cF^{-k} \quad \text{if} \quad k \to \infty \tag{3.54}$$

with

$$c = 2.6327\ldots, \quad F = 4.669202\ldots.$$

This behavior was already in 1975 noted by Feigenbaum, who found also — what is most important — that for a very large class of maps, the constant F has the same value. Honoring Feigenbaums pioneering work, the constant F is now named the universal Feigenbaum constant. Let us continue now the bifurcation analysis and proceed to the region $r_\infty < r < 4$. We find there a chaotic behavior of the orbits except for an infinite number of small windows of r-values for which there exist a stable m-cycle. The first such cycles which appear beyond r_∞ are of even period. Next, odd cycles appear as e.g. a 3-cycle at $r = 3.828427$ which stays stable up to $r = 3.841499$. Then a 6-cycle follows and one observes further period doublings. Outside the window there are no stable periodic orbits although there is an infinite number of unstable cycles. The most chaotic case is reached at $r = 4$. A quantitative way to describe the chaotic behavior in the region $r > r_\infty$ is the Lyapunov exponent defined in Section 2.5. In our simple case of 1-d maps the Lyapunov exponent is defined as the time average of the logarithmic slope

$$\lambda = \langle \phi(x(t)) \rangle_t \tag{3.55}$$

where

$$\phi(x) = \log|df(x)/dx|. \tag{3.56}$$

Following an orbit of the system we may write

$$\lambda = \lim_{n\to\infty} \cdot \frac{1}{n} \sum_{k=1}^{n} \log|df/dx|_{x=x_k}. \tag{3.57}$$

Except for a measure of value zero, λ is independent of the starting point. According to the definition of chaos in Section 2.5 we say in the case $\lambda > 0$ that the orbit is chaotic. Stable orbits as fixed points and limit cycles are characterized by $\lambda < 0$. Lyapunov exponents for the logistic map in dependence on the r-parameter were calculated for the interval where chaos is observed in several papers (see e.g. Ebeling, Steuer & Titchener, 2001).

There is no need here to discuss the bifurcation behavior of the tent map Eq. (3.50) in the same detail. Let us just mention that the interesting interval is here $1 < r < 2$. So far we have seen only an interesting bifurcation picture but no connection to the problems of statistical thermodynamics and stochastics, which are all connected with certain probabilities. We shall show now, how probabilities and thermodynamic quantities come into play. Let us consider for example again the logistic map in the case that all the orbits of the system are chaotic at the given value of r (e.g. $r = 4$). We will show that in this case probability may be introduced in a natural way (Lauwerier, 1986). The invariant distribution $p(x)$ is defined by the normalized probability $p(x)dx$ of finding an image $T : x$ in the interval $(x, x + dx)$. The normalization condition reads

$$\int p(x)dx = 1. \tag{3.58}$$

For simplicity we consider only unimodal maps on the interval $0 \leq x \leq 1$; then any x has at most two pre-images y and z. The probability to find x in the interval $(x, x + dx)$ should equal the sum of finding the pre-images y and z in the intervals $(y, y + dy)$ and $(z, z + dz)$. In this way we find

$$p(x)dx = p(y)dy + p(z)dz. \tag{3.59}$$

Introducing now

$$x = f(y), \qquad x = f(z) \tag{3.60}$$

and

$$dx/dy = f'(y), \qquad dx/dz = f'(z) \tag{3.61}$$

we get the functional relation

$$p(x) = \frac{p(y)}{f'(y)} + \frac{p(z)}{f'(z)}. \tag{3.62}$$

We call this the (stationary) Frobenius–Perron equation. In the general case the analytic solution of the Frobenius–Perron relation is not known. However a solvable case is just the logistic map with $r = 4$ and

$$y = \frac{1}{2} - \frac{1}{2}\sqrt{1 - x}, \qquad z = \frac{1}{2} + \frac{1}{2}\sqrt{1 - x}. \tag{3.63}$$

$$|f'(y)| = |f'(z)| = 4\sqrt{1 - x}.$$

In this case the Frobenius–Perron equation assumes the form

$$p(x) = \frac{p(\frac{1}{2} - \frac{1}{2}\sqrt{1 - x})}{4\sqrt{1 - x}} + \frac{p(\frac{1}{2} + \frac{1}{2}\sqrt{1 - x})}{4\sqrt{1 - x}}. \tag{3.64}$$

One can check that the normalized solution is given by

$$p(x) = \frac{1}{\pi\sqrt{x(1 - x)}} \qquad \text{if} \qquad 0 \leq x \leq 1. \tag{3.65}$$

This probability distribution has integrable poles at $x = 0$ and $x = 1$ and a minimum at $x = 0.5$. For the tent map the Frobenius–Perron equation reads

$$p(x) = \frac{1}{r}\left[p\left(\frac{x}{r}\right) + p\left(1 - \frac{x}{r}\right)\right]. \qquad (3.66)$$

This equation is solved analytically for $r = 2$ by the homogeneous distribution

$$p(x) = 1 \quad \text{if} \quad 0 \leq x \leq 1. \qquad (3.67)$$

In the general case $p(x)$ cannot be found analytically however there is no problem in finding it by numerical procedures.

For example we can solve Eq. (3.62) by successive iterations starting from certain guess, e.g. the equal distribution which we introduce at the right hand side, calculate the left side etc. Following a theorem of Lasota and Yorke, $p(x)$ is continuous if $f(x)$ is everywhere expanding. In other cases $p(x)$ might be the sum of a continuous function and Dirac δ-functions. Having obtained the probability distribution we get the Shannon entropy of the distribution by integration

$$H = \int dx\, p(x) \cdot \ln(1/p(x)).$$

This gives in the case of the logistic map at $r = 4$ the entropy

$$H = \ln\left(\frac{\pi}{4}\right). \qquad (3.68)$$

Further we obtain for the tent map at $r = 2$ the entropy

$$H = 0. \qquad (3.69)$$

The probability distributions given above are is the stationary (invariant) distribution. They correspond to the invariant probability measures introduced in Section 3.2 in connection with the term ergodicity. We may also discuss the time evolution for the distribution $p(x)$ which is described by the time-dependent Frobenius–Perron equation. In conclusion we undeline the remarkable result, that a purely deterministic, but chaotic dynamics, may define a smooth probability distribution.

3.4 Hamiltonian dynamics: The Liouville equation

As a first elementary example we consider a mechanical system with the linear Hamiltonian (p, q are the action and the angle)

$$H = \beta p - \alpha q. \qquad (3.70)$$

We assume periodic boundary conditions on the surface of the two-dimensional unit square $0 < q < 1, 0 < p < 1$. The equations of motion are given by

$$\dot{p} = \alpha; \qquad \dot{q} = \beta. \qquad (3.71)$$

The solutions are are easily given

$$p(t) = p(0) + \alpha t; \qquad q(t) = q(0) + \beta t. \qquad (3.72)$$

By eliminating the time we see that the phase trajectory is given on the unit square by

$$p(t) = p(0) + \frac{\alpha}{\beta}(q(t) - q(0)). \qquad (3.73)$$

If (α/β) is a rational number,

$$(\alpha/\beta) = m/n, \qquad (3.74)$$

then the trajectory will be periodic and will repeat itself after a period. If α/β is irrational, then the trajectory will be dense in the unit square (but will not necessarily fill it completely). One can show that our system is ergodic, i.e. for any phase function $F(q,p)$ the relation

$$\langle F(q,p)\rangle_t = \langle F(q,p)\rangle_s = \int dq dp F(q,p) \qquad (3.75)$$

holds. In other words, the time average is equal to an ensemble average and there exists an invariant density which is $\rho = 1$. The proof of Eq. (3.75) is simple (Reichl, 1987). This simple example shows, that there exist Hamiltonian systems which observe the general formalism, which requires the existence of probability densities.

Let us consider now the classical mechanics of systems with a more realistic Hamiltonian. We will assume that the Hamiltonian is defined on a space of f coordinates $q = q_1, \ldots, q_f$ and f momenta $p = p_1, \ldots, p_f$. For a large class of systems the Hamiltonian is the sum of a momentum-dependent kinetic energy and a coordinate-dependent potential energy

$$H(q_1, \ldots, q_f, p_1, \ldots, p_f) = T(p_1, \ldots, p_f) + U(q_1, \ldots, q_f). \qquad (3.76)$$

By integration of the Hamiltonian equations (3.30) at given initial state $q_i(t)$, $p_i(t)$, $(i = 1, \ldots, f)$ we may calculate the future state at $t + \delta t$ in a unique way. One of the most important results of modern physics is, that in spite of the deterministic connection between initial and future states, limited predictability occurs. This is due to the fact that most complex Hamiltonian systems are chaotic. However before we discuss this question in more details, let us first look at the simpler question of the reversibility of mechanical motion. Mechanical motions as e.g. the orbits of planets may go in both directions. Forward and backward movement are both allowed, the motion is reversible. On the other hand, macroscopic motions, as one of a comet falling down to earth are irreversible, they cannot go in backward direction. The reversibility of the mechanical motion is formally due to the invariance of the Hamilton equations with respect to the so-called T-transformation, which models the reversal of motion. Let us assume now that $q(t)$ and $p(t)$ are solutions of the

Hamilton equations. The T-transformation leading to reversal of motion at time $t = 0$ is given by ($i = 1, 2, \ldots, f$)

$$q_i(t) \to q_i'(t) = q_i(-t),$$
$$p_i(t) \to p_i'(t) = -p_i(-t). \qquad (3.77)$$

One can show easily, that the $p_i'(t)$ and the $q_i'(t)$ are solutions of the Hamilton equation, i.e. they correspond to allowed motions. A similar argument is true for the quantum-mechanical motion, where the T-transformation is given by

$$\psi(q_1, \ldots, q_f, t) \to \psi'(q_1, \ldots, q_f, t) = \psi^*(q_1, \ldots, q_f, -t). \qquad (3.78)$$

Here ψ is the wave function and ψ^* its complex conjugate. The wave function should satisfy the Schrödinger equation

$$\partial_t \psi(q_1, \ldots, q_f, t) = \mathbf{H}\psi(q_1, \ldots, q_f, t) \qquad (3.79)$$

where $\mathbf{H}$ is the Hamilton-operator. One can easily show now, that ψ' is also a solution of the Schrödinger equation, i.e. it represents a possible motion of the system. Since the times of Boltzmann, Loschmidt, Poincare and Zermelo there is a never ending discussion about the origin of the breaking of time symmetry observed in macroscopic physics (see e.g. Linde, 1984; Prigogine & Stengers, 1988; Prigogine, 1989; Ebeling, et al., 1990). Our point of view is in brief, that the observed irreversibility might be a property of the expanding world in which we are living. The second law is a basic property of this, our actual Universe. A priori we cannot exclude the possibility that in contracting phases of the Universe or in other Universes (if such exist) the second law is not valid. Actually all our observations refer to the expanding Universe surrounding us. Merely for philosophical reasons we share Boltzmann's and Einstein's view, that globally the world is uniform in space and time. Stationarity on big scales is one of its basic properties. Boltzmann expressed his views in the following sentences (quotation from Brush, 1965):
"*The second law of thermodynamics can be proved from the mechanical theory if one assumes, that the present state of the universe, or at least that part that surrounds us, started to evolve from an improbable state and is still in a relatively improbable state. This is a reasonable assumption to make, since it enables us to the facts of experience, and one should not expect to be able to deduce it from anything more fundamental. ... For the universe as a whole, there is no distinction between the "backwards" and "forwards" directions of time, but for the worlds on which living beings exist, and which are therefore in relatively improbable states, the direction of time will be determined by the direction of increasing entropy, proceeding from less to more probable states*". We believe, that Boltzmann, who was a really deep thinker, was right in the general respect, his views were just limited by the knowledge of his time. Nowadays our knowledge about the fundamental laws of dynamics is no more limited to the classical mechanics. Modern physics is based on quantum mechanics, general relativity theory on the big scales and quantum field theories

on the small scales. Let us imagine how Boltzmann would rephrase his ideas in our days, nearly 100 years after his reply to Zermelo's paper. Probably Boltzmann would start from general relativity, quantum field theories, relativistic thermodynamics and modern cosmological theory (Neugebauer, 1980; Linde, 1984). Guided by his general view about stationarity he would like the model of the closed Universe which has oscillating solutions (Linde, 1984). Modifying the standard picture about regular oscillations, Boltzmann would possibly assume stochastic oscillations. He would not insist on the purely thermal character of the oscillations but would admit as well vacuum fluctuations. Maybe he would say that our Universe is subject to some colored noise with a basic period of about 1–100 billions of years. Still he would insist on his hypothesis: "*Among these worlds the state probability increases as often as it decreases. For the Universe as a whole the two directions of time are indistinguishable, just as in space there is no up or down*". We believe that the laws of macroscopic physics are deeply affected by the expansion of our Universe. Expanding space soaks up radiation and acts as a huge thermodynamic sink for all radiation. At present the whole Metagalaxis is filled with a sea of thermal photons having a density of about 500 cm^{-3} and a temperature of about 2.7 K. This so-called background radiation acts as a thermal heat bath which influences all motions of particles in an irreversible way.

Following Boltzmann's view that our world is basically probabilistic, let us consider now the question, how probabilities may be introduced into a many-particle classical mechanical system and what is the dynamics of these probabilities. We postulate that the system may be characterized at time t by a probability density

$$\rho(q_1, \ldots, q_f, p_1, \ldots, p_f, t) \tag{3.80}$$

such that the ensemble average (3.47) of any given phase function F is defined. Trying to find an equation which determines the time evolution of the probability density we note first, that the reversibility of the mechanical motion requires

$$\frac{d\rho}{dt} = 0. \tag{3.81}$$

Here d/dt denotes the substantial derivation corresponding to a coordinate system moving with the phase point. An observer moving on the flow in the phase space will see a constant probability, otherwise he could differ between the past and the future. The so-called Liouville equation (3.81) is equivalent to the well-known invariance of the phase volume with respect to motion. Let us transform now Eq. (3.81) by using the Hamilton equations (3.30). We get

$$\frac{d\rho}{dt} = \frac{\partial \rho}{\partial t} + \sum_r \left(\frac{\partial \rho}{\partial q_r} \dot{q}_r + \frac{\partial \rho}{\partial p_r} \dot{p}_r \right) = 0. \tag{3.82}$$

This is the explicit form of the Liouville equation which we may write formally, employing the Poisson brackets, as

$$\frac{\partial \rho}{\partial t} + [H, \rho] = 0. \tag{3.83}$$

There exists a different way to derive the Liouville equation starting from the property of the Hamiltonian flow to be free of sources (see Chapter 2). This way we get

$$\frac{\partial \rho}{\partial t} + \sum_r \left(\frac{\partial \rho \dot{q}_r}{\partial q_r} + \frac{\partial \rho \dot{p}_r}{\partial p_r} \right) = 0. \tag{3.84}$$

By using the fact that the divergence of Hamilton flows is zero

$$\rho \sum_r \left(\frac{\partial \dot{q}_r}{\partial q_r} + \frac{\partial \dot{p}_r}{\partial p_r} \right) = 0. \tag{3.85}$$

we arrive again at the Liouville equation Eq. (3.83).

The Liouville equation is still reversible; strictly speaking it is not a kinetic equation. We may compare it with the Frobenius–Perron equation introduced in Section 3.3. Stationary solutions of Eq. (3.83) are easily found. We have to observe however several requirements, a probability has to fulfill, as smoothness and integrability.

A rather general solution is

$$\rho(q_1, \ldots, q_f, p_1, \ldots, p_f) = F(H(q_1, \ldots, q_f, p_1, \ldots, p_f)) \tag{3.86}$$

with a free function F which is arbitrary up to certain requirements. One example of a stationary solution of this type is

$$\rho(q_1, \ldots, q_f, p_1, \ldots, p_f) = C \exp[-\beta H(q_1, \ldots, q_f, p_1, \ldots, p_f)] \tag{3.87}$$

where C and β are positive constants. Later this distribution will get the name "canonical". Further any function of constants of motion is a stationary solution. Namely, in the case we can find s constants of the motion $I_1, \ldots, I_s$ which are single valued differential (analytic) functions. Furthermore these s functions must be independent of each other and exist globally, i.e. for all allowed values of the coordinates and momenta. Then, again for a rather general class of functions F we find a whole class of solutions in the form

$$\rho(q_1, \ldots, q_f, p_1, \ldots, p_f) = F(I_1, \ldots, I_s). \tag{3.88}$$

The problem with the Liouville equation is, that it has so many solutions. In this way we come back to Boltzmann's hypothesis stating that possibly the Hamiltonian is the only single valued analytic integral of motion, i.e. the solutions of type (3.87) expressing Gibbs' canonical distributions would be sufficiently general.

Why should this be true? The key ideas are based on Poincare's work from 1890, about the instability of many-body motions. The further development of this work we owe especially to Birkhoff, Hopf, Krylov, Born, Kolmogorov, Arnold, Moser, Sinai, Chirikov, Zaslavskii and others. We have explained the concept of instability and Lyapunov exponents already in the previous chapter. Let us repeat the main ideas in brief: The states and the trajectories of dynamical systems are never exactly known and are subject to stochastic perturbations. Therefore the stability of motion with respect to small changes is of large interest. The stability of trajectories $x(t) = [q_1(t), \ldots, q_f(t), p_1(t), \ldots, p_f(t)]$ is studied by investigating besides the original trajectories $x(t; x_0, t)$ which starts at x_0, t_0 also a second one $x(t; x_0 + \delta x_0, t_0)$, which at the initial time t_0 starts at $x_0 + \delta x_0$ where δx_0 is a small shift vector. The motion is called globally stable if for all t and any $\epsilon > 0$ there exists always an $\eta(\epsilon, t_0)$ such that for $|\delta x(t_0)| < \eta$ follows $|\delta x(t)| < \epsilon$ for any $t > t_0$. If such an η does not exist, the motion is called unstable.

From a more detailed analysis of the instability we obtain the spectrum of eigenvalues of singular points, the spectrum of Floquet exponents of periodic orbits and the spectrum of Lyapunov exponents. These exponents are all related to the properties of the Jacobian J defined earlier by Eq. (2.89) of the dynamical system.

For Hamilton systems the trace of the Jacobian is zero

$$\operatorname{Tr} J(x) = 0. \tag{3.89}$$

Correspondingly the eigenvalues are either real, consisting of symmetrically located pairs or conjugated imaginary. This means that we will not find asymptotically stable singular points. All singular points will be of saddle type or centers. An analog statement can be given for the stability of periodic orbits. The sum of the Lyapunov-exponents λ_i is always zero

$$\sum_i \lambda_i = 0 \tag{3.90}$$

what corresponds to the conservative character of Hamiltonian systems. If at least one (the largest) Lyapunov exponent is positive, the motion is chaotic. Generally we expect that the real parts of the spectrum have positive and negative contributions

$$\lambda_1 \geq \cdots \geq \lambda_j > 0 > \lambda_{j+1} \geq \cdots \geq \lambda_n. \tag{3.91}$$

Then the sum of the positive Lyapunov exponents is in most cases equal to the Kolmogorov entropy (Pesin, 1977; Ledrappier & Young, 1985). The Kolmogorov entropy is closely connected with the problem of the predictability of motions (Eckmann & Ruelle, 1985). Originally Kolmogorov introduced this quantity on the basis of the many-point entropies of the Shannon theory (Kolmogorov, 1958). Here we use only the simplified version based on the λ-spectrum. Let us define the Pesin-entropy by (Pesin, 1977; Ebeling, Steuer & Titchener, 2001)

$$K_\lambda = \sum_i^+ \lambda_i, \qquad (\lambda_i > 0). \tag{3.92}$$

The Pesin entropy is identical to the Kolmogorov entropy for a big class of interesting systems (Pesin identity) (Pesin, 1977, Eckmann & Ruelle, 1985). The dynamics is characterized as unstable if $K_\lambda > 0$. If $K_\lambda > 0$ for a certain region of the phase space we say that this region is stochastic. In this case predictability is quite limited. Trajectories tend to diverge at least in certain directions, what makes long term predictions impossible. Small uncertainties at zero time will arrive at very large values very soon.

One of the most important results of the modern theory of Hamiltonian systems is, that most many-body systems have stochastic regions (Krylov, 1950, 1979; Arnold and Avez, 1968; Sinai, 1970, 1972; Chirikov, 1979; Lichtenberg and Lieberman, 1983; Zaslavskij, 1984; Arnold, 1987). Systems with positive K-entropy are called now K-systems or K-flows. The property of being a K-flow includes mixing and ergodicity. The opposite however is not true. Sinai (1970, 1972) has shown that systems of $N > 2$ hard spheres in a box with hard walls are K-systems. This makes rather probable that all systems of particles with rather hard repulsion are also K-systems. We consider this to be one of the most important results of modern statistical mechanics.

3.5 Markov models

In the previous two sections we have considered several examples of chaotic dynamical systems leading to stationary probability densities, corresponding to invariant measures (Lasota & Mackey, 1985). For a simple example we shall demonstrate now, how Markov models for the dynamics may be derived. Such Markov models correspond to irreversible kinetic equations for the process to be described. Following the work of Nicolis, Piasecki and McKernan (1992) we study first the tent map (3.50) (Nicolis, Martinez & Tirapegui, 1991; Nicolis, Piasecki and McKernan, 1992; McKernan, 1993; Nicolis and Gaspard, 1993)

$$\begin{aligned} x(t+1) &= rx(t) &&\text{if} && x \leq 0.5\,, \\ x(t+1) &= r[1 - x(t)] &&\text{if} && x > 0.5\,. \end{aligned} \tag{3.93}$$

The time-dependent version of the Frobenius–Perron equation for the tent map is obtained by a generalization of Eq. (3.66). Considering the balance of probabilities at time t and $t+1$ we obtain

$$p(x, t+1) = \frac{1}{r}\left(p\left(\frac{x}{r}, t\right) + p\left(1 - \frac{x}{r}, t\right)\right). \tag{3.94}$$

Let us consider now the special parameter value $r = 2$; then the map is fully chaotic and the time evolution is given by:

$$p(x, t+1) = \frac{1}{2}\left(p\left(\frac{x}{2}, t\right) + p\left(1 - \frac{x}{2}, t\right)\right). \tag{3.95}$$

As we can verify by substitution, the stationary distribution is given by

$$p_0(x) = 1 \quad \text{if} \quad 0 < x < 1. \tag{3.96}$$

In other words, the equal distribution satisfies the stationary Frobenius–Perron equation at $r = 2$. The mean uncertainty corresponding to the equal distribution (3.96) is $H = 0$. Any other normalized distribution has a lower value of the mean uncertainty. This is exactly the behavior we expect from the point of view of thermodynamics. We note, that for continuous distributions the mean uncertainty is not always positive definite. Another more serious problem with the Frobenius–Perron equation (3.94) is, that an initial distribution must not necessarily converge to the stationary distribution. In other words, we have no irreversibility of the evolution, no Markov property. The solution of this problem comes from the rather old idea of coarse-graining introduced already by Gibbs. The introduction coarse-grained descriptions leads us and to evolution equations with Markov character. So far we have considered a fine (microscopic) description of our dynamic system based on the exact state $x(t)$ at any time t or the corresponding distribution $p(x, t)$. A coarse description does not specify the state exactly but only with respect to certain intervals. Let us introduce a partition of the state space by

$$\boldsymbol{P} : [C_1, \ldots, C_\lambda], \quad C_i \cup C_j = 0 \quad \text{if} \quad i \neq j. \tag{3.97}$$

$\boldsymbol{P}$ is called a partition of the interval $[0, 1]$. Now we restrict our description by giving only the number of the interval in which the exact state $x(t)$ is located. On this coarse-grained level the state is one of the λ discrete possibilities. The corresponding dynamics is a hopping process between the intervals. Since the hopping is a discrete process, the description of the dynamics has necessarily to use stochastic methods. We note at this place that for for discrete processes deterministic descriptions do not exist. This is an important point. For the case of continuous state spaces we have a free choice between deterministic and probabilistic descriptions, both are strictly equivalent. After introducing coarse-graining no deterministic description exists anymore and the stochastic description is a must. Let us introduce $P(i, t)$ as the probability of finding the system at the time t on the level (the interval) i. In exact term this probability is defined by

$$P(i, t) = \int_{C_i} dx \cdot p(x, t). \tag{3.98}$$

In accordance with our general reasoning we assume now that the evolution of the probability is a Markov process defined by a stochastic matrix $\boldsymbol{W} = [W_{ij}]$ and the equations

$$P(i, t+1) = \sum_j W_{ij} P(j, t) \tag{3.99}$$

with

$$\sum_i W_{ij} = 1.$$

The problem with the rough descriptions is, that the Markov picture might not exist or be incompatible with Eqs. (3.97)–(3.99). In any case a general proof of the existence of Markov descriptions seems to be difficult. A partition of the original phase space which leads to a Markov description is called a *Markov partition*. Evidently no general prescription is known, how to find Markov partitions for an arbitrary given dynamics. However several rules are known, which might be helpful in finding Markov partitions for given deterministic dynamics.

An example where this procedure works was demonstrated by Nicolis et al. (1991, 1992). We will discuss this example here without giving a full prove. Let us now consider again the tent map with $r = 0.5(1+\sqrt{5})$. For this map the dynamics is chaotic and has an attractor located in the interval $[r(1-r/2), r/2]$. This means, in the coarse of the time evolution, any initial state will be attracted by this interval. In the coarse grained description a 2-partition is generated by means of the maximum of the tent at 0.5:

$$C_1 = [0, 0.5), \qquad C_2 = [0.5, 1]. \tag{3.100}$$

The resulting states, denoted e.g. by "L" and "R" may be viewed as the letters of an alphabet. Then following Nicolis et al. (1992) the stochastic matrix $\boldsymbol{W}$ is exactly given by

$$W_{11} = 0 \qquad W_{12} = 1/r^2 \tag{3.101}$$

$$W_{21} = 1 \qquad W_{22} = 1/r \tag{3.102}$$

We can easily verify the Markov properties

$$W_{11} + W_{21} = 1, \qquad W_{12} + W_{22} = 1.$$

Further we verify, that the stationary (invariant) distribution has the components

$$\begin{aligned} P(1) &= 1/(1+r) = 0.27639\ldots, \\ P(2) &= r/(1+r) = 0.72360\ldots. \end{aligned} \tag{3.103}$$

A direct proof of these relations may be given by comparison by carrying out the integrations in Eq. (3.99), this way we may confirm that Eq. (3.102) and Eq. (3.103) are true. Some generalization of this description is possible for the 4-partition

$$C_1 = [r(1-r/2), 1/2]; \qquad C_2 = [1/2, r/2]; \qquad (3.104)$$

$$C_3 = [0, r(1-r/2)]; \qquad C_4 = [r/2, 1]. \qquad (3.105)$$

After some transitory dynamics, the attractor is reached, which cannot be left, i.e. the transitions $1 \to 3$, $1 \to 4$, $2 \to 3$, $2 \to 4$ are impossible. The corresponding matrix elements should disappear.

$$W_{13} = 0 \qquad W_{14} = 0 \qquad W_{23} = 0 \qquad W_{24} = 0.$$

The matrix elements $W_{11}, W_{21}, W_{12}, W_{22}$ remain unchanged. The message is, that there might be several Markov descriptions for a given deterministic dynamics.

Another example which can be treated this way is the logistic map at $r = 4$, where also a family of Markov partitions is known. For instance, the points of the unstable periodic orbits $x = 0.345\ldots$; $x = 0.905\ldots$ define a three-cell Markov partition. The resulting 3 states, e.g. "O", "L" and "M" by be considered again as an alphabet. As shown by Nicolis et al. (1989) the corresponding probability matrix has the elements

$$W_{11} = 1/2 \qquad W_{12} = 0 \qquad W_{13} = 1 \qquad (3.106)$$

$$W_{21} = 1/2 \qquad W_{22} = 1/2 \qquad W_{23} = 0 \qquad (3.107)$$

$$W_{31} = 0 \qquad W_{32} = 1/2 \qquad W_{33} = 0. \qquad (3.108)$$

Many other examples for generating Markov partitions are known. However the deep problem remains open, what are in general the conditions for making a transition to Markov descriptions. Evidently Markov-like descriptions are the condition sine qua non for the formulation of a "statistical mechanics" including "kinetic equations" and the irreversible transition to an "equilibrium thermodynamics". In conclusion we may state that nonlinear systems in the chaotic regime are leading to probabilistic and thermodynamic descriptions in a quite natural way. In other words, nonlinearity, chaos, and thermodynamics are closely linked together. This point will occur to be fundamental for all the problems discussed in this book. This was the reason to explain this connection for simple examples already at the beginning of this book.

Our basic hypothesis is, even without having a general prove of this statement, that the macroscopic systems which are the objects of "statistical physics" have similar properties as the simple chaotic maps studied in this Chapter.

Chapter 4
Entropy and Equilibrium Distributions

4.1 The Boltzmann–Planck principle

According to our general concepts, Statistical Physics is the bridge between the microscopic and macroscopic levels of description. Basic tools are probability and entropy. We have shown in the last Chapter how these concepts come into play. Probability may be introduced axiomatically as we did in Chapter 2 as an appropriate concept for the description of mesoscopic or macroscopic systems. However, and this is even more interesting in the present context, it may arise (see Sections 3.3–3.5) in a natural way if the dynamics of systems is chaotic. Having probabilities we may calculate entropies as first done for physical systems by Boltzmann (see Section 3.1). Entropy concepts were used already several times in the previous Chapters. In the introduction we gave, based on the historical point of view, a brief discussion of the probabilistic physical entropy concept developed in the pioneering work of Boltzmann, Planck and Gibbs. As we mentioned already, strictly speaking this concept has still an earlier root in game theory. Already in the 18th century DeMoivre used the expression $\log(1/p_i)$ as a measure for the uncertainty of predictions in the context of describing the outcome of games. The mathematical concept, *entropy as mean uncertainty*, was worked out later by Shannon, who formed the basis of the modern information theory. Shannon's information theory has nowadays very many applications reaching from technology, to medicine and economy. The concept of entropy used in the stochastic theory is also based on Shannon's entropy.

In order to explain these ideas in brief, we consider a system with discrete states numbered with $i = 1, \ldots, s$ which are associated with the probabilities p_i. The states i are standing here for certain states of order parameters of the system. Then the Shannon entropy is defined as the mean uncertainty per state

$$\mathcal{H} = \langle \log(1/p_i) \rangle = -\sum_i p_i \log p_i. \tag{4.1}$$

This quantity is always between zero and one:

$$0 \leq \mathcal{H} \leq 1. \tag{4.2}$$

The mean uncertainty is zero, if all the probability is concentrated on just one state

$$\mathcal{H} = 0 \quad \text{if} \quad p_k = 1, \quad p_i = 0 \quad \text{if} \quad i \neq k. \tag{4.3}$$

On the other hand the uncertainty is maximal, if the probability is equally distributed on the states

$$\mathcal{H} = 1 \quad \text{if} \quad p_i = 1/s \quad i = 1, \ldots, s. \tag{4.4}$$

The equal distribution corresponds to maximal uncertainty. This property will play an important role in the considerations in the last section of this Chapter. Another important fact is that $\mathcal{H}$ has the property of additivity. In order to prove this we consider two weakly coupled systems 1 and 2 with the states i and j respectively and with

$$p_{ij}^{12} = p_i^1 \cdot p_j^2. \tag{4.5}$$

Using this relation we can show quickly that the mean uncertainty is an additive function

$$\mathcal{H}^{12} = \mathcal{H}^1 + \mathcal{H}^2. \tag{4.6}$$

This way, the mean uncertainty has quite similar properties as the entropy and we arrive at the hypothesis that both quantities are closely related, may be even proportional. Still we have to find out, what are the conditions for a proportionality.

The expression for the mean uncertainty may be generalized to continuous state spaces. Let us assume that x is a set of n order parameters on the dynamic-stochastic level of description. If $p(x)$ denotes the probability density for this set of order parameters which describe the macroscopic state, the mean uncertainty (informational entropy) of the distribution (the $\mathcal{H}$-function) is defined by

$$\mathcal{H} = -const \int dx \; p(x) \cdot \log p(x). \tag{4.7}$$

In the case of discrete variables $i = 1, 2, \ldots, s$ we come back to the classical Shannon expression with a sum instead of the integral. As well known these are the basic formulas of information theory. We shall come back to this later several times.

In order to come from the general expression for the mean uncertainty (informational entropy) to Boltzmann's physical entropy we identify the state space with the phase space of one molecule q, p: Then the Boltzmann's mean uncertainty of states in phase space is given by

$$H_B = -\int \frac{dpdq}{h^3} f(\boldsymbol{p}, \boldsymbol{q}, t) \cdot \ln f(\boldsymbol{p}, \boldsymbol{q}, t). \tag{4.8}$$

Here $f(\boldsymbol{p}, \boldsymbol{q}, t)$ is the one-particle distribution function. Further we introduced a constant h with the dimension of an action. This was h^3 has the same dimension as

$dpdq$ and this makes the whole integral dimensionless. Here is a point where classical theory has to borrow results from quantum theory. Heisenberg's uncertainty relation teaches us, that there is no way to measure location and momentum at the same time with an accuracy better than h (Planck's constant). This makes h^3 the natural choice for the minimal cell in Boltzmann's theory. We compare now our results with the Boltzmann formula for the entropy of ideal gases. As shown in the previous chapter, the Boltzmann entropy of ideal gases is given by

$$S_B = -k_B \int \frac{dpdq}{h^3} f(\boldsymbol{p},\boldsymbol{q},t) \cdot \ln f(\boldsymbol{p},\boldsymbol{q},t), \qquad (4.9)$$

with the normalization

$$N = \int \frac{dpdq}{h^3} f(\boldsymbol{p},\boldsymbol{q},t). \qquad (4.10)$$

We see that Boltzmann's entropy is proportional to the uncertainty of molecular states:

$$S_B = k_B H_B \qquad (4.11)$$

where k_B is the universal Boltzmann constant.

For the generalization to interacting systems we must realize in accordance with Chapter 3, that the new state space is the phase space of all the N molecules in the system. For this general case we introduce after Gibbs the normalized probability density ρ in the $6N$-dimensional phase space. Assuming that Shannon's state space is the phase space of all the molecules forming the system, the Shannon entropy of the system is given by the N-particle probability ρ. Therefore Gibbs' mean uncertainty H_G is the phase space entropy for the distribution of the molecules in the total phase space

$$H_G = -\int \frac{d\boldsymbol{p}^N d\boldsymbol{q}^N}{h^{3N}} \rho(\boldsymbol{q}^N,\boldsymbol{p}^N) \cdot \ln \rho(\boldsymbol{q}^N,\boldsymbol{p}^N). \qquad (4.12)$$

By multiplication with Boltzmann's constant we get the statistical Gibbs entropy

$$S_G = k_B H_G \qquad (4.13)$$

or explicitly

$$S_G = -k_B \int \frac{dpdq}{h^{3N}} \rho(\boldsymbol{q}^N,\boldsymbol{p}^N) \cdot \ln \rho(\boldsymbol{q}^N,\boldsymbol{p}^N). \qquad (4.14)$$

Now the basic theorems of statistical thermodynamics tell us, that in the case of ideal gases the Boltzmann entropy equals the thermodynamic entropy, i.e.

$$S_B = S. \qquad (4.15)$$

For interacting systems we postulate that the thermodynamic entropy corresponds to the Gibbs entropy

$$S_G = S. \tag{4.16}$$

In this way — in some sense — the thermodynamic entropy may be considered in a specific (formal) sense as a special case of the Shannon entropy. It is (up to a constant) just the mean uncertainty of the location of the molecules in the phase space. The close relation between the thermodynamic entropy and the Shannon entropy is the solid basis for the embedding of the information concept into the theoretical physics (Brillouin, 1956; Grandy & Schick, 1991).

The Gibbs expression includes all interaction effects which in general lead to a decrease of the value of the entropy in comparison to the ideal gas (Ebeling & Klimontovich, 1984). In principle, Eq. (4.11) works for nonequilibrium states as well, however this is true only *cum grano salis* as we will see later. In a different but closely related approach developed by Boltzmann and Planck the entropy of a macrostate was defined as the logarithm of the thermodynamic probability

$$S_{BP} = k_B \ln W. \tag{4.17}$$

which is defined as the total number of equal probable microstates corresponding to the given macro state. This fundamental formula is carved on Boltzmann's gravestone in the *"Zentralfriedhof"* cemetery in Vienna.

For the special case of equilibrium systems with fixed energy E the Gibbs relations Eqs. (4.11) and (4.13) reduce to the Boltzmann–Planck formula (4.14).

Let us mention further on that after Einstein one may invert relation (4.14) what gives us the probability that the nonequilibrium state occurs as the result of a spontaneous fluctuation

$$W(y_1, y_2, \ldots, y_n) = const \exp[-\delta S(y_1, y_2, \ldots, y_n)/k_B] \tag{4.18}$$

where δS is the lowering of entropy. We will come back to this relation in Chapter 6 and use it as a basis for developing Einstein's theory of fluctuations.

Another interesting aspect aspect of Eq. (4.15) is its relation to considerations in Chapter 2 on measures of distance from equilibrium. The inspection of relation (4.15) shows directly that δS is a kind of measure of the distance from equilibrium (at $E = const$) since equilibrium is the most probable state.

In Shannon's approach the basic role play some order parameters x and the corresponding probabilities $p(x)$ or p_i. In order to find a closer connection to the Boltzmann–Gibbs approach, we have to consider the order parameters as certain functions of the microscopic variables.

$$x = x(q_1, \ldots, p_{3N}). \tag{4.19}$$

Let us assume now that Gibbs' probability density may be represented as the product of the probability density in the order parameter space and the conditional probability (formula of Bayes)

$$\rho(p, q) = \int p(x) \cdot \rho(p, q|x) dx. \tag{4.20}$$

Then a brief calculation yields

$$S_G = k_B H + S_b \tag{4.21}$$

with

$$S_b = \int p(x) S(x) dx, \tag{4.22}$$

$$S(x) = -k_B \int \frac{dpdq}{h^3} \rho(p, q|x) \log \rho(p, q|x). \tag{4.23}$$

Here $S(x)$ is the conditional statistical entropy for a given value of the order parameter x. In this way we have shown that, up to a factor, the Shannon entropy of an order parameter is a fully legitimate part of the Gibbs entropy. As Eq. (4.21) shows, the contribution $k_B H$ constitutes the statistical entropy contained in the order parameter distribution. In general this is a very small part of the total statistical entropy, the overwhelming part comes from the term S_b. The part collected in S_B reflects the entropy contained in the microscopic state, it is not available as information. Let us give an example: The Gibbs entropy of a switch with two states is the sum

$$S_G = k_B \log 2 + S_b \tag{4.24}$$

where S_b is the usual (bound) entropy in one of the two positions. The two contributions to the total entropy are interchangeable in the sense discussed already by Szilard, Brillouin and many other workers (Denbigh & Denbigh, 1985). Information (i.e. macroscopic order parameter entropy) may be changed into thermodynamic entropy (i.e. entropy bound in microscopic motions). The second law is valid only for the sum of both parts, the order parameter entropy and the microscopic entropy.

4.2 Isolated systems: The microcanonical distribution

From the point of view of statistical physics, as we have shown already in Section 4.1, entropy is deeply connected with the mean uncertainty of the microscopic state in the phase space. Let us study now in more detail systems which are energetically isolated from the surrounding in the sense that the energy is closely concentrated around a given value E:

$$E - (1/2)\delta E \leq H \leq E + (1/2)\delta E. \tag{4.25}$$

The part of the phase space enclosed by this relation is called the energy shell. In isolated systems the available part of the phase space is the volume $\Omega(E)$ of this energy shell enclosing the energy surface $H(q,p) = E$. Boltzmann's postulate is, that under conditions equilibrium, any point on the energy shell (more precise the neighborhood of the point) is visited equally frequent. This so-called ergodic hypothesis is physically plausible but very difficult to prove for concrete systems (Sinai, 1970, 1972, 1977). Equivalent is the assumption that the probability is constant on the shell

$$\rho(q,p) = \frac{1}{\Omega(E)}. \qquad (4.26)$$

Equal probability on the shell is equivalent to maximum of the entropy. Therefore thermodynamic equilibrium corresponds to maximal entropy. This property which appears here as a consequence of certain aspects of the dynamics on the shell (ergodicity) will be turned in the last section to a first principle of statistical physics.

If the system is isolated (i.e. located on an energy shell) but not in equilibrium only certain part of the energy shell will be available. In the course of relaxation to equilibrium the probability is spreading over the whole energy shell filling it finally with constant density. Equilibrium means equal probability, and as we have seen and shall explain in more detail later, least information about the state on the shell.

We shall come back now to our basic question about the meaning of entropy. Usually entropy is considered as a measure of disorder, but the entropy is like the face of Janus, it allows other interpretations. The standard one is based on the study of phase space occupation. The number of states with equal probability corresponds to the volume of the available phase space $\Omega(A)$. Therefore the entropy is given by

$$S = k_B \ln \Omega^*(A). \qquad (4.27)$$

Here A is the set of all macroscopic conditions $\Omega(A)$ the corresponding phase space and

$$\Omega^*(A) = \frac{\Omega(A)}{h^{3N} N!}. \qquad (4.28)$$

Here the mysterious factor $N!$ was introduced as a consequence of the indisguishability of the microscopic particles. In the nonequilibrium states the energy shell is not filled with equal density, but shows regions with increased density (attractor regions). Let $S(E,t)$ be the entropy at time t. Then we may define an effective volume of the occupied part of the energy shell by

$$S(E,t) = k \ln \Omega^*_{\text{eff}}(E,t), \qquad (4.29)$$

$$\Omega^*_{\text{eff}}(E,t) = \exp[S(E,t)/k_B]. \qquad (4.30)$$

In this way, the relaxation on the energy shell may be interpreted as a monotonous increase of the effective occupied phase volume. This is connected with a devaluation of the energy, a point of view discussed already in Section 2.3.

4.3 Gibbs distributions for closed and for open systems

Starting from the microcanonical distribution we intend now to derive the equilibrium canonical distributions for the microscopic variables of the Γ-space. First we consider the standard cases:

(i) Closed systems, allowing exchange of energy with a second system.
(ii) Open systems which involve additionally particle exchange.

An isolated system with energy E and volume V is divided in two subsystems, further called bath and system. Both subsystem should be macroscopic bodies and, therefore, allow the introduction of intensive thermodynamic variables. They are both in thermal equilibrium which implies the equality of their temperatures. We label the bath volume by V_b and the volume of the system by V, respectively and evidently follows

$$V_0 = V_b + V. \tag{4.31}$$

In the Hamilton function the interaction part though it is essential for the relaxation to the equilibrium state will be neglected. Therefore choosing the analogous labeling it reads

$$E_0 = H_b + H + H_{int} \simeq H_b(q_1^b, \ldots p_{3M}^b) + H(q_1, \ldots, p_{3N}). \tag{4.32}$$

Here $q_1, \ldots, q_{3N}$ and $p_1, \ldots, p_{3N}$ are the microscopic variables of the system and $q_1^b, \ldots, q_{3M}^b, p_1^b, \ldots, p_{3M}^b$ those of the bath, respectively. First we look for the distribution of the microscopic variables of the system if the exchange of particles is forbidden. Suppose the energy of the system is fixed at a value $H = E$. From Eq. (4.32) immediately follows that the bath should possess the energy $H_b = E_b = E_0 - E$. The probability to find the microscopic variables of the considered system in states with energy E is found by collection of the probability of the bath to be in states $H = E_b$. It means

$$\rho(E) = \exp[-S_0(E_0, V_0)/k_B] \int d\Gamma_b \delta(H_b + E - E_0). \tag{4.33}$$

The integral is taken over the bath variables $q_1^b \ldots, p_{3N}^b$. Otherwise this integral gives just the thermodynamic weight of the numbers of microscopic (quantum) states of the bath which realize the volume V and the energy E_b

$$\Omega(E_b, V_b) = \int d\Gamma_b \delta(H_b + E - E_0). \tag{4.34}$$

Since the bath is considered as a macroscopic body we are able to find its thermodynamic entropy

$$S_b(E_b, V_b) = k_B \ln \Omega(E_B, V_B). \tag{4.35}$$

By inversion of this formula we are left with

$$\rho(E) = \exp[(S_b(E_0 - E, V_0 - V) - S_0(E_0, V_0))/k_B]. \tag{4.36}$$

We reformulate the expression in the exponent by using the definition of the free energy $F = E - TS$. Hence we derive

$$S_b(E_0 - E, V_0 - V) - S(E_0, V_0) = \frac{1}{T}(F_0 - F_b E) = \frac{1}{T}(F(T, V) - E) \tag{4.37}$$

where $F(T, V)$ is the free energy of the considered system. In deriving (4.37) we made use of the equality of the temperatures. Subsequently we interpret $E = H(q_1, \ldots, p_{3N})$ as the Hamilton function of the microscopic variables of the system and find the canonical distribution

$$\rho(q_1, \ldots, p_{3N}) = \exp\left[\frac{F(T, V) - H(q_1, \ldots, p_{3N})}{k_B T}\right]. \tag{4.38}$$

This way we derived now Gibbs' canonical distribution, already known to us from Chapter 3, from the microcanonical distribution.

The bridge between thermodynamics and statistics follows from the normalization condition of this distribution. Accordingly to

$$\int \frac{d\Gamma}{h^{3N}} \rho = 1 \tag{4.39}$$

the free energy of a system with volume V and embedded in a thermal bath with temperature T is defined by the statistical sum $Q(T, V)$

$$F(T, V) = -k_B T \ln Q(T, V) = -k_B T \ln \int \frac{d\Gamma}{h^{3N}} \exp(-\beta H). \tag{4.40}$$

The mean energy follows from the relation

$$E = \langle H \rangle = -\frac{\partial}{\partial \beta} \ln Q(T, V) = k_B T \frac{\partial}{\partial T} \ln Q(T, V) \tag{4.41}$$

and the dispersion is defined by

$$\langle H^2 \rangle - (\langle H \rangle)^2 = \frac{\partial^2}{\partial \beta^2} \ln Q(T, V) = -\frac{\partial}{\partial \beta} \langle H \rangle = k_B T^2 C_V. \tag{4.42}$$

This way all important thermodynamical quantities are given by the partition function $Q(T, V)$.

Quite similar we can proceed if particle exchange between the bath and the system is possible. The additional condition concerns with the conservation of the particle number of the bath and the system.

$$N_0 = N_b + N. \tag{4.43}$$

The thermodynamic entropy of the bath is analogous to equation (4.37) but depends now also on N. Therefore in the exponent of the probability distribution the following difference of the entropies occurs

$$\Delta S = S_b(E_0 - E, V_0 - V, N_0 - N) - S(E_0, V_0, N_0). \tag{4.44}$$

Introducing the thermodynamic potential $\Xi = -pV$ it becomes

$$\Delta S = \frac{1}{T}(\Xi + \mu N - E) \tag{4.45}$$

and we end with

$$\rho_N(q_1, \ldots, p_{3N}) = \exp\left[\frac{\Xi + N\mu - H_N}{k_B T}\right]. \tag{4.46}$$

By using again a normalization condition which reads now

$$1 = \sum_N \int \frac{dq_1 \ldots dp_{3N}}{N! h^{3N}} \rho_N(q_1, \ldots, p_{3N}) \tag{4.47}$$

we find a relation between thermodynamics and statistical quantities

$$\Xi = -kT \ln \sum_N \int \frac{d\Gamma}{N! h^{3N}} \exp[-\frac{H_N}{k_B T}]. \tag{4.48}$$

In conclusion let us shortly comment on the general procedure how the canonical distributions were derived from the microcanonical one. First of all the fixed energy of the system defined by the microscopic state was interpreted as an extensive thermodynamic variable of the system, further we neglected the interaction energy. Secondly the exponents in the distributions (4.38) and (4.46) correspond to that value of the overall entropy when the temperatures of the bath and the system have been equilibrated. The overall entropy is maximized under the constraints E, V, N for the considered system and E_0, V_0, N_0 for bath plus system. Our result could be derived also from the minimal reversible work which has to be applied isoenergetically to generate the state of our system embedded in the bath.

$$\Delta S = -\frac{1}{T} R_{\min}. \tag{4.49}$$

This minimal work obviously depends on the actual thermodynamic embedding of the considered system. In the case of a closed system where the volume and temperature are constant $R_{\min} = F(T, V) - H$. Fixing the pressure instead of the volume we would have

$$R_{\min} = G(T, p) - H = F(T, V) + pV - H \tag{4.50}$$

with $G(T, p)$ being the Gibbs free energy.

4.4 The Gibbs–Jaynes maximum entropy principle

Summarizing and looking again at the previous derivations of Gibbs ensembles, one might be not too happy with the logical built-up of the theory. We started with the Boltzmann–Planck expression for the entropy of systems with equally probable states, applying this formula to macroscopically isolated systems with fixed energy. In fact we postulated that constant probability on the energy shell is given *a priori*. Also we did not care much about the mathematical difficulties to prove "ergodicity" for concrete systems. In a following step we derived, by using some embedding procedure, the probability distributions for other situations, as e.g. systems in a heat bath.

The great follower of Gibbs' work E. T. Jaynes criticizes this approach with the following remarks (Jaynes, 1985): "*A moment's thought makes it clear how useless for this purpose is our conventional textbook statistical mechanics, where the basic connections between micro and macro are sought in ergodic theorems. These suppose that the microstate will eventually pass near every one compatible with the total macroscopic energy; if so, then the then the long-time behavior of a system must be determined by its energy. What we see about us does not suggest this.*"
In order to find an alternative, and possibly more elegant procedure, let us turn the question around, starting now from Gibbs original work. We quote again a sentence of Jaynes (1985): "*Why is that knowledge of microphenomena does not seem sufficient to understand macrophenomena? Is there an extra general principle needed for this? Our message is that such a general principle is indeed needed and already exists, having been given by J. Willard Gibbs 110 years ago .... A macrostate has a crucially important further property — entropy — that is not determined by the microstate.*"

We will show now that all Gibbsian ensembles may be derived in a unique way just from one principle, the *Gibbs–Jaynes maximum entropy principle*. We start as Gibbs and Jaynes from a very general variational principle. In order to explain this variational principle we start with the following abstract problem: We consider a macroscopic systems, given incomplete information A. Then we postulate that holds the

Gibbs–Jaynes principle:

If an incomplete information A is given about a macroscopic system, the best prediction we can make of other quantities are those obtained from the "ensemble" ρ that has maximum information entropy H while agreeing with A. By "agreeing" with A we mean that the average $\langle A \rangle$ calculated with ρ corresponds to the given information A.

Let us sketch briefly how this general principle works in the case that $\rho = \rho(q,p)$ is a probability density in the phase space and $A = [A_1, \ldots, A_m]$ stands for a set of real functions on the phase space. The most important case is that the $A_k(q,p)$

are constants of the motion (energy, angular momentum, particle numbers etc.). In order to find the probability density under the constraints

$$A'_k = \langle A_k \rangle = \int dq dp A_k(q,p) \rho(q,p) \tag{4.51}$$

we maximize the information entropy

$$\mathcal{H} = -\int dq dp \rho(q,p) \ln \rho(q,p) \tag{4.52}$$

under the given constraints. We define an m-component vector $\lambda = [\lambda_1, \ldots, \lambda_m]$ of Lagrange multipliers. Then the probability density that agrees with the given data $\mathbf{A}'$ follows from

$$\delta \left[\mathcal{H} + \sum_k \lambda_k \left(A'_k - \int dq dp A_k(q,p) \rho(q,p) \right) \right] = 0. \tag{4.53}$$

This leads to

$$\rho(q,p) = Z^{-1} \exp\left[-\sum_k \lambda_k A_k(q,p)\right], \tag{4.54}$$

where the normalization factor, the so-called partition function, is given by

$$Z(\lambda_1, \ldots, \lambda_m) = \int dq dp \exp\left[-\sum_k \lambda_k A_k(q,p)\right]. \tag{4.55}$$

The found probability density ρ spreads the probability as uniformly as possible over all microstates subject to the constraints. The Lagrange multipliers are found from the relations

$$A'_k = \langle A_k \rangle = -\frac{\partial}{\partial \lambda_k} \ln Z(\lambda_1, \ldots, \lambda_m). \tag{4.56}$$

The dispersion is given by second derivatives

$$\langle (A_k - A'_k)^2 \rangle = \frac{\partial^2}{\partial \lambda_k^2} \ln Z(\lambda_1, \ldots, \lambda_m) = -\frac{\partial}{\partial \lambda_k} \langle A_k \rangle. \tag{4.57}$$

We see that the linear "ansatz" (4.54) implies that the mean values and the dispersion are closely connected. Let us turn now to more concrete examples:

- Microcanonical ensemble:
  The case that there are no constraints, except the fixation of the system on an energy shell. This leads to the ensemble (on the shell)

$$\rho(q,p) = const. \tag{4.58}$$

- Canonical ensemble:
  The case that the mean of the energy is given $A'_1 = E = \langle H \rangle$ leads to the ensemble

  $$\rho(q,p) = Z^{-1} \exp[-\beta H(q,p)]. \tag{4.59}$$

- Grand canonical ensemble:
  The case that besides the mean energy $A'_1 = E$ also the s mean particle numbers are given $A'_2 = \langle N_1 \rangle, \ldots, A'_{s+1} = \langle N_s \rangle$ are given leads to the ensemble

  $$\rho(q,p; N_1, \ldots, N_s) = Z^{-1} \exp[-\beta H(q,p) - \lambda_1 N_1 - \cdots - \lambda_s N_s]. \tag{4.60}$$

  After identifying β with the reciprocal temperature and λ_k/β with the chemical potentials, we are back at the formulae derived in the last section.

- Canonical ensemble for rotating bodies:
  We consider here the case of rotating bodies consisting of N particles in internal equilibrium. We assume that the angular velocity of the body is ω and the angular momentum is $\mathbf{L}$. In a coordinate system which rotates with the body we find the Hamiltonian

  $$H_r = H - \boldsymbol{\Omega} \cdot \mathbf{L}. \tag{4.61}$$

The assumption that in the rotating system the system behaves like a standard canonical ensemble we find (Landau & Lifshits, 1990)

$$\rho(q,p) = Z_r^{-1} \exp[-\beta H_r(q,p)]. \tag{4.62}$$

This leads in the original system of coordinated to the distribution

$$\rho(q,p) = Z^{-1} \exp\left[-\beta \left(H(q,p) - \boldsymbol{\Omega} \cdot \mathbf{L}(\mathbf{q},\mathbf{p})\right)\right]. \tag{4.63}$$

This distribution may be obtained directly from Jaynes method by assuming that $\mathbf{L}$ plays the role of an additional observable which is an integral of motion and $\boldsymbol{\Omega}$ is the corresponding Langrange parameter, connected with the mean value of the angular momentum by

$$\langle \mathbf{L} \rangle = -\beta^{-1} \frac{\partial}{\partial \boldsymbol{\Omega}} \ln Z(\beta, \boldsymbol{\Omega}). \tag{4.64}$$

This way we have shown that the Gibbs–Jaynes maximum entropy principle is indeed very powerful, it contains all known distributions for equilibrium situations as special cases. In fact, the principle provides much more useful information (Levine & Tribus, 1978). A few other applications, including non-equilibrium situatations, will be demonstrated below. In non-equilibrium, in general, the linear "ansatz"

(4.54) is no more sufficient, since the means and the dispersion may be independent variables. In order to admit such situations we have to use quadratic functions in the exponent of the distribution function. Examples will be demonstrated in the next Chapter.

Chapter 5

Fluctuations and Linear Irreversible Processes

5.1 Einstein's theory of fluctuations

In this chapter we will present several aspects of the statistical theory of fluctuations and the theory of linear irreversible processes (Haase, 1963; Prigogine, 1967, Keller, 1977; DeGroot & Mazur, 1984, Schwabl, 2000, Schimansky-Geier & Talkner, 2002). This theory is a well developed part of the statistical physics which holds true close to equilibrium. The main result of the statistical theory near to equilibrium is the foundation of a deep connection between fluctuations and dissipation, both phenomena seemingly not related one to the other. One of the main topics is the study of the interaction of the degree of freedom under consideration with the surrounding. In this relation the concrete way of interaction with the particles or dynamic modes of the surrounding is not of interest for the description of the relevant variables. This interaction causes dissipation, i.e. the distribution of concentrated energy to many degrees of freedom and causes deviations from the equilibrium states, fluctuations. Thus as will be shown, the correlation functions and spectra of the fluctuations stand in close relation to the response function on external forces, to the dissipative and transport coefficients as long as the linear approximation around an equilibrium states holds. Unfortunately the extension of this beautiful theory to all systems far from equilibrium is impossible.

The *Boltzmann–Planck principle* developed in Section 4.1 was generalized by *Einstein* and applied to fluctuations in one of his seminal papers published in 1905. Suppose that x should be a fluctuating quantity of a thermodynamic system. Taking the general view of statistical physics, we will assume that x is an explicit function of the microscopic variables.

$$x = x(q_1, \ldots, p_f). \tag{5.1}$$

Suppose further on that we deal with an isolated thermodynamic system, i.e. E, N, V are (macroscopic) constants. We want to find the probability distribution $w(x)$ of the fluctuating value x under the given isoenergetic conditions. The key for the solution of this problem is the statistical definition of the thermodynamic entropy due to Boltzmann and Planck (see Section 4.1). Under the assumption of thermodynamic equilibrium at fixed E, V, N we will introduce a conditional entropy $S(x|E, V, N)$ for the microscopic states where a certain value of x is realized. Geometrically the form $x = const$ is some subset of the hypersurface $E = const$. Therefore, this entropy is determined by the number of microscopic states or in other words by the thermodynamic weight $\Omega(x|E, V, N)$ of the states which correspond to the value of x:

$$S(x|E, V, N) = k_B \ln \Omega(x|E, V, N). \tag{5.2}$$

The probability distribution to find a certain value of x is then defined by

$$\omega(x|E, V, N) = \frac{\Omega(x|E, V, N)}{\Omega(E, V, N)} \tag{5.3}$$

or respectively according to (6.1)

$$\omega(x) = \exp\left[-\frac{1}{k_B T}\left(S(E, V, N) - S(x|E, V, N)\right)\right]. \tag{5.4}$$

Therefore we reduced the information on the thermodynamic system to the knowledge of the distribution of x. The main idea in Einstein's approach consists in the calculation of the entropy difference δS for two states based on thermodynamic relations, i.e. the Gibbsian fundamental equation. Additionally since (6.4) is a normalized distribution we do not have to know the full thermodynamic entropy $S(E, N, V)$, but just relative changes. Further on we set $S(x|E, V, N) = S(x)$. Then obviously, is true that

$$\omega(x) = \frac{\exp[S(x)/k_B]}{\int dx \exp[S(x)/k_B]}. \tag{5.5}$$

The same result may be obtained by application of a projection technique. By introducing a δ-function we collect the microscopic probability corresponding to the value x. The conditional entropy (6.2) then reads

$$S(x|E, N, V) = k_B \ln \int d\Gamma \delta(x - x(q_1, \ldots, p_{3N})) . \delta(H - E). \tag{5.6}$$

The probability distribution follows respectively

$$\omega(x) = \exp[-S(E, V, N)/k_B] \cdot \int d\Gamma \delta(x - x(q_1, \ldots, p_{3N})) \delta(H(q_1, \ldots, p_{3N}) - E). \tag{5.7}$$

This expression is equivalent to the distribution (6.4). The present approach can also be applied to microscopic distributions under different thermodynamic constraints. For example, if x is a fluctuating value of an isothermic-isochoric system the projection procedure yields

$$\omega(x|T,V) = \exp\left[\frac{F(t,V)}{k_BT} - \frac{F(x|T,V)}{k_BT}\right] \quad (5.8)$$

where

$$F(x|T,V) = -kT \ln \int d\Gamma \delta(x - x(q_1,\ldots,p_{3N})) \exp\left[-H(q_1,\ldots,p_{3N}/k_BT\right] \quad (5.9)$$

is the conditional free energy for the isothermic-isochoric system under the condition that x is fixed.

Let us list some general properties of the probability distribution near equilibrium. First of all looking for distributions of x around the equilibrium value x_0 the thermodynamic potentials have extremal properties with respect to x. From the second law follows $S \to$ max, the maximum is reached in equilibrium, i.e. if x reaches the value x_0. Therefore an expansion around a stable equilibrium state gives

$$S(x) = S(x_0) - \frac{1}{2}k_B\beta(x - x_0)^2 \quad (5.10)$$

with

$$\beta = -\frac{1}{k_B}\left(\frac{\partial^2 S}{\partial x^2}\right) > 0. \quad (5.11)$$

This way we obtain that the fluctuations around equilibrium states are Gaussian.

$$\omega(x) = \sqrt{\frac{\beta}{2\pi}} \exp\left[-\frac{1}{2}\beta(x - x_0)^2\right]. \quad (5.12)$$

The standard deviations are determined by the second derivatives of the entropy

$$\langle (x - x_0)^2 \rangle = 1/\beta. \quad (5.13)$$

For other thermodynamic embeddings we obtain in an analogous way Gaussian distributions where β is the positive second derivative of the corresponding thermodynamic potential. As a concrete example we derive the distribution of the fluctuating position of a mechanical spring. It is surrounded by a gas and both the spring and the gas should be in equilibrium under the condition that the overall energy is constant. The change of the entropy of the overall system (gas and spring) can be calculated using the concept of the reversible "*Ersatzprozess*". It yields for small deviations

$$dS = \left(\frac{\partial S}{\partial E}\right)_V dE = -\frac{1}{T}dW_{\min} \quad (5.14)$$

where $W_{\min}$ is the required minimal reversible work which has to be applied to bring the spring out off equilibrium. In our case, if x_0 is the equilibrium position and χ

the elasticity

$$dS = -\frac{1}{T}\chi(x-x_0)dx\,. \tag{5.15}$$

Therefore we arrive at

$$\omega(x) = \sqrt{\frac{\chi}{2\pi k_B T}}\exp\left[-\frac{\chi(x-x_0)^2}{2k_B T}\right]. \tag{5.16}$$

This formula played an important for the experimental determination of the Boltzmann constant k_B.

5.2 Fluctuations of many variables

In the case that there are several fluctuating variables $x_1, \ldots, x_s$, the variable x may be treated as a vector $\boldsymbol{x}$ with components $x_i, (i = 1 \ldots n)$ (Klimontovich, 1982, 1986). The resulting Gaussian distribution near the equilibrium state $\boldsymbol{x}_0$ reads

$$\omega(\boldsymbol{x}) = \sqrt{\frac{\det \beta}{(2\pi)^n}}\exp\left[-\frac{1}{2}\sum \beta_{ij}(x_i - x_{i0})(x_j - x_{j0})\right] \tag{5.17}$$

with the matrix

$$\beta_{ij} = -\frac{1}{k_B}\left(\frac{\partial^2 S}{\partial x_i x_j}\right) \tag{5.18}$$

and the second moments

$$\langle (x_i - x_{i0})(x_j - x_{j0})\rangle = (\beta^{-1})_{ij}\,. \tag{5.19}$$

As an application we determine now the fluctuations of the thermodynamic variables T, V, N etc. of a subvolume which is embedded into a thermal bath with temperature T. Both bath and subvolume together should be adiabatically isolated, i.e. their common energy, volume and particle number E, V, N are constants. Two standard problems are:

(i) V is fluctuating and N is fixed, or
(ii) N is fluctuating and V is fixed.

Here N, V denote the particle number and volume of the subvolume respectively. We have to calculate the entropy change of an occurring fluctuation taking into account the thermodynamic constraints. If the subvolume is a macroscopic body, this entropy change consists of two parts coming from the subvolume and from the bath, respectively

$$\Delta S_{\text{total}} = \Delta S + \Delta S_b\,. \tag{5.20}$$

First we let V be fluctuating and the particle number of the subvolume N be conserved. Conservation of the overall volume and energy makes the entropy change of the bath to be a function of the values for the subvolume

$$\Delta S_b = \frac{\Delta E_b + p_b \Delta V_b}{T_0} = -\frac{\Delta E + p_0 \Delta V}{T_0}. \tag{5.21}$$

Further we expand ΔE in a series for small deviations of the entropy and the volume from its equilibrium values.

$$\Delta E = T_0 \Delta S - p_0 \Delta V + \frac{1}{2}(\Delta T \Delta S - \Delta p \Delta V). \tag{5.22}$$

Combining these relations we find in quadratic approximation the probability for occurring fluctuations of the thermodynamic variables

$$W = A \exp\left[\frac{\Delta p \Delta V - \Delta T \Delta S}{2k_B T_0}\right]. \tag{5.23}$$

Analogously we get for a fixed subvolume and a fluctuating particle number

$$W = A \exp\left[-\frac{\Delta \mu \Delta N + \Delta T \Delta S}{2k_B T_0}\right]. \tag{5.24}$$

We mention that in both cases the values in the exponents correspond to the negative minimal work which has to be applied to bring the subvolume to the nonequilibrium state with $\Delta T, \Delta S$ etc. Now we remember that the deviations from equilibrium are not independent one from each other near to equilibrium. In both expressions only two variable are independent and will govern the behavior of the remaining variables by the caloric and thermic state equation. Selecting in (6.23) ΔT and ΔV as independent variables it determines the behavior of ΔS and Δp

$$\Delta S = \left(\frac{\partial S}{\partial T}\right)_V \Delta T + \left(\frac{\partial S}{\partial V}\right)_T \Delta p, \tag{5.25}$$

$$\Delta p = \left(\frac{\partial p}{\partial T}\right)_V \Delta T + \left(\frac{\partial p}{\partial V}\right)_T \Delta V. \tag{5.26}$$

This way we get the formula

$$W(\Delta T, \Delta V) = A \exp\left[-\frac{C_V}{2k_B T_0^2}(\Delta T)^2 + \frac{1}{2k_B T}\left(\frac{\partial p_0}{\partial V_0}\right)_{T_0}(\Delta V)^2\right]. \tag{5.27}$$

Several interesting questions can be discussed on the basis of this expression:

First of all we see that in the linear approximation near equilibrium, extensive and intensive thermodynamic variables are decoupled. This follows from

$$\langle \Delta T \Delta V \rangle = 0 \tag{5.28}$$

and equivalently for the combination of the other variables. Further on the standard deviations of intensive variables scale with

$$\langle (\Delta T)^2 \rangle = \frac{k_B T_0^2}{C - V} \simeq \frac{1}{V_0}. \tag{5.29}$$

For extensive variables the same scaling is found for the mean square deviations of relative quantities, e.g.

$$\frac{\langle (\Delta V)^2 \rangle}{V_0^2} = k_B T_0 \frac{K_T}{V_0}. \tag{5.30}$$

Here K_T denotes the relative isothermic expansion coefficient

$$K_T = -\frac{1}{V_0} \left(\frac{\partial V_0}{\partial p_0} \right)_{T_0}. \tag{5.31}$$

A third point we want to discuss is the question of thermodynamic stability. Positivity of C_V and K_T is a consequence from thermodynamic inequalities resulting from the second law. Allowing additionally fluctuation of the thermodynamic variables equilibrium is defined as the maxima of the corresponding distributions. For the particle number fluctuations we find starting from (5.24)

$$\langle (\Delta N)^2 \rangle = k_B T_0 \left(\frac{\partial N_0}{\partial \mu_0} \right)_{T_0, V_0}. \tag{5.32}$$

This relation is in close relation to the stability of thermodynamic phases upon the variation of particle numbers.

5.3 Onsager's theory of linear relaxation processes

According to Einstein's view, any macroscopic quantity x may be considered as a fluctuating variable, which is determined by certain probability distribution $\omega(x)$. The mean value is given as the first moment of the probability distribution

$$x_0 = \langle x \rangle = \int x \cdot \omega(x) dx. \tag{5.33}$$

In a stationary state we may shift the origin and assume $x_0 = 0$, without loss of generality. Let us assume now that the stationary state, the target of our investigation, is the state of thermodynamic equilibrium. Then $x_0 = 0$ corresponds to equilibrium and any value of $x(t)$ different from zero is strictly speaking a nonequilibrium state. According to the 2nd Law, there exists a Lyapunov function and therefore the equilibrium state is an attractor of the dynamics (see Section 2.2). The equilibrium state corresponds to a maximum of the entropy. This means in the present situation:

$$S(x = 0) = \max; \tag{5.34}$$

$$\left(\frac{\partial S}{\partial x} \right)_{x=0} = 0; \quad \left(\frac{\partial^2 S}{\partial x^2} \right)_{x=0} \leq 0. \tag{5.35}$$

According to Onsagers's view, the relaxation dynamics of the variable x is determined by the first derivative of the entropy, which is different from zero outside equilibrium. Starting from a deviation from the equilibrium (an entropy value below the maximum) the spontaneous irreversible processes should drive the entropy to increase

$$\frac{d}{dt}S(x) = \frac{\partial S}{\partial x} \cdot \frac{dx}{dt} \geq 0. \tag{5.36}$$

In this expression two factors appear, which were interpreted by Onsager in a quite ingenious way. According to Onsager the derivative

$$X = -\frac{\partial S}{\partial x} \tag{5.37}$$

is considered as the driving force of the relaxation to equilibrium. In irreversible thermodynamics this term is called in analogy to mechanics the *thermodynamic force*, the analogy means that the (negative) entropy takes over the role of a potential. The second term

$$J = -\frac{dx}{dt} \tag{5.38}$$

is considered as the *thermodynamic flux* or *thermodynamic flow*. In a seminal paper, concerned with the question of the relaxation of nonequilibrium states to equilibrium, Onsager (1931) postulated a linear relation between the thermodynamic force and the flux

$$J = LX. \tag{5.39}$$

The idea behind is, that the thermodynamics force is the cause of the thermodynamic flow and both should disappear at the same time. The coefficient L is called *Onsager's phenomenological coefficient*, or *Onsager's kinetic coefficient*. From the 2nd Law follows that the Onsager-coefficients are strictly positive.

$$P = \frac{d}{dt}S(x) = J \cdot X = L \cdot X^2 \geq 0. \tag{5.40}$$

Onsager's postulate about a linear connection between thermodynamic forces and fluxes is the origin of the development of the thermodynamics of linear dissipative system, called also *linear irreversible thermodynamics*. A remarkable property of the linear theory is the bilinearity of the entropy production

$$P = J \cdot X. \tag{5.41}$$

Referring now to the fluctuation theory Eq. (6.10) we find for the neighborhood of the equilibrium state the relation

$$X = -\frac{\partial S}{\partial x} = k_B \beta x. \tag{5.42}$$

Using the previous equations we get finally the following linear relaxation dynamics

$$\dot{x} = -k_B L \beta x. \tag{5.43}$$

With the abbreviation

$$\lambda = L k_B \beta \tag{5.44}$$

being the so-called *relaxation coefficient* of the quantity x we get finally

$$\dot{x} = -\lambda x. \tag{5.45}$$

This linear kinetic equation describes the relaxation of a thermodynamic system brought initially out of equilibrium. Starting with the initial state $x(0)$ the dynamics of the variable $x(t)$ is

$$x(t) = x(0) \exp[-\lambda t]. \tag{5.46}$$

We see that $t_0 = \lambda^{-1}$ plays the role of the decay time of the initial deviation from equilibrium. On the other hand this coefficient which is responsible for the relaxation to equilibrium is in close relation to the fluctuation properties of the considered system. Indeed Eq. (6.44) connects a kinetic property λ with a fluctuation quantity β. In this way we arrived for the first time at a so-called fluctuation-dissipation relation. In fact, Onsager assumed that deviations from equilibrium and fluctuations around the equilibrium observe the same kinetics.

Quite similar we might proceed if we are dealing with several thermodynamic values. Indeed taking the entropy in dependence on $x_i, (i = 1, \ldots, N)$ the entropy production reads

$$S(x_1, \ldots, x_n) = S_{\max} - \frac{1}{2} \beta_{ij} x_i x_j. \tag{5.47}$$

We agree, here and further on, to sum over repeating indices (Einstein's convention).

Following the Onsager ideas described above we get for the thermodynamic forces and fluxes the relations

$$X_i = -k_B \beta_{ij} x_j, \tag{5.48}$$

$$J_i = -\dot{x}_i. \tag{5.49}$$

The generalized linear Onsager-ansatz reads

$$J_i = L_{ij} X_j. \tag{5.50}$$

Again the 2nd law requires positivity of the entropy production. This requires

$$L_{ij} X_i X_j \geq 0 \tag{5.51}$$

for any value of X_i and disappearance only for $X_i = 0, i = 1, \ldots, n$. This corresponds to the requirement of positive definiteness of the matrix L_{ij}. By inserting Eqs. (5.48) and (5.49) into Eq. (5.50) we get

$$\dot{x}_i = -k_B L_{ij} X_j, \tag{5.52}$$

and introducing the matrix of relaxation coefficients of the linear processes near equilibrium states we end up with

$$\dot{x}_i = -\lambda_{ij} x_j, \tag{5.53}$$

$$\lambda_{ij} = k_B L_{ik} \cdot \beta_{kj}. \tag{5.54}$$

Since the matrix β_{ij} determines the dispersion of the stationary fluctuations, we have found again a close relation between fluctuations and dissipation, i.e. we have got a fluctuation-dissipation relation for a set of fluctuating and relaxing variables.

5.4 Correlations and spectra of stationary processes near equilibrium

This section is devoted to the time correlation functions and their spectrum (Klimontovich, 1982, 1986; Schimansky-Geier & Talkner, 2002). The time correlation functions will be defined here as averages over the stationary probability distribution $w(x)$. We consider only stationary processes and the corresponding stationary probability distributions. As a consequence of stationarity all characteristic functions depending on two times t, t' are functions of their time difference $t - t'$ only. We define the correlation function of a variable $x(t)$ as the mean over the product with the same function taken at a later time.

$$C(\tau) = \langle x(t) x(t+\tau) \rangle_t = \int dx\, w(x) x(t) x(t+\tau); \qquad \tau > 0. \tag{5.55}$$

There are two equivalent ways of definition:

(i) the time average over a long (infinite) time interval,
(ii) an ensemble average based on certain probability distribution $w(x)$. For the case of many fluctuating variables $x_i(t), (i = 1, \ldots, n)$ the time correlation function is defined as

$$C_{ij}(\tau) = \langle x_i(t) x_j(t+\tau) \rangle_t = \int dx_1, \ldots, dx_n w(x_1, \ldots, x_n) x_i(t) x_j(t+\tau) \tag{5.56}$$

where $\tau > 0$ and $w(x_1, \ldots, x_n)$ stands for the simultaneous probability distribution of all x_i-variables.

We will study now several general properties of the time correlation functions.

- The first (evident) property is:
  The time correlation functions should vanish for infinitely large time and be equal to the covariance coefficient for small time

$$\lim_{t\to\infty} C_{ij}(\tau) = 0; \qquad C_{ij}(\tau = 0) = \langle x_i x_j \rangle. \tag{5.57}$$

- A second property follows immediately from the stationarity: Since $C_{ij}(\tau)$ does not depend on the actual time we find by substituting $t \to t' - \tau$

$$C_{ij}(\tau) = \langle x_i(t) x_j(t+\tau) \rangle = \langle x_i(t'-\tau) x_j(t') \rangle = C_{ji}(-\tau). \tag{5.58}$$

We introduce now the Fourier-component of the time correlation function

$$S_{ij}(\omega) = \int_{-\infty}^{\infty} d\tau C_{ij}(\tau) \exp[i\omega\tau]. \tag{5.59}$$

This matrix function is called the spectrum of the fluctuating values. It can be calculated directly from the time correlation function via Eq. (5.59). Another way is by the analysis of the dynamics of the fluctuating values. We introduce first the Fourier-components of $x_i(t)$

$$x_{i\omega} = \int_{-\infty}^{+\infty} dt x_i(t) \exp[i\omega t]. \tag{5.60}$$

We multiply this expression with $x_{j\omega'}$ and average over the stationary distribution

$$\langle x_{i\omega} x_{j\omega'} \rangle = \int\!\!\int_{-\infty}^{+\infty} dt dt' \langle x_i(t) x_j(t') \rangle \exp[i(\omega+\omega')t + i\omega'(t'-t)]. \tag{5.61}$$

Due to the stationarity the correlator $\langle x_i(t) x_j(t') \rangle$ depends on the difference $t' - t$

$$\langle x_{i\omega} x_{j\omega'} \rangle = 2\pi\delta(\omega+\omega') S_{ij}(\omega). \tag{5.62}$$

This is a form of the *Wiener–Khintchin-theorem* which connects the averaged product of the modes of a fluctuating stationary system with the spectrum of the fluctuating values. Later on we will make use of this equation.

In the last section we derived equations for the linear relaxation dynamics of macroscopic variables. Onsager stated that these equations are valid for the relaxation of fluctuations too. Indeed, in the derivation of the relaxation dynamics we never made a statement whether the initial nonequilibrium state has been prepared, as a result of an external force like usually by considering the dynamics of mean values or as the result of a spontaneous fluctuation. This way Onsager postulated the validity of these equations for the regression of fluctuating variables. As a consequence we may calculate the dynamic characteristics of fluctuations like the time correlation function from the relaxation kinetics. There exist several approaches to prove the Onsager postulate (Klimontovich, 1982, 1986, 1995; Landau & Lifshits, 1990), we take it here as a quite evident hypothesis.

In order to calculate the correlation functions we start from the relaxation equations for the variables $x_i(t)$ which reads in the simplest case of one component $\dot{x}(t) = -\lambda x(t)$ (see previous section). We assume now that this relation is valid also for a deviation caused by a spontaneous fluctuation. We multiply the relaxation equation with the initial value $x(0)$ and find

$$\frac{d}{dt}(x(t)x(0)) = -\lambda(x(t)x(0)). \qquad (5.63)$$

After averaging with respect to an ensemble of realizations we get a kinetic equation for the time correlation function

$$\frac{d}{d\tau}C(\tau) = -\lambda C(\tau) \qquad (5.64)$$

with the initial conditions

$$C(\tau = 0) = \langle x^2 \rangle = \int x w dx = \beta^{-1}. \qquad (5.65)$$

By integrating (5.64) we find the explicit expression for the correlation function

$$C(\tau) = \frac{1}{\beta}\exp[-\lambda|\tau|]. \qquad (5.66)$$

The generalization to several fluctuating variables is straightforward. The application of Onsager's regression hypothesis leads to the kinetic equations

$$\frac{d}{d\tau}C_{ij}(\tau) = -\lambda_{ik}C_{kj} \qquad (5.67)$$

with the initial condition

$$C_{ij}(0) = \langle x_i x_j \rangle. \qquad (5.68)$$

The most elegant method to solve these equations are one-sided Fourier-transforms (Klimontovich, 1984). We represent the time correlation functions as

$$S_{ij}^+(\omega) = \int_0^\infty d\tau C_{ij}(\tau)\exp[i\omega\tau]. \qquad (5.69)$$

The negative part of the spectrum is just the complex conjugate

$$S_{ij}^-(\omega) = \int_{-\infty}^0 d\tau C_{ij}(\tau)\exp[i\omega\tau] = [S_{ij}^+(\omega)]^*. \qquad (5.70)$$

Taking into account the initial conditions we find for the positive part

$$(-i\omega\delta_{ik} + \lambda_{ik})S_{kj}^+(\omega) = \langle x_i x_j \rangle. \qquad (5.71)$$

From here on we consider C_{ij} and λ_{ik}, and $S_{ij}^\pm$ as elements of matrices $\mathbf{C}$ and Λ and $\mathbf{S}^\pm$. We moreover introduce the matrix $\mathbf{B} = \mathbf{C}^{-1}(0)$, and use the notation $\mathbf{I}$ for the unit matrix with elements δ_{ij}.

By adding positive and negative parts we get the complete spectrum

$$\mathbf{S}(\omega) = (-i\omega\mathbf{I} + \mathbf{\Lambda})^{-1}\mathbf{B}^{-1} + \mathbf{B}^{-1}(-i\omega\mathbf{I} + \mathbf{\Lambda})^{-1}. \tag{5.72}$$

Here we have used the matrix inversion of Eq. (5.71) and the symmetry relation $C_{ij}(\tau) = C_{ji}(-\tau)$, valid for stationary processes. The correlation function $C_{ij}(\tau)$ follows from the inverse Fourier transform. In order to illustrate this procedure we take as a simple example the motion of a heavy particle in a viscous liquid. In the next chapters we will study this case of Brownian motion in more detail. We assume for the velocity of the particle the equation of motion

$$\dot{v} = -\gamma_0 v. \tag{5.73}$$

For the stationary correlation function we get

$$\frac{d}{d\tau}\langle v(t+\tau)v(t)\rangle = -\gamma_0 \langle v(t+\tau)v(t)\rangle \tag{5.74}$$

with the initial condition

$$\langle v(t)^2 \rangle = \frac{k_B T}{m}. \tag{5.75}$$

Solving these equations we find the correlation function

$$\langle v(t+\tau)v(t)\rangle = \frac{k_B T}{m} \exp[-\gamma|\tau|] \tag{5.76}$$

with the spectrum

$$S_{vv}(\omega) = \frac{k_B T}{m} \frac{2\gamma_0}{\gamma_0^2 + \omega^2}. \tag{5.77}$$

Sometimes this frequency distribution is called a red spectrum, since the maximum of intensity is at low ω. A more rich structures of the spectrum is obtained if the particle is additionally under the influence of a harmonic force. Then the dynamical equations read

$$\dot{x} = v; \qquad \dot{v} = -\gamma_0 v - \omega_0^2 x \tag{5.78}$$

and the corresponding system of equations for the correlation functions is

$$\dot{C}_{xv}(\tau) = C_{vv}(\tau); \qquad \dot{C}_{vv} = -\gamma_0 C_{vv} - \omega_0^2 C_{xv}. \tag{5.79}$$

Following the approach described above we find the spectrum

$$S_{vv}(\omega) = \frac{k_B T}{m} \frac{2\gamma_0}{\gamma_0^2 + (\omega - \omega_0^2/\omega)^2}. \tag{5.80}$$

This distribution has a peak at $\omega \simeq \omega_0$, a so-called resonance. Details and applications will be discussed in the next Chapters.

5.5 Symmetry relations and generalizations

At the end of this chapter we want to come back to the Onsager matrix of phenomenological coefficients L_{ij}, relating the thermodynamic fluxes and forces near to equilibrium. Empirically it was found that the matrix is symmetrical with respect to exchange of the indices and in a few cases one observes antisymmetry

$$L_{ji} = \pm L_{ij}. \tag{5.81}$$

This macroscopic property means that if a force X_j will induce the thermodynamic flux J_i we will find also a flux J_j generated by the force X_i. This circumstance, was also found already in Onsager's paper.

For the proof of this relation we use properties of the correlation function, in particular $C_{ij}(\tau) = C_{ji}(-\tau)$. The second fact which has to be taken into account is that the fluctuating values are functions of the microscopic variables. Therefore it holds

$$x_i(t) = \epsilon_i x_j(-t) \tag{5.82}$$

which expresses the reversibility of the microscopic motion. Here ϵ is the parity coefficient which is $+1$ for even variables and -1 for odd ones. It yields that

$$C_{ij}(\tau) = \langle x_i(t) x_j(t+\tau) \rangle = \epsilon_i \epsilon_j \langle x_i(-t) x_j(-t-\tau) \rangle = \epsilon_i \epsilon_j C_{ij}(-\tau). \tag{5.83}$$

Taking into account the stationarity condition we find

$$C_{ij}(\tau) = \epsilon_i \epsilon_j C_{ji}(\tau). \tag{5.84}$$

As a consequence of the general regression dynamics we find the following dynamical equations for the correlation functions

$$\frac{d}{d\tau} C_{ij}(\tau) = -\lambda_{ik} C_{kj}(\tau). \tag{5.85}$$

Hence we obtain

$$\lambda_{ik} C_{kj}(\tau) = \epsilon_i \epsilon_j \lambda_{jk} C_{ki}(\tau). \tag{5.86}$$

For $\tau = 0$ the correlation function reduces to the standard deviation with $C_{ij}(0) = [\beta_{ij}]^{-1}$; using further the definition of the relaxation coefficients λ_{ij} we get finally the famous *Onsager-Casimir symmetry relation*

$$L_{ji} = \epsilon_i \epsilon_j L_{ij}. \tag{5.87}$$

This relation is one of the fundamentals of linear irreversible thermodynamics; it is well confirmed by many experiments.

The approach presented so far in this Chapter is restricted to irreversible processes close to equilibrium. The extension to far from equilibrium situations is extremely difficult. There exist many approaches to solve this problem. We mention the early work of Onsager and Machlup (1953), of Machlup and Onsager (1953), of Ginzburg and Landau (1965), and of Glansdorff and Prigogine (1971). An advanced theory of nonlinear irreversible processes is due to the work of the late Rouslan Stratonovich (Stratonovich, 1994). We mention also the extended thermodynamics and other new approaches (Muschik, 1988; Jou, Casas-Vazques & Lebon, 1993; Ebeling & Muschik, 1993; Luzzi, Vasconcellos & Ramos, 2000). Special attention deserves a new approach (Grmela & Öttinger, 1997), named GENERIC (general equations for the nonequilibrium reversible-irreversible coupling), which seems to contain most of the theories mentioned above as special cases (Öttinger, 2005). We will give here only one of the basic ideas. The general time-evolution equation postulated by Grmela and Öttinger is of the following structure:

$$\frac{dx}{dt} = L\frac{\delta E}{\delta x} + M\frac{\delta S}{\delta x}. \tag{5.88}$$

Here x represents a set of independent variables, in many cases x will depend on continuous position-dependent fields, such as mass, momentum, and energy densities. Further, E and S are the total energy and entropy expressed in terms of the state variables, and L and M are certain linear operators or matrices. The application of a linear operator may include integrations over continuous labels and then $\delta/\delta x$ typically implies functional rather than partial derivatives.

There is no room here to go into the details of these more or less fruitful but not exhaustive methods. The whole field is still in development.

Chapter 6
Nonequilibrium Distribution Functions

6.1 A simple example — driven Brownian particles

The general theory of nonequilibrium systems is still in the first stages of development. At present we have a well developed theory of ideal gases going back to *Boltzmann*. We presented the main ideas of this rather old theory already in the 3rd Chapter. Further we have a nice theory for small (linear) deviations from equilibrium, which essentially goes back to *Onsager*, this theory was presented in Chapter 5. A theory of the same generality as the Gibbs theory of equilibrium systems does not exist. In particular the formulation of the statistical mechanics of far-from equilibrium systems is an extremely difficult task which is full of surprises and interesting applications to many interdisciplinary fields (Haken, 1973, 1983; Klimontovich, 1995: Zubarev, 1976: Schimansky-Geier & Pöschel, 1997). On the other hand we may treat many special examples, several of them will presented in this and in the following sections.

In order to bring a system to nonequilibrium we need some driving force. Historically the first case of a treatment of driven systems is connected with the theory of sound developed in the 19th century by Helmholtz and Rayleigh. The treatment of these early models will lead us later to other interesting applications as the theory of driven Brownian particles. This is a unification of the model of Brownian motion due to Einstein, Smoluchowski and Langevin with the model of (acoustic) oscillators driven by negative friction developed by Rayleigh, Helmholtz, van der Pol and many other workers. We will use here the definitions and results introduced in Sections 2.4 and 2.5.

Let us study at the beginning a force-free Brownian particles on a line ($d = 1$) under equilibrium conditions. Using the phenomenological method we find the Maxwellian distribution, which is a special case of the canonical distribution function defined in Sections 4.3–4.4:

$$f_0(\mathbf{v}) = \rho(H) = C \exp\left[-\frac{mv^2}{2k_B T}\right]. \tag{6.1}$$

There is an alternative method to derive this distribution based on the Langevin–Fokker–Planck method. We start from the Langevin equation (see Section 2.5)

$$\frac{dv}{dt} = -\gamma_0 v + (2D_v)^{1/2}\xi(t) \tag{6.2}$$

which describes standard Brownian motion in equilibrium systems (γ_0 — friction coefficient, D_v — diffusion constant for the velocities). The corresponding Fokker–Planck equation reads (see Section 2.5)

$$\frac{\partial P(v,t)}{\partial t} = \frac{\partial}{\partial v}\left[\gamma_0 v P(v,t) + D_v \frac{\partial P(v,t)}{\partial v}\right] \tag{6.3}$$

This equation is solved by the Gaussian distribution

$$f_0(v) = C \exp\left[-\frac{\gamma_0 v^2}{2D_v}\right]. \tag{6.4}$$

The Gaussian corresponds to the equilibrium Maxwell distribution just in the case that the so-called *Einstein relation*

$$D_v = \frac{k_B T \gamma_0}{m} \tag{6.5}$$

holds. This case will be studied in more detail in Chapters 6–9. Here we are more interested in the transition to non-equilibrium situations when Eqs. (6.4) and (6.5) are not observed. In this case the system might be driven away from equilibrium; then we speak about active Brownian motion. The corresponding Langevin equation of motion reads ($m = 1$):

$$\frac{dv}{dt} = F(v) + (2D(v))^{1/2}\xi(t). \tag{6.6}$$

Here $F(v)$ is the dissipative force acting on the particle and $D(v)$ a diffusion function. The simplest case

$$F(v) = -\gamma_0 v; \qquad D(v) = D_v = const \tag{6.7}$$

leads as back to standard Brownian motion. Now we will consider nonequilibrium situations corresponding to nonlinear expressions for the force. Models of nonlinear dissipative forces were first considered for a one-dimensional oscillation problem by Rayleigh (1894) who studied in his "*Theory of Sound*" the case

$$F(v) = (a - bv^2)v; \qquad D(v) = 0 \tag{6.8}$$

with $b > 0$. Then the equation of motion has for $a < 0$ the only stationary state $v = 0$ and for $a > 0$ it possesses two stationary states

$$v = \pm v_0; \qquad v_0 = \sqrt{\frac{a}{b}}. \tag{6.9}$$

The point $a = 0$ corresponds to a kinetic phase transition which possesses quite interesting properties (Klimontovich, 1982, 1986; Haken, 1983; Horsthemke & Levefer, 1984). Klimontovich (1982, 1986) studied this system in much detail including a Langevin source with constant noise. Assuming $m = 1$ the Hamiltonian is $H = v^2/2$. We see that the dissipative force is fully determined by the Hamiltonian

$$F(v) = -(a - 2bH)v. \tag{6.10}$$

Such systems are called "canonical dissipative". This rather interesting class of systems will be analyzed in more detail in the next section. The Fokker–Planck equation reads now

$$\frac{\partial P(v,t)}{\partial t} = \frac{\partial}{\partial v} \left[(2bH - a)vP(v,t) + D_v \frac{\partial P(v,t)}{\partial v} \right]. \tag{6.11}$$

This equation has the following solution which represents a stationary distribution function (Klimontovich, 1982, 1986)

$$f_0(v) = \rho(H) = C \exp\left[-\frac{aH - bH^2}{D_v}\right]. \tag{6.12}$$

For $a < 0$ this distribution is quite similar to a Maxwellian. For $a > 0$ the system is driven away from equilibrium. In this case the velocity distribution is bistable and has two maxima in the velocity space. The maximum of the energy distribution corresponds to

$$H_0 = \frac{v_0^2}{2} = \frac{a}{2b}; \qquad v_0^2 = \frac{a}{b}.$$

This corresponds to a system of particles which move all either with velocity v_0 to the right or with $-v_0$ to the left. The picture is like a hydrodynamic flow to the right or to the left. Lateron we will discuss interesting biological applications to the movement of swarms (see Chapter 12).

The theory given above may be easily generalized to self-oscillating systems of a special type, discussed briefly already in Section 2.4, which have the equation of motion (Klimontovich, 1982, 1986)

$$\frac{dv}{dt} + \omega_0^2 x = (a - bH)v + (2D_v)^{1/2}\xi(t). \tag{6.13}$$

Here the Hamiltonian is given by ($m = 1$):

$$H = \frac{1}{2}v^2 + \frac{1}{2}\omega_0^2 x^2. \tag{6.14}$$

Again the friction force is determined by a Hamiltonian, corresponding to the so-called canonical-dissipative structure. Let us first discuss the stationary state for $D = 0$. An exact solution of the dynamic equations is

$$x(t) = (v_0/\omega_0)\sin(\omega_0 t + \delta); \qquad v(t) = v_0 \cos(\omega_0 t + \delta). \qquad (6.15)$$

The stationarity requires

$$H = H_0 = v_0^2 = a/b. \qquad (6.16)$$

This is a sustained oscillation with stationary amplitude $x_0 = v_0/\omega_0$. Due to the canonical-dissipative character we may find again an exact solution for the stationary Fokker–Planck equation, which is similar to Eq. (6.12). We find

$$f_0(v) = \rho(H) = C \exp\left[-\frac{aH - bH^2/2}{D_v}\right]. \qquad (6.17)$$

Here we have to observe the different expression for the Hamiltonian H and consequently also for the normalization C. For the passive case $a < 0$ we may neglect the nonlinearity putting $b = 0$ and find simply a Maxwellian.

$$f_0(v) = \rho(H) = C \exp\left[-\frac{-|a|H}{D_v}\right]. \qquad (6.18)$$

For $a = 0$ we observe in the deterministic system a bifurcation which leads to the auto-oscillating regime. The distribution function for the transition point reads:

$$f_0(v) = \rho(H) = C \exp\left[-\frac{bH^2}{D_v}\right]. \qquad (6.19)$$

We notice the large dispersion which is characteristic for all phase transitions. The first moments of this distribution are (Klimontovich, 1982, 1986)

$$\langle H \rangle = \frac{2}{\sqrt{\pi}}\sqrt{\frac{D_v}{2b}}; \qquad \langle H^2 \rangle = \frac{D_v}{b}. \qquad (6.20)$$

In the case of driven motion $a > 0$, that means for the regime of developed oscillations, the distribution reads

$$f_0(v) = \rho(H) = C \exp\left[-\frac{b(H - a/b)^2}{2D_v}\right]. \qquad (6.21)$$

where the constant follows again from normalization (Klimontovich, 1982, 1986).

$$C = \sqrt{\frac{2b}{\pi D}}\left[1 + \Phi\left(\frac{a}{\sqrt{2bD_v}}\right)\right]^{-1}.$$

Here $\Phi(x)$ is the standard error function. We can easily check that the curve of maximal probability has the shape of a crater. In the limit of small noise ($D_v \to 0$) the curve of maximal probability is exactly above the deterministic limit cycle. This is, of course, a necessary condition which any correct solution of the Fokker–Planck equation has to fulfill. Further we note, that the exponent $(H - a/b)^2$ correspond to the Lyapunov function of the system identified in Section 2.4. We found this way a model system which is exactly solvable in equilibrium and for any distance from equilibrium. Admittedly the force function which we used $F = (a - bH)v$ is not very realistic from the physical point of view, however it may be shown, that the more realistic Rayleigh force $F = (a - bv^2)v$ introduced for modeling sound oscillations as well as the van der Pol force $F = (a - bx^2)$ introduced for modeling electric oscillations may be converted in good approximation to our canonical-dissipative force $F = (a - bH)v$. This may be shown by using the procedure of phase-averaging (Klimontovich, 1982, 1986). We may use our model system as a good testing ground for more advanced methods of nonequilibrium theory. A special conclusion is, that the stationary distribution functions may depend not only on invariants as in our case H, but also on higher powers of the invariants, as in our case H^2.

6.2 Canonical-dissipative systems

As we have seen in the previous section, there exist systems, were the dissipative forces are determined by the Hamiltonian, which allows for exact solutions. Now we will generalize this concept and treat a whole class of solvable systems. This idea is mainly based on works of *Haken and Graham* (Haken, 1973, Graham, 1981). We will show that at least for this special class of far from equilibrium systems, the so-called canonical-dissipative systems, a general ensemble theory similar to Gibbs approach may be developed (Graham, 1981; Ebeling, 1981, 2000, 2002; Feistel and Ebeling, 1989). The theory of canonical-dissipative systems is the result of an extension of the statistical physics of Hamiltonian systems to a special type of dissipative systems where conservative and dissipative elements of the dynamics are both determined only by invariants of the mechanical motion. This theory is in close relation to the simple example of driven Brownian motion and auto-oscillating systems, presented in the first section of this Chapter. Later we will show a close relation to a recently developed theory of active Brownian particles (Schweitzer et al., 1998; Ebeling et al., 1999; Erdmann et al., 2000; Schweitzer et al., 2001; Schweitzer, 2003). The main ideas of the theory of active Brownian motion will be explained in Chapter 12.

We start the development of the theory of canonical-dissipative systems with a rather general study of the phase space dynamics of a driven many-particle system with f degrees of freedom $i = 1, \ldots, f$. Assuming that the Hamiltonian is given by $H(q_1 \ldots q_f p_1 \ldots p_f)$ the mechanical motion is given by Hamilton equations. The solutions are trajectories on the plane $H = E = const$. The constant energy $E = H(t = 0)$ is given by the initial conditions, which are (in certain limits) arbitrary. We construct now a canonical-dissipative system with the same Hamiltonian (Haken

1973; Graham 1981; Ebeling, 2000; Schweitzer et al., 2001)

$$\frac{dp_i}{dt} = -\frac{\partial H}{\partial q_i} - g(H)\frac{\partial H}{\partial p_i}. \tag{6.22}$$

We will assume that the dissipation function $g(H)$ is nondecreasing. Equation (6.22) defines a canonical-dissipative system which does not conserve the energy. In regions of the phase space where $g(H)$ is positive, the energy decays and in regions where $g(H)$ is negative, the energy increases. The simplest possibility is constant friction $g(H) = \gamma_0 > 0$ which corresponds to a decay of the energy to the ground state. Of more interest is the case when the dissipative function has a root $g(E_0) = 0$ at a given energy E_0. Then the states with $H < E_0$ are supported with energy, and from states with $H > E_0$ energy is extracted. Therefore any given initial state with $H(0) < E_0$ will increase its energy up to reaching the shell $H(t) = E_0$ and any given initial state with $H(0) > E_0$ will decrease its energy up to the moment when the shell $H(t) = E_0$ is reached. Therefore $H = E_0$ is an attractor of the dynamics, any solution of Eq. (6.22) converges to the surface $H = E_0$. On the surface $H = E_0$ itself the solution corresponds to a solution of the original Hamiltonian equations for $H = E_0$. The simplest dissipation function with the wanted properties is a linear function

$$g(H) = c(H - E_0). \tag{6.23}$$

The speed of the relaxation process is proportional to c^{-1}. The linear dissipative function (6.23) has found applications to Toda chains (Makarov et al., 1999). More general dissipative functions were considered in the theory of active Brownian motions (Schweitzer et al., 1998; Ebeling et al., 1999; Erdmann et al., 2000). We mention that all noninteracting systems $H = H(\mathbf{p}^2)$ with $g = g(\mathbf{p}^2)$ are canonical-dissipative. The attractor of the dissipative system (5.22) is located on the surface $H = E_0$. This does not mean that the full $(2f - 1)$-dimensional surface is the attractor of the system. Such a statement is correct only for the case $f = 1$, which has been considered in the last section, further this statement may be true also for systems which are ergodic on the surface $H = E_0$. In the general case the attractor may be any subset of lower dimension, possibly even a fractal structure.

Let us consider for example the case of one particle moving in an external field with radial symmetry

$$H = \frac{\mathbf{p}^2}{2m} + U(r); \qquad \mathbf{p} = m\mathbf{v}. \tag{6.24}$$

Then the equation of motion reads

$$\frac{d\mathbf{p}}{dt} = -\frac{\partial H}{\partial \mathbf{r}} - g(H)\mathbf{v}. \tag{6.25}$$

The corresponding equation for the angular momentum $\boldsymbol{L} = \boldsymbol{r} \times \boldsymbol{p}$ reads

$$\frac{d\boldsymbol{L}}{dt} = -\frac{1}{m}g(H)\boldsymbol{L}.\tag{6.26}$$

We see that on surfaces with $g(H) = 0$ obligatory $\boldsymbol{L} = \boldsymbol{L_0} = const$ holds. Of special interest are cases where this constant is different from zero. In other words, the system shows rotations. The concrete value of $\boldsymbol{L_0}$ may be obtained by explicit solutions of the equations of motion on the surface $H = E_0$.

A more general class of canonical-dissipative systems is obtained, if beside the Hamiltonian also other invariants of motion are introduced into the driving functions. Let us assume that the driving functions depend on $H = I_0$ and also on some other invariants of motion $I_0, I_1, I_2, \ldots, I_s$ for example

- $I_1 = \boldsymbol{P}$ — total momentum of the system,
- $I_2 = \boldsymbol{L}$ — total angular momentum of the system. etc.

For the equation of motion we postulate

$$\frac{dp_i}{dt} = -\frac{\partial H}{\partial q_i} - \frac{\partial G(I_0, I_1, I_2, \ldots)}{\partial p_i}.\tag{6.27}$$

We include now an external white noise source restricting now our study to the case were the dynamics is determined by H. The Langevin equations read

$$\frac{dp_i}{dt} = -\frac{\partial H}{\partial q_i} - g(H)\frac{\partial H}{\partial p_i} + (2D(H))^{1/2}\xi(t).\tag{6.28}$$

The essential assumption is, that noise and dissipation depend only on H. The corresponding Fokker–Planck equation reads

$$\frac{\partial \rho}{\partial t} + \sum \frac{\partial H}{\partial p_i}\frac{\partial \rho}{\partial q_i} - \sum \frac{\partial H}{\partial q_i}\frac{\partial \rho}{\partial p_i}\tag{6.29}$$

$$= \sum \frac{\partial}{\partial p_i}\left[g(H)\frac{\partial H}{\partial p_i}\rho + D(H)\frac{\partial \rho}{\partial p_i}\right].\tag{6.30}$$

An exact stationary solution is

$$\rho_0(q_1 \ldots q_f p_1 \ldots p_f) = Q^{-1}\exp\left(-\int_0^H dH'\,\frac{g(H')}{D(H')}\right).\tag{6.31}$$

The derivative of ρ_0 vanishes if $g(H = E_0) = 0$. This means the probability is maximal at the surface $H = E_0$.

For the special case of a linear dissipation function we find a stationary solution

$$\rho_0(q_1 \ldots q_f p_1 \ldots p_f) = Q^{-1}\exp\left(\frac{cH(2E_0 - H)}{2D}\right) = Q_1^{-1}\exp\left(\frac{-c(H - E_0)^2}{2D}\right).\tag{6.32}$$

The problem with these distributions is, that they might be formally exact but nevertheless lacking physical meaning. In particular, distributions of type (6.31) do not admit translational or rotational flows, since

$$\langle p_i \rangle = 0; \qquad \langle L_i \rangle = 0 \qquad (6.33)$$

holds for symmetry reasons. Let us consider for example a mass point rotating in a central field on a plane. Then $H = E_0$ is a sphere in a 4-dimensional space. Since the angular momentum has only two possible directions (up or down), corresponding to right or left rotations, the system is unable to fill the whole surface $H = E_0$ but just a part of it, compatible with these two possibilities of left or right rotations. The easiest way to admit rotations is to include, as for the equilibrium case explained in Section 4.4, the invariant $\boldsymbol{\Omega} \cdot \boldsymbol{L}$ into the distribution

$$\rho_0(q_1 \ldots q_f p_1 \ldots p_f) = Z^{-1}(\boldsymbol{\Omega}') \exp\left(-\int_0^H dH' \frac{g(H')}{D(H')} + \boldsymbol{\Omega}' \cdot \boldsymbol{L}\right). \qquad (6.34)$$

This distribution admits rotations due to the different symmetry character. The mean value of the angular momentum is given by

$$\langle \boldsymbol{L} \rangle = \frac{\partial \ln Z(\boldsymbol{\Omega}')}{\partial \boldsymbol{\Omega}'}. \qquad (6.35)$$

By replacing the mean value by the deterministic value $\boldsymbol{L}_0$

$$\boldsymbol{L}_0 = \frac{\partial \ln Z(\boldsymbol{\Omega}')}{\partial \boldsymbol{\Omega}'} \qquad (6.36)$$

we may get a good approximate solution, reflecting the most important physical properties of our dynamical system. More general forms of the distribution will be discussed in Section 5.5. The existence of exact solutions for the probability distributions admits to derive the thermodynamic functions as the mean energy and the entropy. The system has further a Lyapunov functional K which is provided by the Kullback entropy which is a nonincreasing function

$$K[\rho, \rho_0] = \int dq_1 \ldots dq_f dp_1 \ldots dp_f \rho \log[\rho/\rho_0]. \qquad (6.37)$$

This theorem governs the approach to the stationary state.

6.3 Microcanonical non-equilibrium ensembles

As shown above, canonical-dissipative forces drive the system to certain subspaces of the energy surface, where the total momentum or the angular momentum are fixed. In many cases the system is ergodic on these surfaces; this question has to be checked separately for any special case. Assuming that ergodicity (quasi-ergodicity) is given we may postulate that in the long run the measure of the trajectories is equally distributed on certain shells around the surfaces

$$H(q_1 \ldots q_f p_1 \ldots p_f) = E_0, \tag{6.38}$$

$$I_k(q_1 \ldots q_f p_1 \ldots p_f) = I_k, k = 2, 3, \ldots, s. \tag{6.39}$$

The idea about the ergodicity of the trajectories leads us to microcanonical ensembles. There exist many examples of physical systems which are well described by microcanonical ensembles (Gross, 2001). Here we consider systems which are uniform in space but far from equilibrium. As examples may serve laser plasmas, homogeneous turbulent fluids and systems of active Brownian particles. These systems have in common, that they are driven by energy supply to far from equilibrium states. We construct now a special non-equilibrium ensemble which is characterized by a constant probability density on a energy shell

$$E_0 - \frac{1}{2}\delta E \leq H(q_1 \ldots q_f p_1 \ldots p_f) \leq E_0 + \frac{1}{2}\delta E. \tag{6.40}$$

This is a rather restrictive assumption, in particular it means that due to the symmetry of the system all mean fluxes are zero. The mean energy of the ensemble is fixed $\langle H \rangle = E_0 = U$ and the entropy is given by the volume of the energy shell according to Boltzmann's formula. Since our system is not in thermal equilibrium, not all thermodynamic relations, e.g. the relation between energy and temperature, are valid. The most typical properties of an equilibrium ensemble is, that the mean energy is proportional to the noise level T and that the mean quadratic deviation $\langle (\delta H)^2 \rangle$ is proportional to T^2. In the equilibrium all energy comes from the thermal fluctuations. In nonequilibrium the energy of the nonlinear excitations and the noise energy are decoupled. In other words, the energy $\langle H \rangle = U \simeq E_0$ is given mostly by the properties of the energy source and is nearly independent of the noise level which we will denote from now by D. On the other hand the energy fluctuations depend strongly on the noise strength D and are nearly independent on the energy of the excitations

$$\langle \delta E^2 \rangle = \left(\langle H^2 \rangle - \langle H \rangle^2 \right) \simeq D. \tag{6.41}$$

We see immediately that the canonical distribution is not compatible with these properties. By constructing other exponential distribution functions from the maximum entropy principle we have to observe the correct behavior of the mean and the dispersion (Ebeling & Röpke, 2004). The most simple example of a distribution with the right properties is the Gaussian Eq. (6.32). As an example may serve a nonequilibrium gas where all particles move with the same modulus of the velocity $v_i = v_0^2$, a so-called isokinetic ensemble (Hoover, 2001). The direction of the velocities fluctuates stochastically. A physical system which behaves in such a way is a fluid in the state of uniform turbulence. Other examples are strongly excited laser plasmas and active Brownian particles.

In a generalization of the approach described above we assume that beside the energy also some other invariants of motion as e.g. the momentum or the angular momentum are conserved. Then the generalized microcanonical ensemble assumes that the probability density is constant on the shells around the prescribed invariants

$$I_k - \frac{1}{2}\delta I_k \leq I_k(q_1 \ldots q_f p_1 \ldots p_f) \leq I_k + \frac{1}{2}\delta I_k \tag{6.42}$$

for $k = 0, 1, \ldots, s$. This means that the density is concentrated on certain submanifolds of the energy shell $k = 0$. We note that the invariants are not necessarily smooth functions. Fluxes may be prescribed as far as they are expressed by invariants of motion, e.g a macroscopic flow may be prescribed by the total momentum $\boldsymbol{I}_1 = \boldsymbol{P}$. The entropy is given by the Boltzmann formula

$$S = k_B \log \Omega(I_0, I_1, \ldots, I_s). \tag{6.43}$$

where Ω expresses the volume of the manifold defined by the fixed invariants of motion.

6.4 Systems driven by energy depots

We started this Chapter with an example of a driven system far from equilibrium which goes back to to Klimontovich (1982, 1986, 1995). The simple Klimontovich system, which allows a full treatment, is a unification of the model of Brownian motion due to Einstein, Smoluchowski and Langevin with the model of (acoustic) oscillators driven by negative friction which originally is due to Rayleigh and Helmholtz. A "minus" of this nice model is it purely phenomenological character. We did not touch at all the question, where the forces, driving the system out of equilibrium, may come from. Let us start from a Langevin equation for a particle with coordinate $\boldsymbol{r}$ and velocity $\boldsymbol{v}$:

$$m\frac{d\boldsymbol{v}}{dt} + \nabla U(\boldsymbol{r}) = \boldsymbol{F}(\boldsymbol{v}) + m(2D_v)^{1/2}\boldsymbol{\xi}(t). \tag{6.44}$$

Now we will derive the dissipative force $\boldsymbol{F}(\boldsymbol{v})$ in a more physical way from energy depot models. There are many possibilities to couple a system to energy reservoirs. We will start with the treatment of two different variants. The first variant of the depot model (SET-model) developed in (Schweitzer et al., 1998, Ebeling et al., 1999) is based on the ansatz

$$\boldsymbol{F}(\boldsymbol{v}) = m\boldsymbol{v}(de - \gamma_0) \tag{6.45}$$

where e is the energy content of a depot and d a conversion parameter. The first term expresses an acceleration in the direction of $\boldsymbol{v}$. The second term $m\gamma_0\boldsymbol{v}$ is the usual passive friction, which by assumption is connected with the noise by an Einstein relation $D_v = \gamma_0 k_B T/m$. We assume further that the Brownian particles are able to take up energy with the rate q, which can be stored in the depot e. This internal energy can be converted into kinetic energy with a momentum dependent rate $d\boldsymbol{v}^2$, which results in the acceleration in the direction of movement. The internal energy dissipates with the rate ce, The balance of the depot energy then reads

$$\frac{de}{dt} = q - ce - de\boldsymbol{v}^2. \tag{6.46}$$

The energy balance of the particle follows by multiplying the Langevin equation with the velocity v, we get

$$\frac{dH}{dt} = dem\boldsymbol{v}^2 - m\gamma_0 \boldsymbol{v}^2 + m\boldsymbol{v}\sqrt{2D_v}\cdot\boldsymbol{\xi}(t). \tag{6.47}$$

Assuming $q > 0$ and requiring that the internal energy depot relaxes fast compared to the motion of the particle we get in adiabatic approximation for the *depot model*

$$\boldsymbol{F} = m\boldsymbol{v}\left(\frac{dq}{c+d\boldsymbol{v}^2} - \gamma_0\right). \tag{6.48}$$

For sufficiently large values of q and d the friction function may have a zero at

$$v_0^2 = \frac{q}{\gamma_0} - \frac{c}{d} = \frac{c}{d}\zeta,$$

$$\zeta = \frac{qd}{c\gamma_0} - 1. \tag{6.49}$$

For positive values of the bifurcation parameters ζ and small velocities $|\boldsymbol{v}| < v_0$, we observe input of free energy, the system is driven (self-propelling). Another convenient way of writing the friction function is

$$\gamma(v^2) = \gamma_0\left(1 - \frac{\delta}{1+v^2/v_1^2}\right) \tag{6.50}$$

with $v_1^2 = c/d$ and $\delta = \zeta + 1$. This friction force was first studied in (Schweitzer et al., 1998; Ebeling et al., 1999; Erdmann et al., 2000). The parameter $v_1 > 0$ is connected to internal dissipation and δ controls the conversion of the energy taken up from the external field into kinetic energy. Actually, the parameter δ is now the essential bifurcation parameter of this model. The parameter value $\delta = 0$ corresponds to equilibrium, the region $0 < \delta < 1$ stands for nonlinear passive friction and $\delta > 1$ corresponds to active friction. The value of the transition (bifurcation) from one to the other regime is $\delta = 1$. For the passive regime $0 < \delta < 1$ the friction function has one zero point at $v = 0$ which is the only attractor of the deterministic motion. In Fig. 6.1 we represented the velocity-dependent friction for different values of driving. Without noise all particles come to rest at $v = 0$. For

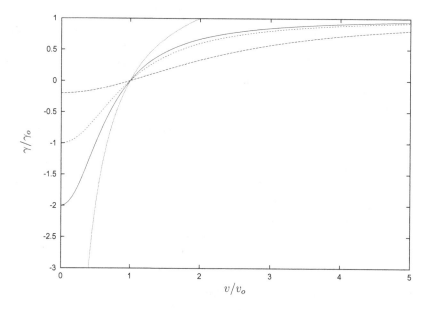

Fig. 6.1 Friction function for several values of the driving strength.

$\zeta > 1$ the point $v = 0$ becomes unstable but we have now two additional zeros at

$$v = v_0 = \pm v_1 \sqrt{\delta - 1}. \tag{6.51}$$

These two velocities are the new attractors of the free deterministic motion if $\delta > 1$. In Fig. 6.2 we have plotted the friction force for the two values $\delta = 0$ (equilibrium) and $\delta = 2$ (strong driving). The figure includes the representation of a useful piecewise linear approximation of the friction force for $\delta > 1$ which reads

$$\boldsymbol{F} = -m\gamma_1 \boldsymbol{v} \left[1 - \frac{v_0}{|v|} \right],$$

$$\gamma_1 = 2\gamma_0 \frac{\delta - 1}{\delta}. \tag{6.52}$$

This is a linear approximation to the friction force near to the two stable velocities $v = \pm v_0$. We see also that between the zero points the force function may be well approximated by a cubic law

$$\boldsymbol{F} = m\boldsymbol{v}[a - bv^2] \tag{6.53}$$

with $v_0^2 = a/b$. This simple law was introduced already in the 19th century by Lord Rayleigh in his "*Theory of sound*" (Rayleigh, 1893). One-dimensional stochastic systems with a friction function corresponding to Rayleigh's law were discussed in detail by Klimontovich (1983, 1995).

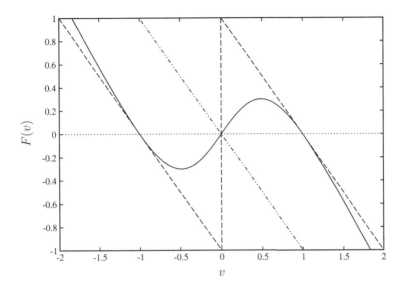

Fig. 6.2 Dissipative force for the parameter values $\delta = 0$ (equilibrium) and $\delta = 2$ (strong driving) and a piecewise linear approximation for $\delta = 2$.

The piecewise linear friction function introduced above as an approximation is similar to the friction law found empirically by Schienbein and Gruler from experiments with cells (Schienbein & Gruler, 1993; Schienbein et al., 1994). The Schienbein–Gruler function reads

$$F = -m\gamma_0 v \left[1 - \frac{v_0}{|v|}\right]. \tag{6.54}$$

The asymptotics of the depot model (SET-model) is in agreement with the Schienbein–Gruler model; in both cases we have the asymptotics $-\gamma_0 v$. In the zero point v_0 the derivative in the zero point is in the SET-model $-\gamma_1$ and in the Schienbein Gruler law it is fixed by γ_0.

Let us come now to the question of the distribution functions. The systems of active Brownian particles introduced above are canonical dissipative systems in the sense discussed in the previous section. Therefore the distribution function can be given exactly. On order to prove this we write

$$\boldsymbol{F} = -m\boldsymbol{v}\gamma(v^2). \tag{6.55}$$

Since v^2 is proportional to the kinetic energy — and there is no potential energy so far — this system is indeed canonical-dissipative. Based on the canonical-dissipative character we find in the general case the distribution function

$$\rho = C \exp\left(-\frac{m}{2D_v} \int \gamma(v^2) dv^2\right). \tag{6.56}$$

Let us discuss now more concrete cases. The distribution for the depot model (SET-model) reads:

$$\rho_0(v_x, v_y) = C \exp\left[-\frac{mv^2}{2kT} + \frac{q}{2D_v}\log\left(1 + \frac{d}{c}v^2\right)\right]. \tag{6.57}$$

Approximately this may be written as

$$\rho_0(v_x, v_y) = C \exp\left[-\frac{qd^2}{4D_v c^2}(v^2 - v_0^2)^2\right]. \tag{6.58}$$

For piecewise linear force laws, including the Schienbein–Gruler law, the distribution is of particular simplicity

$$\rho(v_x, v_y) = C \exp\left(-\frac{\gamma_1}{2D_v}(|\boldsymbol{v}| - v_0)^2\right). \tag{6.59}$$

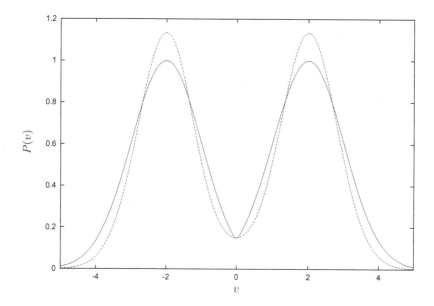

Fig. 6.3 Distribution function of the velocity for the depot model (dashed line) and for the piecewise linear approximation (full line). (Parameters: $D_v = \gamma_0 = c = d = 1, q = 5, v_0 = 2$.)

The piecewise linear force law may be used as an approximation for arbitrary friction laws of type (6.55). Since the modulus of $\boldsymbol{v}^2$ is most time around the value v_0^2 the linearization is well justified. Due to the simplicity of piecewise linear friction laws we are able to find also several more general solutions. In particular the *mean square displacement* may be calculated (Schienbein et al., 1994; Mikhailov & Meinköhn,

1997; Erdmann et al., 2000)

$$\langle (\boldsymbol{r}(t) - \boldsymbol{r}(0))^2 \rangle = \frac{2v_0^4}{D_v}t + \frac{v_0^6}{D_v^2}\left[\exp\left(-\frac{2D_v t}{v_0^2}\right) - 1\right]. \qquad (6.60)$$

The analytical expressions for the stationary velocity distribution and for the mean squared displacement are in good agreement with computer simulations (Schweitzer et al., 2001) and with measurements on the active movements of granulocytes (Schienbein & Gruler, 1993; Schienbein et al., 1994). We notice also the close relation of this theory to systems with isokinetic thermostats (Hoover et al., 1987; Hoover, 2001). Driving by velocity-dependent dissipative forces may have similar effects as driving by isokinetic thermostats as will be shown in the next section. In the adiabatic approximation all the models discussed above have similar properties, in particular they are of canonical-dissipative form.

Summarizing the results of this and the previous sections we underline that these studies were limited to a rather special type of nonequilibrium systems which are pumped from external sources with free energy. We started from the Hamiltonian theory of mechanical systems and constructed forces to drive them away from equilibrium. Then in order to extend the known methods of statistical physics for conservative systems to dissipative systems we developed a general theory of canonical-dissipative systems. Special canonical-dissipative system were constructed, which solution converges to the solution of the conservative system with given energy, or other prescribed invariants of motion. In this way we were able to generate nonequilibrium states characterized by certain prescribed invariants of mechanical motion. We postulated special distributions which are analogues of the microcanonical ensemble in equilibrium. Further we found solutions of the Fokker–Planck equation which may be considered as analogues of the canonical equilibrium ensembles. We proposed to call these distributions canonical-dissipative ensembles. By the help of the explicit nonequilibrium solutions we were able to construct for canonical-dissipative systems a complete statistical thermodynamics including an evolution criterion for the nonequilibrium.

In Chapter 12 we will study applications of the theory to problems of active Brownian motion. We will show that this class on systems is of interest for modeling biological motion (Gruler & Schienbein, 1993, Schweitzer, 2002) and traffic (Helbing, 1997, 2001). Finally we want to say, that canonical-dissipative systems are a rather artificial class of models. In reality, canonical-dissipative systems, strictly according to their definition, do not exist. However there are many complex systems which have several properties in common with canonical-dissipative systems as

- the mechanical character of the motion, i.e. the existence of space and momentum, of quasi-Newtonian dynamics etc.,
- the support of the dynamics with free energy from internal or external reservoirs and the existence of mechanisms which convert the reservoir energy into acceleration.

If so, then the question arises, how systems near to having canonical-dissipative character can be treated by some kind of perturbation theory (see e.g. Haken, 1973; Ebeling, 1981).

6.5 Systems of particles coupled to Nose–Hoover thermostats

The nonequilibrium systems which we considered so far, were driven by negative friction, the stochastic effects were generated by an embedding into a thermal reservoir. This thermostat was assumed to be in equilibrium, what determined the passive friction and the noise. In connection with molecular-dynamical simulations, Nose' (1991) and Hoover (1988, 1998, 1999, 2001) developed a completely different class of thermostats, which found many applications (Hoover & Posch, 1996, Klages et al., 1999; Rateitschak et al., 2000). Because of the great interest in these systems, we will discuss now this new kind of imbedding.

The equations of motion for a Nose–Hoover thermostated particle on a plane confined in a parabolic well read (Hoover, 1998; 2001)

$$\frac{d}{dt}\mathbf{r} = \mathbf{v}\,;\quad \frac{d}{dt}\mathbf{v} = -\gamma(t)\mathbf{v} - a\mathbf{r} \tag{6.61}$$

here γ is a variable friction

$$\frac{d}{dt}\gamma = \frac{1}{\tau^2}\left(\frac{v^2}{2T} - 1\right). \tag{6.62}$$

In the following we will use instead of the temperature T of the thermostat a characteristic velocity $v_0^2 = 2T$. By multiplication with v we get the energy balance

$$\frac{d}{dt}\left(\frac{v^2}{2} + \frac{ar^2}{2}\right) = -\gamma v^2. \tag{6.63}$$

The solutions for v^2 and γ are typically periodic functions. Let us now first consider the case $a = 0$. In Fig. 6.4 we have shown periodic trajectories in the plane v^2 vs. γ. The shape of the trajectories is a hint to the existence of invariants of motion. By subtracting both equations and using the relation

$$\gamma = -\frac{d}{2dt}\log(v^2) \tag{6.64}$$

we find the following invariant of motion

$$\frac{d}{dt}\left(\frac{v^2}{2} - \frac{v_0^2}{2}\log\frac{v^2}{v_0^2} + \frac{\tau^2}{2}\gamma^2\right) = 0. \tag{6.65}$$

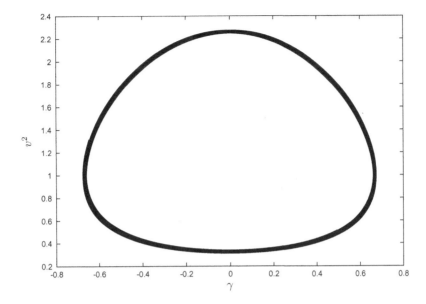

Fig. 6.4 Typical trajectories of free particles with a Nose–Hoover thermostat on the plane v^2 against γ.

Therefore the trajectories are located on the surface

$$I_0 = \frac{v^2}{2} - \frac{v_0^2}{2} \log \frac{v^2}{v_0^2} + \frac{v_0^2}{2}\tau^2\gamma^2 = const. \tag{6.66}$$

The invariant of motion I_0 has the dimension of an energy (in our units $m = 1$). In Fig. 6.5 we have demonstrated the invariant I_0 and $v^2(t)$ as functions of time. There seems to be a relation between I_0 and Dettmann's Hamiltonian (Hoover, 2001). In Fig. 6.6 the invariant is represented in the plane v^2 against γ for a noisy integration. The distribution function for a canonical ensemble is given by

$$f_0(\boldsymbol{v}) = C \exp(-\beta I_0(v, \gamma)). \tag{6.67}$$

In an approximation where $\beta v_0^2 \ll 1$ we find

$$f_0(\boldsymbol{v}, \gamma) = C \exp\left(-\beta \frac{v^2}{2} - \beta\tau^2 \frac{\gamma^2}{2}\right). \tag{6.68}$$

In the opposite case where $\beta v_0^2 \gg 1$ we get

$$f_0(\boldsymbol{v}, \gamma) = C \exp\left(-\beta \frac{1}{4v_0^2}(v^2 - v_0^2)^2 - \beta\tau^2 \frac{\gamma^2}{2}\right). \tag{6.69}$$

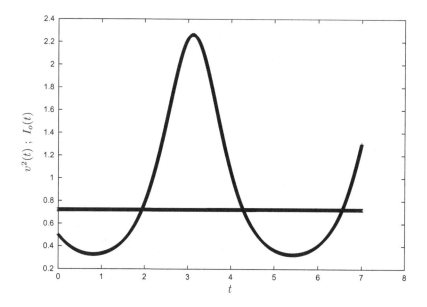

Fig. 6.5 Free particles in the NH-thermostat. We show the time dependence of $v^2(t)$ and the invariant I_0 according to a numerical integration.

Finally we consider also the case of many particle $i = 1, 2, \ldots, N$ driven by thermostats through a time-dependent friction function

$$\frac{d}{dt}\boldsymbol{r}_i = \boldsymbol{v}_i; \quad \frac{d}{dt}\boldsymbol{v}_i = -\gamma_i(t)\boldsymbol{v}_i - a\boldsymbol{r}_i + \sqrt{2D}\xi_i(t). \tag{6.70}$$

In this case we may distinguish between two types of thermostats:
(i) individual thermostats defined by

$$\frac{d}{dt}\gamma_i = \frac{1}{\tau^2}\left(\frac{v_i^2}{2T} - 1\right), \tag{6.71}$$

(ii) collective thermostats with the same friction for all particles

$$\frac{d}{dt}\gamma = \frac{1}{\tau^2}\left(\frac{\sum v_i^2}{2NT} - 1\right). \tag{6.72}$$

Collective of thermostats (case (ii)) correspond in the case $\beta v_0^2 \gg 1$ approximately to a quasi-microcanonical ensemble in the dimension $d = 3N$. As shown by Klages et al. (1999) the distribution functions of this system may show central dips depending on the dimension d. We conclude this section by discussing the relations between Nose–Hoover thermostats and the depot models discussed in the previous section. We introduce first a generalized depot model with dissipative forces $\boldsymbol{F}$ in

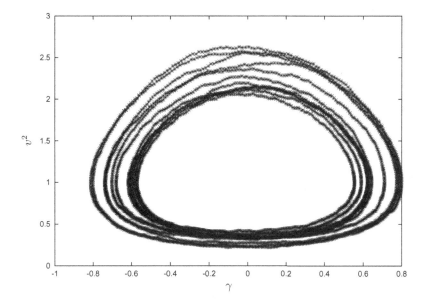

Fig. 6.6 The invariant of free particles in the NH-thermostat from a noisy integration.

the following form:

$$\boldsymbol{F}(e, \boldsymbol{v}) = \boldsymbol{v}(de^n - \gamma_0). \qquad (6.73)$$

where e is the energy content of a depot and d is a conversion parameter. The first term expresses an acceleration in the direction of $\boldsymbol{v}$. The second term $\gamma_0 \boldsymbol{v}$ is the usual passive friction, which by assumption is connected with the noise by an Einstein relation $D = \gamma_0 kT/m$. We assume further that the Brownian particles are able to take up energy, which can be stored in the depot e. This internal energy can be converted into kinetic energy with a momentum dependent rate $de^n v^2$, which results in the acceleration in the direction of movement. The exponent n is free so far. The dissipative flow of energy into the mechanical systems is

$$\boldsymbol{v} \cdot \boldsymbol{F} = v^2(de^n - \gamma_0). \qquad (6.74)$$

If this energy flow comes from a depot with content e then we have the depot dynamics

$$\frac{de}{dt} = s(e) - de^n v^2. \qquad (6.75)$$

Here $s(e)$ is a function describing the support of the depot from an outside source. Now we specify the model: Asuming $n = 1$ and $s(e) = q - ce$ we come back to the depot model studied in Section 5.4. In order to find a Nose–Hoover-type dynamics we assume the exponent is $n = 1/2$ and postulate further the following balance of the depot (Ebeling & Röpke, 2003)

$$\frac{de}{dt} = q\sqrt{e} - ce - d\boldsymbol{v}^2\sqrt{e}. \tag{6.76}$$

Within this model we get in adiabatic approximation assuming $q > 0$ and requiring that the internal energy depot relaxes fast compared to the motion of the particle

$$\boldsymbol{F} = \boldsymbol{v}\left(\frac{d}{c}(q - d\boldsymbol{v}^2) - \gamma_0\right). \tag{6.77}$$

Now, under the condition $((qd/c) - \gamma_0) > 0$, a root $v_0 > 0$ exists and we get a cubic law which corresponds to the Rayleigh friction law. We reformulate now the model in order to show the relation to NH-thermostats. Introducing a time-dependent friction variable $\gamma = \gamma_0 - d\sqrt{e}$ we find the "dissipative force"

$$\boldsymbol{F} = -\gamma(t)\boldsymbol{v}. \tag{6.78}$$

The friction variable satisfies the dynamic equation

$$\frac{d\gamma}{dt} = \frac{d}{2}\left(d\boldsymbol{v}^2 - \frac{c}{d}\gamma - q + \frac{c}{d}\gamma_0\right). \tag{6.79}$$

By introducing here the new variables

$$v_0^2 = \frac{q}{d} - \frac{c\gamma_0}{d^2} \tag{6.80}$$

and

$$\tau^2 = \frac{2}{d^2 v_0^2}, \qquad \xi = \frac{c}{2} \tag{6.81}$$

we get the dynamics

$$\frac{d\gamma}{dt} = \frac{1}{\tau^2}\left(\frac{v^2}{v_0^2} - 1\right) - \xi\gamma, \tag{6.82}$$

where ξ is a dissipative constant. Assuming a special time evolution for ξ, we arrive at a dynamics discussed in Section 7.10.3 of Hoover's book (Hoover, 2001). In the limit $\xi = 0$, i.e. $c = 0$ our system reduces to a conservative Nose–Hoover dynamics (Hoover, 1988, 1998, 2001; Klages et al., 1999). In the case $\xi = const > 0$ we find a dissipative Nose–Hoover dynamics. The corresponding Langevin equation contains a stochastic force. In order to simplify we assume that only the passive friction generates noise

$$D = \gamma_0 kT, \tag{6.83}$$

$$\langle \xi_i(t) \rangle = 0; \qquad \langle \xi_i(t)\xi_j(t') \rangle = \delta(t - t')\delta_{ij}. \tag{6.84}$$

In the simplest case of force-free particles the kinetic energy is a conserved quantity: $v^2 = const$. This means, the system is canonical-dissipative. Consequently we are able to find exact solutions for the probability distribution following Sections 5.3–5.4. For the dissipative Nose–Hoover dynamics we get the stationary distribution

$$f_0(v_1, v_2) = C \exp\left[-\alpha(v^2 - v_0^2)^2\right] \quad (6.85)$$

with $\alpha = d/(4cD)$. We notice the close relation to Lorentz gas distributions (Klages et al., 1999; Rateitschak et al., 2000). In conclusion we may say, that driving by velocity-dependent dissipative forces may have similar effects as driving by isokinetic thermostats.

6.6 Nonequilibrium distributions from information-theoretical methods

In this section we will show, how non-equilibrium distribution functions may be constructed by means of the information-theoretical methods which were introduced in Section 4.4. This method which basically was developed by Gibbs and Jaynes (Jaynes, 1957; 1985). was further developed and applied to nonequilibrium situations by Zubarev (Subarev, 1976; Zubarev et al., 1996, 1997)). This information-theoretical method is of phenomenological character and connected with the maximum entropy approach. There is no reason to restrict the method to the equilibrium case. As stated by Jaynes (1957, 1985) the method should work under much more general conditions. While in equilibrium we prescribe certain invariants of motion as e.g. energy, particle numbers etc. (see Section 4.4.) we have to look in nonequilibrium for more general observables. As we shall see, under non-equilibrium conditions, we should consider also macroscopic constraints due to the boundary conditions, as e.g. prescribed flows, angular momenta etc.. We consider quantities which may be represented as averages over the distribution and are prescribed in average by the conditions. This does not mean that these quantities are fixed, only the averages are prescribed and fluctuations around the means which are described by the distribution are possible. Then one constructs distribution functions which are compatible with the given averages. The free parameters in distribution function are found by maximizing the entropy of the distributions.

In the following we will pay special attention to the construction of non-equilibrium ensembles for systems with rotational modes, by using the Zubarev-formalism (Subarev, 1976; Zubarev, Morozov and Röpke, 1996, 1997). We will show in detail how the distributions are constructed if the angular integrals of motion are prescribed. Let us start now to explain the Jaynes–Zubarev approach to non-equilibrium systems. We consider an N-particle system with the Hamiltonian:

$$H_S = \sum_i^N \frac{p_i^2}{2m} + \frac{1}{2}\sum_{i \neq j} V(\mathbf{r}_i - \mathbf{r}_j). \quad (6.86)$$

Including an embedding in surroundings, a heat bath, we have two additional terms in the total Hamiltonian:

$$H_{\text{total}} = H_S + H_B + H_{SB} \tag{6.87}$$

here the bath is modeled by $H_B = H_B(Q, P)$, and the coupling by $H_{SB} = \sum_i^N V(r_i, Q)$. Special cases are the isolated system with $H_{SB} = 0$ and external fields $V^{\text{ext}}(r_i)$ modeling traps.

Let us discuss now the form of the distribution functions. In non-equilibrium, in general, the linear "ansatz" used in the previous Chapter (4.54) is no more sufficient, since the means and the dispersion may be independent variables. In order to admit such situations we have to use quadratic function in the exponent of the distribution function. Let us assume that the mean values of the observables A_k and A_k^2 are given. In order to find the probability density under the constraints

$$A_k' = \langle A_k \rangle = \int dq dp A_k(q, p) \rho(q, p), \tag{6.88}$$

$$A_k'' = \langle A_k^2 \rangle = \int dq dp A_k^2(q, p) \rho(q, p), \tag{6.89}$$

we maximize the information entropy

$$\mathcal{H} = -\int dq dp \rho(q, p) \ln \rho(q, p) \tag{6.90}$$

under the given constraints. We define two m-component vectors $\lambda' = [\lambda_1', \ldots, \lambda_m']$ and $\lambda'' = [\lambda_1'', \ldots, \lambda_m'']$ of Lagrange-multipliers. Then the probability density that agrees with the given data $\mathbf{A}'$ follows from

$$\delta[\mathcal{H} + \sum_k \lambda_k' \left(A_k' - \int dq dp A_k(q, p) \rho(q, p) \right)$$

$$+ \sum_k \lambda_k'' (A_k'' - \int dq dp A_k^2(q, p) \rho(q, p))] = 0. \tag{6.91}$$

This leads to

$$\rho(q, p) = Z^{-1} \exp\left[-\sum_k (\lambda_k' A_k(q, p) + \lambda_k'' A_k^2(q, p)) \right], \tag{6.92}$$

where the normalization factor, the so-called partition function, is now given by

$$Z(\lambda_1, \ldots, \lambda_m) = \int dq dp \exp\left[-\sum_k (\lambda_k' A_k(q, p) + \lambda_k'' A_k^2(q, p)) \right]. \tag{6.93}$$

The choice of the parameters should reflect the conditions (6.88) and (6.89).

The choice of the observables A_k describing the physical problem is not clear from the beginning. Candidates are the invariants of the dynamics and quantities which are prescribed by the boundary conditions. Among the invariant quantities the angular momentum is of special interest.

We will consider now this class of observables in more detail. The interest in a theory of rotational systems is due to the important role of rotations in nature. For example we mention the typical rotations of astrophysical objects as stars and planets, in macroscopic, mesoscopic and in atomic systems. In general one can say that most confined or selfconfined physical systems show rotational modes. In recent times finite Coulomb systems as quantum dots are in the center of interest (Bonitz et al., 2002; Dunkel et al., 2004).

In our context the N-particle angular momentum is given as the sum of the momenta of the particles:

$$\boldsymbol{L}_N = \sum_i^N \boldsymbol{L}_i, \quad \boldsymbol{L}_i = \boldsymbol{r}_i \times \boldsymbol{p}_i. \tag{6.94}$$

Further $\boldsymbol{L}_N^2$ and higher powers of the angular momenta sometimes are of interest (see e.g. the theory of atoms and molecules). The time evolution of the angular momentum is given by:

$$\frac{d}{dt}\boldsymbol{L}_N = \frac{i}{\hbar}[H, \boldsymbol{L}_N] = \frac{i}{\hbar}[H_{SB}, \boldsymbol{L}_N]. \tag{6.95}$$

We will assume central forces, so that any change of the angular momentum is due to the "bath". If the coupling H_{SB} is absent or has rotational symmetry then the total angular momentum is a conserved quantity. Alternatively we may prescribe the angular momentum by boundary conditions. The total angular momentum $\boldsymbol{L}_N$ is a conserved quantity, if the coupling H_{SB} is absent or has rotational symmetry. Alternatively we may prescribe it by boundary conditions. Let us assume that the average

$$\langle \boldsymbol{L}_N^n \rangle = \int d\Gamma \rho \boldsymbol{L}_N^n = const \tag{6.96}$$

for any given n is fixed. Then, we will postulate, that the distribution is a function of H_S and of $\boldsymbol{L}_N$

$$\rho = \rho(H_S, N, \boldsymbol{L}_N). \tag{6.97}$$

The concrete form of the distribution is assumed to be determined by the expectation values. This may be a canonical distribution, but also any other forms of functional dependence are possible. Two cases are of special interest:

(1) For rotating bodies of N particles which are only weakly coupled to an external heat bath and which otherwise are in internal equilibrium, the distribution is of particular simplicity. Assuming that conserved quantities H_N, N and $\boldsymbol{L}_N$ are the observables and that the corresponding Lagrange

parameters are β, μ and $\boldsymbol{\omega}$, we find according to Section 4.4. an extended canonical Gibbs distribution (Landau & Lifshits, 1990).

$$\rho_{\text{eq}} = \frac{1}{Z_{\text{eq}}} \exp\left[-\beta(H_S - \mu N - \boldsymbol{\omega} \cdot \boldsymbol{L}_N)\right], \qquad (6.98)$$

here $\boldsymbol{\omega}$ is the angular velocity.

(2) If the torque due to the surrounding bath is weak, then $\boldsymbol{L}_N$ is nearly conserved, it is a long-living mode of the system. In particular this is the case if the cluster and the surrounding have nearly spherical symmetry. Then $\boldsymbol{L}_N, \boldsymbol{L}_N^2$, etc. are quasi-conserved and should be included in the relevant distribution ρ_{rel} (Zubarev et al., 1996, 1997; Ebeling & Röpke, 2004).

We may assume that several n-order moments of the angular momentum are given at time t as e.g.

$$\langle \boldsymbol{L}_N \rangle^t = \int d\Gamma \cdot \rho_{\text{rel}}(t) \boldsymbol{L}_N,$$
$$\langle \boldsymbol{L}_N^2 \rangle^t = \int d\Gamma \cdot \rho_{\text{rel}}(t) \boldsymbol{L}_N^2. \qquad (6.99)$$

We assume further that the relevant distribution ρ_{rel} is again a generalized Gibbs distribution but including higher momenta

$$\rho_{\text{rel}}(t) = \frac{1}{Z_{\text{rel}}(t)} \exp\left[-\beta\left(H_S - \mu N - \sum_n \boldsymbol{\Omega}_n(t) \boldsymbol{L}_N^n\right)\right]. \qquad (6.100)$$

In this distribution in addition to β, μ, the free parameters $\boldsymbol{\Omega}_n$ appear, which characterize nonequilibrium quantities. The following step according to Jaynes–Zubarev is to maximize the Gibbs entropy

$$S = \int d\Gamma \rho_{\text{rel}} \cdot \ln \rho_{\text{rel}} \qquad (6.101)$$

under the given constraints (6.96). In this procedure the new parameters appear as the Lagrange multipliers $\boldsymbol{\omega}_{(n)}(t)$ and may be determined by the prescribed averages (6.96) as shown in detail e.g. in (Jaynes, 1957; Subarev, 1976; Zubarev et al., 1996). Of special importance for our problems is the case $n = 2$, i.e. the square of the angular momentum is given.

If the torque is weak, then the relaxation time of the angular momenta is quite long. An estimate of the relaxation time may be found also by linear response theory, which yields the expression (Zubarev et al., 1996)

$$\tau_L \propto \int_0^\infty dt \left\langle \left(\frac{d\boldsymbol{L}_N}{dt}\right)_t \left(\frac{d\boldsymbol{L}_N}{dt}\right)_{t=0} \right\rangle.$$

The equilibrium correlation function may be evaluated by simulations or by analytical methods like perturbation expansions.

Among the quantities which may be prescribed by the surrounding, the kinetic energy is of special interest. Due to specific boundary conditions in some situations the individual kinetic energies are prescribed up to certain accuracy. For example, the interaction with a laser beam or with a surrounding plasma may fix the kinetic energies of the individual particles T_i around some value given by the laser intensity

$$T_i = \frac{m_i}{2} v_i^2 \simeq T_0.$$

In a recent papers Trigger and Zagorodny have shown that the charged grains in a dusty plasma have a velocity distribution which is peaked around a characteristic velocity (kinetic energy) of the grains (Trigger and Zagorodny, 2002, 2003; Trigger, 2003).

As already mentioned, sometimes often the experimental conditions are such that the particles are driven to to a prescribed kinetic energy

$$v_i^2(t) \to v_0^2.$$

An example are charged grains in dusty plasmas (Trigger & Zagorodny, 2002, 2003, Trigger, 2003) or Coulomb clusters driven by a strong laser field. Due to the interaction with the surrounding, in the examples the "bath" is given by the plasma ions or the radiation field, the particles are accelerated to certain kinetic energy. In th is case we postulate that the distribution is of the following form

$$\rho_N(\boldsymbol{v}_1,\ldots,\boldsymbol{v}_N) = \Pi_{i=1}^N \exp\left[-\alpha_v(\boldsymbol{v}_i^2 - v_0^2)^2\right]. \qquad (6.102)$$

According to this distribution the most probable squared velocities (kinetic energies) are at the values v_0^2. Here α_v is an appropriate parameter characterizing the dispersion. We denote this type of ensemble "isokinetic ensemble". This notation is borrowed from molecular dynamics, where Gaussian isokinetic ensembles play a big role (Hoover, 1988, 1998, 1999; Rateitschak et al., 2000).

A generalization are conditions fixing the individual energies H_i of the particles on the given value H_0. This leads to the Gaussian distributions

$$\rho_N(r_1,\ldots,r_N,v_1,\ldots,v_N) = \Pi_{i=1}^N \exp[-\alpha_H(H_i - H_0)^2]. \qquad (6.103)$$

In radially symmetric systems the driving to v_0^2 or H_0 implies for a special class of systems (an example will be given below) that $\boldsymbol{L}_i$ is also fixed to certain L_{i0}. This is due to internal connections between the invariants of motion. The existence of rotations means, that the symmetry with respect to all directions in space is broken, i.e correspondingly the distribution functions should be modified. Therefore we will consider also the distributions

$$\rho_N(L_1,\ldots,L_N) = \Pi_{i=1}^N \exp\left[-\alpha_L(\boldsymbol{L}_i - L_{i0})^2\right]. \qquad (6.104)$$

A different "ansatz" is based on the invariant $H - \mathbf{\Omega} \cdot \mathbf{L}$ which is a kind of internal Hamiltonian and reads

$$\rho_0(q_1 \ldots q_f p_1 \ldots p_f) = Z^{-1}(\Omega) \exp(-\alpha_1 (H - E_0)^2 - \alpha_2 (H - \mathbf{\Omega} \cdot \mathbf{L})). \quad (6.105)$$

This way we may prescribe the most probable value of the total energy and of the internal energy separately by the choice of α_1, α_2.

More general forms of the distributions are Gaussian distributions centered around the prescribed invariants of motion

$$\rho_0(q_1 \ldots q_f p_1 \ldots p_f) = const. \, \Pi_{k=0}^s \exp\left(-c_k \frac{(I_k(q_1 \ldots q_f, p_1 \ldots p_f) - I_k)^2)}{2D}\right). \quad (6.106)$$

However as we have shown above, there is no guaranty that the distributions constructed in this way already have the correct physical symmetry. This has to be checked in every concrete case. In the limit of very strong driving the probability reduces to a kind of microcanonical ensemble corresponding to the shell defined by $I_0, I_1, \ldots, I_s$ with a constant probability density on the shell.

Let us discuss now in brief the physical meaning and symmetry properties of probability distributions which express prescribed angular momenta. Where is the

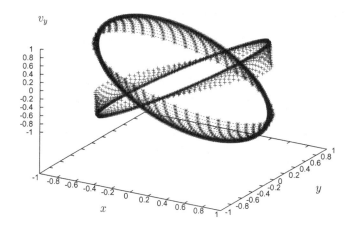

Fig. 6.7 Typical distribution of a rotating 2-d particle with fixed angular momentum $\mathbf{L}^2 = L_0^2$. The two possible values of the angular momentum select two tires on the sphere $H = const$. We see a projection of the 4-dimensional phase on the $x - y - v_y$-space.

probability concentrated in the phase space? We consider as a simple case of a particle rotating on a 2-dimensional plane, i.e. we have a 4-dimensional phase space (a physical realization was considered already in Section 5.1 and will be studied in more detail in Chapter 12). Then fixation of the most probable value of the total energy $H = E_0$ defines the surface of a sphere (or a cylinder if the kinetic energy is prescribed) in the 4-dimensional space:

$$\frac{m}{2}(v_1^2 + v_2^2) + \frac{m}{2}\omega_0^2(x_1^2 + x_2^2) = E_0 .$$

In order to characterize the two rotating states we introduce two non-negative invariants of motion

$$J_+ = H - \omega_0 L = \frac{m}{2}(v_1 + \omega_0 x_2)^2 + \frac{m}{2}(v_2 - \omega_0 x_1)^2 \qquad (6.107)$$

and

$$J_- = H + \omega_0 L = \frac{m}{2}(v_1 - \omega_0 x_2)^2 + \frac{m}{2}(v_2 + \omega_0 x_1)^2 . \qquad (6.108)$$

The condition $J_+ = 0$ characterizes rotations with positive angular momentum ($L \simeq H_0/\omega_0$) and the condition $J_- = 0$ characterizes rotations with negative angular momentum ($L \simeq -H_0/\omega$). Geometrically $J_\pm = 0$ define two surfaces in the 4-dimensional space which intersect the sphere $H = H_0$ in two circles representing the trajectories of the rotating states (see Fig. 6.7). For the corresponding probabilities we may assume the Gaussian distribution with two free parameters

$$\rho(\mathbf{r}, \mathbf{v}) = C \exp\left[-\alpha_H (H - H_0)^2\right] \left[\exp(-\mu J_+) + \exp(-\mu J_-)\right] . \qquad (6.109)$$

We note that for the special choice $\mu = 2\alpha_H H_0$ we may bring this to the more special form

$$\rho(\mathbf{r}, \mathbf{v}) = C' \exp\left[-\frac{\mu}{2H_0}(H^2 - H_0^2)\right] \left[\exp(+\mu\omega_0 L) + \exp(-\mu\omega_0 L)\right] . \qquad (6.110)$$

This way the probability will be concentrated on two tires in the phase space.

Let us summarize our findings: We used in this section the method of generalized Gibbs distributions (Gibbs, 1902; Landau & Lifshits, 1990) and in particular the Zubarev method for constructing relevant distributions (Subarev, 1976; Zubarev et al., 1996, 1997). These methods are of phenomenological character and connected with the maximum entropy approach. We repeat the basic assumptions: Due to the boundary conditions certain averages as e.g. here the angular momentum L_N or L_N^2 are prescribed. This does not mean that the quantities are fixed, only the averages or the most probable values are prescribed and fluctuations around these values which are described by the distribution are possible. The distribution function is found by maximizing the entropy of distributions which fulfill the given constraints.

In the Chapters 7–12 we will develop more special microscopically oriented methods for the construction of nonequilibrium distributions for several types of Brownian motion.

Chapter 7
Brownian Motion

7.1 Einstein's concept of Brownian motion

In what follows we discuss in detail the corresponding approaches and results in order to grasp their strengths and their interconnections. In the case of Brownian motion all the approaches work equally well and are essentially equivalent, however, each of them has its own area of application and its own domain of validity outside of the case of diffusion processes: they apply to different cases of the so-called anomalous diffusion as well. Therefore we do not hesitate to discuss the same issue from different points of view and to present different derivations of the same results. Such a deep understanding always pays off.

Motivated mostly by description of Brownian motion several different but strongly interconnected approaches were devised for the description of phenomena where the stochastic nature of systems plays a considerable role. The ingenious microscopic derivation of the diffusion equation by A. Einstein (which contained, in a nutshell, several different approaches) and the discussion by P. Langevin marked the two main ways of description of fluctuation phenomena, one based on the discussion of the deterministic equations for the probability densities, another one based on the discussion of a particular stochastic realizations of the process. These approaches, refined both from physical and from the mathematically point of view build now the main instrument of description of both equilibrium and nonequilibrium processes on a mesoscopic scale.

Einstein, in his analysis of the situation has connected the motion of suspended particles with diffusion and showed that this diffusive behavior follows from the three postulates. First, the particles considered are assumed not to interact with each other: their trajectories are independent. Second, one assumes that the motion of the particles lacks long-time memory: one can choose such a time interval τ, that the displacements of the particle during two subsequent intervals are independent. Third, the distribution of a particle's displacements s during the subsequent time intervals $\phi(s)$ possesses at least two lower moments. Moreover, for the force-free situation $\phi(s)$ is symmetric. The displacement of the particle can thus be considered as a result of many tiny, independent, equally distributed steps.

The further line of his reasoning is very close to what we will call now a Kramers–Moyal expansion, see Chapter 2. For the simplicity, following Einstein, we analyze a one-dimensional problem:

The concentration of particles n in vicinity of point x is proportional to the probability density $f(x,t)$ to find one particle at this point. Comparing the probabilities at time τ and at time $t+\tau$ we get (due to the independence of the new displacement of the previous position and to the fact that $x(t+\tau) = x(t) + s$)

$$f(x, t+\tau) = \int f(x-s, t)\phi(s)ds. \qquad (7.1)$$

Now, since both τ and s are both small compared to the time- and space-scales of interest, one can expand the function f in Taylor series on both sides of the equation. On the left-hand side it is enough to expand up to the first order in t, on the right-hand side we need the second order in s. We get:

$$f + \left(\frac{\partial f}{\partial t}\right)\tau + ... = f + \left(\frac{\partial f}{\partial x}\right)\int s\phi(s)ds + \frac{1}{2}\left(\frac{\partial^2}{\partial x^2}f\right)\int s^2\phi(s)ds + \cdots . \qquad (7.2)$$

The integral $\int s\phi(s)ds$ vanishes due to the symmetry. In the lowest order we thus get:

$$\frac{\partial f}{\partial t} = \frac{\sigma^2}{2\tau}\frac{\partial^2}{\partial x^2}f, \qquad (7.3)$$

where $\sigma^2 = \int s^2\phi(s)ds$. Here we recognize a diffusion equation, and associate $\sigma^2/2\tau$ with the diffusion coefficient D. The solution of Eq. (7.3) is, clearly, a Gaussian

$$f(x,t) = \frac{1}{\sqrt{4\pi Dt}}\exp\left(-\frac{x^2}{4Dt}\right), \qquad (7.4)$$

so that the root-mean-square displacement of the particle along the x-axis would be $\lambda_x = \sqrt{\langle x^2 \rangle} = \sqrt{2Dt}$, which gives the direct way of experimental measurement of D.

The derivation of the diffusion equation by A. Einstein was the very first step of statistical physics into the new domain of non-equilibrium phenomena. Note however, that it was not the derivation of the diffusion equation, which seemed to Einstein to be the main topic of this work: The discussion of the diffusion of particles in the solution gave the way to determine the Avogadro number N_A through the macroscopic measurements on rather large particles, the measurement performed by Perrin some 3 years later. Such measurements were necessary to provide solid basement for atomistic theory of matter. The corresponding theoretical considerations were summarized in Einstein's PhD thesis "Eine neue Bestimmung der Moleküldimensionen" (New method for determination of molecular sizes) presented on April 30, 1906 to the University of Zürich.

7.2 The Langevin equation: Theme and variations

7.2.1 *The Langevin equation*

In his article published in *Comptes Rendus* in 1908 Paul Langevin proposed another approach to description of Brownian motion, than one of Einstein and Smoluchowski. This one was assumed to be "infinitely simpler" than the Einstein's one and seemed to be based only on the equipartition theorem. Here we repeat the main argumentation of the original work (Langevin, 1908).

Let us consider the motion of the Brownian particle in a fluid. On the average this motion is governed by the Newtonian dynamics under friction, $m\dot{v} = -\gamma v$, where γ is the friction coefficient (for a macroscopic spherical particle this friction follows the Stokes law, so that $\gamma = 6\pi\eta r$, where r is the particle's radius). However, this equation, leading to the continuous decay of the particle's velocity, holds only on the average. In order to describe the erratic motion of the particle, resulting from random, uncompensated impacts of the molecules of surrounding fluid, we have to introduce additional, fluctuating force $\xi(t)$ ("noise"). We assume only that this force has zero mean (so that it does not lead to the overall motion on average), and that it is independent on x, which mirrors the homogeneity of the whole system. We thus write

$$m\dot{v} = -\gamma v + \xi(t). \tag{7.5}$$

Our first task will be to find the mean squared displacement of the particle. Let us now multiply both sides of Eq. (7.5) by $x(t)$ and use the evident fact that $x\dot{v} = x\ddot{x} = \frac{d}{dt}(x\dot{x}) - \dot{x}^2$. We thus get

$$m\frac{d}{dt}(x\dot{x}) = m\dot{x}^2 - \gamma x\dot{x} + x\xi. \tag{7.6}$$

Let us now average this equation over the realizations of the process. Dividing both parts of the equation by m we get:

$$\frac{d}{dt}\langle x\dot{x}\rangle = -\frac{\gamma}{m}\langle x\dot{x}\rangle + \langle \dot{x}^2\rangle + \frac{1}{m}\langle x\xi\rangle. \tag{7.7}$$

The last mean value vanishes due to the independence of x and ξ and to the fact that the mean value of ξ is zero: $\langle x\xi\rangle = \langle x\rangle\langle \xi\rangle = 0$. Moreover, due to the equipartition theorem, the mean squared velocity of the particle in our one-dimensional model is so that $m\langle \dot{x}^2\rangle/2 = kT/2$, i.e.

$$\langle \dot{x}^2\rangle = \frac{kT}{m}. \tag{7.8}$$

Thus, for the mean $\langle x\dot{x}\rangle$ one has

$$\frac{d}{dt}\langle x\dot{x}\rangle = -\frac{\gamma}{m}\langle x\dot{x}\rangle + \frac{kT}{m}. \tag{7.9}$$

Let us now assume that the initial particle's position is taken to be at the origin of coordinates. Then $\langle x(0)\dot{x}(0)\rangle = 0$. Under this initial condition Eq. (7.9) can easily be solved and delivers

$$\langle x(t)\dot{x}(t)\rangle = \int_0^t \exp\left[-\frac{\gamma}{m}(t-t')\right]\frac{kT}{m}dt' = \frac{kT}{\gamma}\left[1-\exp\left(-\frac{\gamma}{m}t\right)\right]. \tag{7.10}$$

As a next step, we note that $\langle x(t)\dot{x}(t)\rangle = \frac{1}{2}\frac{d}{dt}\langle x^2(t)\rangle$, so that the mean squared displacement of the particle can be found by an additional integration of Eq. (7.10):

$$\langle x^2(t)\rangle = 2\int_0^t \langle x(t')\dot{x}(t')\rangle\, dt' = 2\frac{kT}{\gamma}\left[t - \frac{m}{\gamma}\left(1-\exp\left(-\frac{\gamma}{m}t\right)\right)\right]. \tag{7.11}$$

For large time the leading term corresponds to

$$\langle x^2(t)\rangle = 2\frac{kT}{\gamma}t, \tag{7.12}$$

i.e. to the diffusive behavior with the diffusion coefficient $D = kT/\gamma$.

7.2.2 Thermalization of the velocity: The fluctuation-dissipation theorem

From our previous discussion it may seem that no properties of the force except for its symmetry are of importance. However, the assumption of the thermalization of the velocity, so that $\langle v^2 \rangle = kT/m$, poses rather tight conditions on the behavior of the force. In order to show this let us examine the Langevin equation more carefully.

From the Langevin equation we immediately have:

$$\dot{v} = -\frac{\gamma}{m}v + \frac{1}{m}\xi(t). \tag{7.13}$$

This is the simplest linear differential equation of the form we have already seen in the previous paragraph, so that its solution can be put down immediately:

$$v(t) = v(0)\exp\left(-\frac{\gamma}{m}t\right) + \frac{1}{m}\int_0^t dt'\xi(t-t')\exp\left[-\frac{\gamma}{m}t'\right]. \tag{7.14}$$

Now it is easy to calculate the velocity squared:

$$v^2(t) = v^2(0)\exp\left(-2\frac{\gamma}{m}t\right) + \frac{2}{m}v(0)\exp\left(-\frac{\gamma}{m}t\right)\int_0^t dt'\xi(t-t')\exp\left[-\frac{\gamma}{m}t'\right]$$
$$+ \left\{\frac{1}{m}\int_0^t dt'\xi(t-t')\exp\left[-\frac{\gamma}{m}t'\right]\right\}^2. \tag{7.15}$$

Averaging this expression, we see that the first term is not changed since it is deterministic, the second one gives on the average zero since the averaging is the linear operation and as those can be permuted with the integration, and since moreover $\langle \xi(t-t') \rangle = 0$. To average the third term we first rewrite the square of the integral as a repeated integral,

$$\left[\int_0^t \xi(t-t') \exp\left(-\frac{\gamma}{m}t'\right) \right]^2$$

$$= \left[\int_0^t dt' \xi(t-t') \exp\left(-\frac{\gamma}{m}t'\right) \right] \left[\int_0^t dt'' \xi(t-t'') \exp\left(-\frac{\gamma}{m}t''\right) \right]$$

$$= \int_0^t \int_0^t dt' dt'' \exp\left(-\frac{\gamma}{m}t'\right) \exp\left(-\frac{\gamma}{m}t''\right) \xi(t-t')\xi(t-t''). \qquad (7.16)$$

Performing now the averaging we get:

$$v^2(t) = v^2(0) \exp\left(-2\frac{\gamma}{m}t\right) + \frac{1}{m^2} \int_0^t \int_0^t dt' dt'' \exp\left[-\frac{\gamma}{m}(t'+t'')\right] \langle \xi(t-t')\xi(t-t'') \rangle. \qquad (7.17)$$

If we now assume the noise to be a stationary random process, then its correlation function depends only on the difference of the arguments, $\langle \xi(t-t')\xi(t-t'') \rangle = C(t'-t'')$. Thus, we can put down

$$v^2(t) = v^2(0) \exp\left(-2\frac{\gamma}{m}t\right) + \frac{1}{m^2} \int_0^t \int_0^t dt' dt'' \exp\left[-\frac{\gamma}{m}(t'+t'')\right] C(t'-t''). \qquad (7.18)$$

Let us now consider the thermalization process. If the overall integration time is much larger than both the correlation time of the force and the characteristic time m/γ of decay of deterministic motion, the limits of integration can be extended to infinity. We can now change to the sum and to the difference variables: $s = (t'+t'')/2$ and $q = (t'-t'')$ (the absolute value of the Jakobian of this transformation is 1). If the correlation function decays fast compared to the deterministic motion, $\tau_c \ll m/\gamma$, the integrations in s and q may be considered as independent so that

$$\langle v^2(t) \rangle \simeq v^2(0) \exp\left(-2\frac{\gamma}{m}t\right) + \frac{2}{m^2} \int_0^\infty ds \exp\left[-2\frac{\gamma}{m}(s)\right] \int_0^\infty C(q) dq$$

$$= v^2(0) \exp\left(-2\frac{\gamma}{m}t\right) + \frac{C}{m\gamma}. \qquad (7.19)$$

where $C = \int_0^\infty C(t) dt$ is the integral of the correlation function $C(t'-t'')$. The first term decays exponentially with time, so that under thermalization we have

$$\langle v^2 \rangle_{eq} = \frac{C}{m\gamma}. \qquad (7.20)$$

Comparing this result with the equipartition condition $\langle v^2 \rangle_{eq} = kT/m$ we see that $C = kT\gamma$. This result connecting the behavior of the random force to the friction coefficient and thus to the dissipation in the deterministic motion is a special case of the fluctuation-dissipation theorem. Using the previously established connection between γ and the diffusion coefficient D we get $C = D\gamma^2$.

The velocity will thermalize and the Langevin approach will be valid if the integral $\int_0^\infty C(t)dt$ converges. This is violated if the correlation function of the force decays so slow that $\int_0^\infty C(t)dt$ diverges. The overall behavior in this case can be vastly different from one described above and could correspond to the so-called *anomalous diffusion*, see Section 7.6.

7.2.3 The properties of the noise

Dividing the overall interaction between the Brownian particle and the medium into a deterministic part (friction) and a random part (noise), we already dispensed from the idea to fully describe the system's dynamics on very short time scales: both the friction and the random force stem essentially from the molecular impacts and follow very complex dynamics. Looking at the situation at very short time scales we would not recognize any constant friction force. The situation gets very clear when we consider, for example, a particle in a rarefied medium, say, a dust particle in the air. In this case it is clear that the interaction between the particle and the air molecules corresponds to a sequence of separated, very short, practically punctual events of momentum transfer (impacts), and the periods of practically no interaction. The velocities of the gas particles are governed by the Maxwell distribution, so that the components of their momenta are Gaussian. In order to separate between that deterministic and the stochastic component we need to perform a kind of temporal pre-averaging over some physically small time interval Δt. The the friction force appears on the average due to the fact that the particle moving, say, with the positive velocity in x-direction meets more gas molecules hitting it from the right (against the its motion direction) than from the left. The velocity of our particle is finite, and momentum transferred in each impact has the finite second moment. The overall change of the particle's momentum $\Delta p = \sum_i \Delta p(t_i)$ where t_i is the instant of i-th impact. The mean force is then $f = \Delta p / \Delta t$ and equals to the mean momentum transfer per unit time. Subtracting this mean from the actual value of Δp we get the noise. Note that since Δp is a sum of many (presumably) independent random variables possessing the finite second moment, then distribution of Δp at longer times (after many impacts) will tend to a Gaussian, as a consequence of the Central Limit Theorem. Thus, the distribution of the transferred momentum during the time Δt reads

$$P(\Delta p) \simeq \frac{1}{\sqrt{2\pi\sigma(\Delta t)}} \exp\left(-\frac{(\Delta p - f\Delta t)^2}{2\sigma^2(\Delta t)}\right). \tag{7.21}$$

Here f is the deterministic force (which was taken $f = -\gamma v$ in our previous considerations) and $\delta p = \Delta p - f\Delta t$ is the random component of the momentum change during the time Δt. Assuming this change to be due to the random force ξ, we get $\delta p = \int_t^{t+\Delta t} \xi(t')dt'$. Then the dispersion $\sigma^2(\Delta t)$ can be connected with the correlation properties of the noise. Since the noise is assumed to be homogeneous in time we can take $t = 0$ without losing the generality. Using the identity

$$\delta p^2(t) = \left[\int_0^{\Delta t} \xi(t')dt'\right]^2 = \int_0^{\Delta t} \xi(t')dt' \cdot \int_0^{\Delta t} \xi(t'')dt''$$
$$= \int_0^{\Delta t} \int_0^{\Delta t} \xi(t')\xi(t'')dt'dt'' \qquad (7.22)$$

and averaging this result over the realizations of the noise we get

$$\sigma^2(\Delta t) = \langle \delta p^2(\Delta t) \rangle = \int_0^{\Delta t} \int_0^{\Delta t} \langle \xi(t')\xi(t'') \rangle \, dt'dt''. \qquad (7.23)$$

Assuming that $\langle \xi(t')\xi(t'') \rangle = C(|t'-t''|)$ and changing to a new variable $q = t'-t''$ we get

$$\sigma^2(\Delta t) = \int_0^{\Delta t} dt' \int_{-t'}^{t'} C(|q|)dq = 2\int_0^{\Delta t} dt' \int_0^{t'} C(q)dq. \qquad (7.24)$$

Assume now that $\int_0^{t'} C(|q|)dq$ converges so fast that

$$2\int_0^{\Delta t} dt' \int_0^{t'} C(q)dq \simeq 2\int_0^{\Delta t} dt' \int_0^{\infty} C(q)dq = 2C\Delta t \qquad (7.25)$$

(this fast convergence is the mathematical expression of the physical assumption that the forces at different Δt-intervals are uncorrelated). Noting that $C = D\gamma^2$ and that according to the fluctuation-dissipation theorem one has $D = kT/\gamma$ we get that the change of the particle's momentum during the time Δt follows the probability distribution

$$P(\Delta p) \simeq \frac{1}{\sqrt{4\pi kT\gamma \Delta t}} \exp\left(-\frac{(\Delta p - f\Delta t)^2}{4kT\gamma \Delta t}\right). \qquad (7.26)$$

Thus, we see that in many cases it is reasonable to assume that the noise in the Langevin equation has a Gaussian distribution. Moreover, one can take this noise to be δ-correlated, so that

$$\langle \xi(t)\xi(t') \rangle = 2C\delta(t-t') = 2kT\gamma\delta(t-t'), \qquad (7.27)$$

which exactly corresponds to the behavior assumed by Eq. (7.25) for any Δt. Thus, the noise is supposed to be Gaussian and white. Moreover, since the averaged force value is subtracted, $\langle \xi(t) \rangle = 0$. The process corresponding to the integral of the noise

$$W(t) = \int_0^t \xi(t)dt \qquad (7.28)$$

is called a Wiener process. The curve depicting this process on the (t, W)-plane is continuous (in the sense that large jumps are very improbable) but nowhere differentiable. Its fractal dimension is 2. The introduction of a Wiener process gives several mathematical advantages and leads to mathematically firm derivation of many results. Physically one however has be aware, that it corresponds to generalizing the intermediate-scale dynamics to the smallest, microscopic scales, which may be incorrect.

7.3 Taylor–Kubo formula and velocity-velocity correlations

There exists one more method of calculation of the mean squared displacement, which may seem to be even simpler than even the original Langevin's one. We already used this method implicitly in our previous discussions, however, it is reasonable to discuss it here in some detail. Let us consider a particle moving with the velocity $v(t)$. If the velocity of the particle is know, it is easy to calculate its displacement

$$x(t) = \int_0^t v(t')dt' \qquad (7.29)$$

(we again assume that the particle's position at $t = 0$ is $x = 0$). To calculate the mean squared displacement we use the already familiar trick:

$$x^2(t) = \left[\int_0^t v(t')dt'\right]^2 = \int_0^t v(t')dt' \cdot \int_0^t v(t'')dt''$$

$$= \int_0^t \int_0^t v(t')v(t'')dt'dt'' . \qquad (7.30)$$

Averaging this equation we get

$$\langle x^2(t)\rangle = \int_0^t \int_0^t \langle v(t')v(t'')\rangle \, dt'dt''. \qquad (7.31)$$

We now assume the stationarity of the random process $v(t)$ and define the velocity-velocity correlation function $B(\tau) = \langle v(t)v(t+\tau)\rangle$ (note that this function is an even function of τ which follows from its independence on t for a stationary process). We get:

$$\langle x^2(t)\rangle = \int_0^t \int_0^t B(|t' - t''|)dt'dt'' . \qquad (7.32)$$

Let us first assume that the velocity-velocity correlation decays rather fast, so that it is integrable. This is true, for example, for the genuine Langevin case, where it decays exponentially, with the characteristic time $\tau_{br} = m/\gamma$. Let us now change in the double integral, Eq. (7.31), to a new variable $q = t' - t''$:

$$\langle x^2(t) \rangle = \int_0^t dt' \int_{-t'}^{t'} B(|q|)dq = 2 \int_0^t dt' \int_0^{t'} B(|q|)dq. \qquad (7.33)$$

Since integral over q is convergent, for $t \gg \tau_{br}$ the upper limit of integration can be changed for infinity:

$$\langle x^2(t) \rangle = 2 \left[\int_0^\infty B(|q|)dq \right] t, \qquad (7.34)$$

which immediately gives us the value of the diffusion coefficient

$$D = \int_0^\infty \langle v(t)v(0) \rangle \, dt : \qquad (7.35)$$

the diffusion coefficient is nothing else than the integral of the velocity-velocity correlation function. This fact was first discovered by G. I. Taylor (in a related but slightly different context of the passive particles' transport by a random wind, Taylor, 1921) and is typically referred to as a Taylor–Kubo formula.

Let us now turn to the case of a Langevin equation and obtain the value of the diffusion coefficient. To calculate the velocity-velocity correlation function we start from Eqs. (7.13) and (7.14). Averaging the solution of Eq. (7.14) over the realizations of the noise, we get

$$\langle v(t) \rangle_{\text{noise}} = v_0 \exp\left(-\frac{\gamma}{m}t\right) \qquad (7.36)$$

so that

$$\langle v(t)v(0) \rangle_{\text{noise}} = v_0^2 \exp\left(-\frac{\gamma}{m}t\right). \qquad (7.37)$$

However, an additional averaging over the initial velocity v_0^2 is necessary. The equipartition theorem says that $\langle v_0^2 \rangle_{eq} = kT/m$, so that

$$B(t) = \langle v(t')v(t'') \rangle = \frac{kT}{m} \exp\left(-\frac{\gamma}{m}t\right). \qquad (7.38)$$

Evaluating the integral, Eq. (7.35), for our correlation function, Eq. (7.38), we get

$$D = kT/\gamma, \qquad (7.39)$$

an already familiar result.

Note that in order to have a long-time diffusive behavior, the convergence of the integral, Eq. (7.35) to a nonzero value is necessary. The situations when the integral diverges or tends to zero lead to different kinds of anomalous diffusion. Quite a few examples of such processes are known. If, for example, $B(\tau) \propto \tau^{-\alpha}$ for τ large, the corresponding integral diverges if $\alpha \geq -1$, leading to superdiffusion. In this case evaluating of Eq. (7.33) gives $\langle x^2(t) \rangle \propto t^{2-\alpha}$ for $\alpha < 1$ and $\langle x^2(t) \rangle \propto t \ln t$ for $\alpha = 1$ and t large. This last situation is physically relevant. In 1968 Alder and Wainwright performed numerical calculations of the velocity autocorrelation function in a system of hard discs and spheres using molecular dynamics (Alder and Wainwright, 1968, 1969). The process here corresponds to a Brownian motion of one (marked) particle in the fluid of the particles of the same mass. They found out that this correlation function decays at long times essentially as $t^{-d/2}$, with d being the dimension of space, i.e. $d = 2$ for the discs and $d = 3$ for spheres. This work has initiated a vivid discussion.

The reason for such a behavior is that the motion of the Brownian particle is in fact a much more complicated process than one assumed by Langevin. The Langevin's assumption of the noise force which gets uncorrelated after a short period of time (which is exactly the same as the Einstein's second postulate) is motivated by the Boltzmann's molecular chaos hypothesis and may be a reasonable approximation in the case of strongly rarefied gases, but not in dence fluids, where the motions (and therefore impacts) of molecules are correlated. The long-time behavior in this case is determined by such correlated motions which can be interpreted as large-scale hydrodynamic motions in a fluid as a continuous medium. In this case one has to consider the contributions of modes of such motions (such as sound waves, heat waves, etc.), which are rather persistent and therefore lead to slow decorrelation of the particle's velocity. The corresponding explanations and calculations for a case of a particle in a dense fluid may be found in the books of Balescu (1975) and of Zwanzig (2001). We would not reproduce them here due to their rather special character (since one has to start from the analysis of the spectrum of such excitations in a real fluid), however in Section 7.7 we consider a much simpler model, which is analytically solvable, and which shows how the spectrum of eigenmodes of the heat bath influences the motion of the Brownian particle. The long-lasting correlations and memory effects lead to non-Markovian Langevin constructions, typically of the form

$$m\ddot{x} = -U'(x) - \int_0^t dt'\, \dot{x}(t') K(t - t') + \xi(t) \qquad (7.40)$$

where $K(t)$ is the memory-kernel. The noise $\xi(t)$ is typically taken to be Gaussian but not white. The corresponding fluctuation-dissipation relation (the Nyquist theorem) connects the memory kernel and the correlation function of the noise, see Section 7.7.

For us now it is important to note that in 3d the velocity-velocity correlation decays fast enough to be integrable, so that the Einstein's and Langevin's picture of the Browninan motion still holds; the behavior in 2d may however be quite different.

7.4 The overdamped limit

Essentially, for the separation of integrations in s and in q in Eq. (7.19) it is of no importance, which time scale shorter, it is only important that they differ strongly.

The overdamped limit

We now turn to the situation, $m/\gamma \ll \tau_c$, corresponding to the so-called overdamped limit of the Brownian motion.

Let us return to the Langevin equation

$$m\dot{v} = -\gamma v + \xi(t) \tag{7.41}$$

and consider now a very light particle with $m/\gamma \to 0$. In this case the particle's motion is governed by the Aristotelian dynamics

$$\dot{x} = v = \frac{1}{\gamma}\xi(t), \tag{7.42}$$

so that the velocity of the particle follows immediately the acting force. The solution follows from the Taylor's result,

$$\langle x^2(t) \rangle = \frac{1}{\gamma^2} \int_0^t dt' \int_{-t'}^{t'} C(q)dq, \tag{7.43}$$

so that $B(r) = \gamma^{-2} C(\tau)$. Assuming that the integral $\int_0^\infty B(\tau)d\tau$ converges and considering times $t \gg \tau_c$ we get

$$\langle x^2(t) \rangle = 2t \frac{1}{\gamma^2} \int_0^\infty C(q)dq, \tag{7.44}$$

so that $\langle x^2(t) \rangle$ grows diffusively. Comparing Eq. (7.44) with the definition of the diffusion coefficient

$$\langle x^2(t) \rangle = 2Dt \tag{7.45}$$

we get in this case

$$D = \frac{1}{\gamma^2} \int_0^\infty C(r)dr. \tag{7.46}$$

This definition of the diffusion coefficient coincides with the previous one since as we have already seen also in the underdamped situation $\int_0^\infty C(r)dr = D\gamma^2$. We now make a small change in our notation. We namely can dispense from γ in the Eq. (7.42) and introduce a new notation for the noise term:

$$\dot{x} = \sqrt{2D}\eta(t), \tag{7.47}$$

where the noise η now possess the unit integral of its correlation function:

$$\int_{-\infty}^\infty \langle \eta(0)\eta(t) \rangle \, dt = 2 \int_0^\infty \langle \eta(0)\eta(t) \rangle \, dt = 1. \tag{7.48}$$

Since we are interested only in the times which are much larger than the correlation times of the noise, we can take the last practically equal to zero and put down

$$\langle \eta(0)\eta(t) \rangle = \delta(t), \tag{7.49}$$

thus assuming the noise to be *white*. We also note that Eq. (7.47) describes the long-time behavior of a massive Langevin particle, now in the long time limit, i.e. for $t \gg m/\gamma$.

We note that the Langevin motion can also be considered under the action of some external force, say, gravitation, like it is the case in the Perrin's experiment with colloidal particles in water. In this case the initial Langevin equation reads

$$m\dot{v} = f - \gamma v + \xi(t) \tag{7.50}$$

where f is the external force (e.g. $f = -mg$ in the Perrin's situation). Taking the limit $m/\gamma \to 0$ (and changing to our new notation) we get

$$\dot{x} = \mu f + \sqrt{2D}\eta(t) \tag{7.51}$$

with the mobility $\mu = 1/\gamma$, which is the form of the Langevin equation of the widest use. Let us now turn to a simple example.

7.5 Example: The Ornstein–Uhlenbeck process

There are several good reasons to discuss the behavior of an overdamped particle in a harmonic potential. The situation was first discussed by Smoluchowski in 1913. His argumentation however was different from the one we use here, where the process is considered as an example for using the Langevin scheme.

The corresponding Langevin equation reads

$$\dot{x} = -\frac{\kappa}{\gamma}x + \sqrt{2D}\eta(t). \tag{7.52}$$

The random process governed by the Eq. (7.52) is called the Ornstein–Uhlenbeck process (Uhlenbeck and Ornstein, 1930). The corresponding process is of interest since it is one of the few less trivial situations where the closed analytical solution is possible. Let us consider the initial condition problem when the particle starts at $t = 0$ at a given coordinate $x(0) = x_0$. We note that this equation for x coincides (up to the notation) with the one governing the velocity of the particle in the classical Langevin problem, so that the correlation function of the particle's coordinate cam be immediately put down. Parallel to Subsection 7.2.2 we have

$$\langle x(t) \rangle = x_0 \exp\left(-\frac{\kappa}{\gamma}t\right) \tag{7.53}$$

and

$$\langle x(t)x(0) \rangle = \langle x_0^2 \rangle \exp\left(-\frac{\kappa}{\gamma}t\right). \tag{7.54}$$

We note that $\tau = \gamma/\kappa$ has a dimension of time; this is the typical relaxation time of the process.

Let us now consider the Ornstein–Uhlenbeck process from the thermodynamical point of view and discuss some its peculiar properties. Consider a general thermodynamical system under isothermal conditions, and take x to be a relevant thermodynamical variable. According to the Zeroth Law, a system, let evolve freely, will sooner or later achieve an equilibrium. Being perturbed from the equilibrium state, the system returns back to it. The thermodynamical potentials (say, the free energy) have at equilibrium their simple, quadratic minima; the corresponding thermodynamic force $f = -\partial F/\partial x$ is thus linear in the variable x describing the deviation from the equilibrium, $f = -\kappa x$, and our problem is equivalent to one of the behavior of an overdamped harmonic oscillator.

Let us now consider fluctuations of x around its equilibrium value (taken to be $x = 0$). According to the Onsager's regression principle introduced in Chapter 6 (which is a generalization of the primary Langevin's picture to the case of liner thermodynamic theories), small fluctuations decay on the average in the same way as macroscopic deviations from equilibrium. However, we know, that these fluctuations in equilibrium do not vanish: this fact is taken into account by the additional random force. Thus, the Ornstein–Uhlenbeck process (or its multidimensional generalizations) describes fluctuations within the linear nonequilibrium thermodynamic scheme, and is just as important and universal as the scheme itself.

The probability density of the particle's position at time t can for this case be found exactly. The approach is based on the linear property of our Langevin process. Using the fact that

$$\dot{x} = -\frac{\kappa}{\gamma}x + \sqrt{2D}\eta(t) \tag{7.55}$$

we put down the formal solution of this ODE in form:

$$x(t) - x_0 \exp\left(-\frac{\kappa}{\gamma}(t - t_0)\right) = \sqrt{2D}\int_{t_0}^{t} dt' \eta(t - t') \exp\left[-\frac{\kappa}{\gamma}(t' - t_0)\right], \tag{7.56}$$

which differs only in notation from our Eq. (7.14). Its right hand side corresponds to a weighted sum of independent Gaussian variables, and thus is also normally distributed. To see this, let consider the subdivision of the integral on the right-hand side into non-intersecting intervals Δt:

$$\int_{t_0}^{t} dt' \eta(t-t') \exp\left[-\frac{\kappa}{\gamma}(t'-t_0)\right] = \sum_n \int_{t_0+n\Delta t}^{t_0+(n+1)\Delta t} dt' \eta(t-t') \exp\left[-\frac{\kappa}{\gamma}(t'-t_0)\right]$$

$$= \sum_n \exp\left[-\frac{\kappa}{\gamma}t_n^*\right] \int_{t_0+n\Delta t}^{t_0+(n+1)\Delta t} dt' \eta(t-t') \tag{7.57}$$

where, according to the mean value theorem, $t_n^* \in [t_0 + n\Delta t, t_0 + (n+1)\Delta t]$. We already know that all integrals $\int_{t_0+n\Delta t}^{t_0+(n+1)\Delta t} dt' \eta(t-t')$ are Gaussian-distributed and independent and have zero mean. So are also the products of these integrals with the exponential prefactors. Since the sum of independent Gaussian variables also has a Gaussian distribution, we see that the left hand side of Eq. (7.56) is distributed according to a symmetric Gaussian law. From the symmetry it follows that the mean value of the right hand side is zero and thus

$$\langle x(t) \rangle = x_0 \exp(-\tau^{-1} t), \tag{7.58}$$

where the characteristic relaxation time $\tau = \gamma/\kappa$ is introduced. It is not hard to find the dispersion of the distribution of the right-hand side. Using the fact that $\langle \eta(t) \eta(t') \rangle = \delta(t - t')$ we get

$$\sigma^2(t) = 2D \left\langle \left(\int_{t_0}^{t} dt' \eta(t - t') e^{\frac{1}{\tau}(t' - t_0)} \right)^2 \right\rangle$$

$$= 2D \int_{t_0}^{t} \int_{t_0}^{t} dt' dt'' \delta(t - t' - t + t'') e^{\frac{1}{\tau}(t' + t'' - 2t_0)}$$

$$= 2D \int_{t_0}^{t} dt' e^{\frac{2}{\tau}(t' - t_0)} = D\tau \left(1 - e^{-\frac{2(t' - t_0)}{\tau}} \right). \tag{7.59}$$

Thus, the probability $p(x, t | x_0, t_0)$ reads

$$p(x, t | x_0, t_0) = \frac{1}{\sqrt{2\pi D\tau (1 - e^{-2(t-t_0)/\tau})}} \exp\left(-\frac{(x - e^{-(t-t_0)/\tau} x_0)^2}{2D\tau (1 - e^{-2(t-t_0)/\tau})} \right). \tag{7.60}$$

The first two central moments $M_1 = \langle x \rangle$, Eq. (7.58), and $M_2 = \langle (x - \langle x \rangle)^2 \rangle$, Eq. (7.59), relax exponentially to their equilibrium values. After the equilibration (i.e. for the times $t \gg \tau$) the process gets stationary and describes equilibrium fluctuations around the value $x = 0$. The distribution of fluctuations $p(x) = p(x, t | x_0, -\infty)$ is Gaussian,

$$p(x) = \sqrt{\frac{1}{2\pi D\tau}} \exp\left(-\frac{x^2}{2D\tau} \right), \tag{7.61}$$

with the mean squared value of x equal to $D\tau$. Now, let us remember that, according to the Einstein's relation $D = kT/\gamma$ and that $\tau = \gamma/\kappa$. Thus, we have $D\tau = kT/\kappa$, so that $p(x)$ corresponds to the Boltzmann distribution

$$p(x) = \sqrt{\frac{\kappa}{2\pi kT}} \exp\left(-\frac{\kappa x^2}{2kT} \right), \tag{7.62}$$

with the equilibrium value of $\kappa \langle x^2 \rangle = kT$, according to the equipartition theorem.

The Ornstein–Uhlenbeck process is one of a few situations when the probability density can be immediately deduced from the Langevin picture and from the properties of the noise. In all other cases the Fokker–Planck approach providing us with the partial differential equation for this probability density is more appropriate.

7.6 The path integral representation

Let us consider a Brownian motion of the overdamped particle taking place under the action of an external force f leading to the mean drift velocity $v = \mu f$, and describe the position of the particle (at x_0 at time $t = t_0$) after some time Δt. Since in the overdamped limit Eq. (7.51) holds, we get

$$\dot{x} - \mu f = \sqrt{2D}\eta(t). \tag{7.63}$$

Integrating this equation within some temporal interval Δt we get:

$$x - x_0 - \mu f \Delta t = \Delta x(\Delta t) \tag{7.64}$$

where $\Delta x(\Delta t) = \sqrt{2D} \int_0^{\Delta t} \eta(t) dt$ is the displacement due to the action of the random forces. This displacement is a sum of many independent small displacements due to the random impacts (random forces), the distribution of Δx will be Gaussian and reads

$$P(\Delta x) = \frac{1}{\sqrt{4\pi D \Delta t}} \exp\left(-\frac{\Delta x^2}{4D\Delta t}\right). \tag{7.65}$$

The distribution of the coordinate x then reads:

$$P(x) = \frac{1}{\sqrt{4\pi D \Delta t}} \exp\left[-\frac{(x - x_0 - \mu f \Delta t)^2}{4D\Delta t}\right]. \tag{7.66}$$

This observation leads to a useful shorthand notation for the random process governing the behavior of our system.

Let us fix Δt small but finite and consider the positions x_n of a particle at time instants $t_n = n\Delta t$. The distributions of independent differences $x_n - x_{n-1}$ are Gaussian distributions of dispersion $\sqrt{2D\Delta t}$. The joint probability to find a trajectory that passes through the points $x_0, x_1, \ldots, x_N$ is then given by a multivariate Gaussian distribution

$$P(x_0, x_1, \ldots, x_N) = \frac{1}{\left(\sqrt{4\pi D \Delta t}\right)^N} \prod_{n=1}^{N} \exp\left[-\frac{(x_n - x_{n-1} - \mu f \Delta t)^2}{4D\Delta t}\right]$$

$$= \frac{1}{\left(\sqrt{4\pi D \Delta t}\right)^N} \exp\left[-\frac{1}{4D\Delta t} \sum_{n=1}^{N} (x_n - x_{n-1} - \mu f \Delta t)^2\right]$$

$$= \frac{1}{\left(\sqrt{4\pi D \Delta t}\right)^N} \exp\left[-\frac{1}{4D} \sum_{n=1}^{N} \left(\frac{x_n - x_{n-1}}{\Delta t} - \mu f\right)^2 (\Delta t)\right]. \tag{7.67}$$

The probability $P(x_0, x_1, \ldots, x_N)$ is associated with the *path* from x_0 to $x_N = x(t)$ which goes through the intermediate points given, i.e. is a coarse-grained approximation for the probability of a given *realization* of the process. Summing up over all possible realizations starting at x_0 and ending at $x(t)$ we obtain the overall transition probability density from x_0 to $x(t)$

$$p(x,t|x_0,0) = \frac{1}{\left(\sqrt{4\pi D \Delta t}\right)^N} \sum_{\substack{\text{all paths of} \\ N \text{ steps}}} \exp\left[-\frac{\Delta t}{2D} \sum_{n=1}^{N} \left(\frac{x_n - x_{n-1}}{\Delta t} - \mu f\right)^2\right]. \tag{7.68}$$

The sum over paths is essentially an integral over x_i:

$$p(x,t|x_0,0) = \left(\sqrt{2\pi D \Delta t}\right)^{-N} \int dx_1 \ldots dx_{N-1} \exp\left[-\frac{\Delta t}{2D} \sum_{n=1}^{N} \left(\frac{x_n - x_{n-1}}{\Delta t} - \mu f\right)^2\right]. \tag{7.69}$$

On the other hand, $(x_n - x_{n-1})/\Delta t$ is nothing else as a discrete approximation for the velocity, and the sum in the exponential in Eq. (7.68) is an integral sum for $\int_0^t (\dot{x}(t) - \mu f)^2 dt$. Thus, the weight of each realization is given by

$$A \exp\left[-\frac{1}{2D} \int_0^t (\dot{x}(t) - \mu f)^2 dt\right] \tag{7.70}$$

(where A is a normalization constant), and the overall transition probability density is given by

$$p(x,t|x_0,0) = A \int \mathcal{D}x(t) \exp\left[-\frac{1}{2D} \int_0^t (\dot{x}(t') - \mu f)^2 dt'\right] \tag{7.71}$$

where $\mathcal{D}x(t)$ denotes the product of the differentials dx_i at different points along the trajectory: this is a *path-integral*. The prefactor A can be obtained *a posteriori* by normalization. Such an object is especially useful for handling perturbative calculations (like in quantum mechanics), so that, in analogy with quantum mechanics, the exponent $\frac{1}{2D}\int_0^t (\dot{x}(t) - v)^2 dt$ is often called the Lagrangian of a diffusion problem (Kleinert, 2003).

Note that in any case the deterministic trajectory for which $(\dot{x}(t) - v)^2$ (with $v = \mu f$ being the deterministic velocity) vanishes, is the most probable one, so that it can be used for the estimates. For $D \to 0$ when the overall motion tends to a deterministic limit described by the Liouville equation, this trajectory represents the actual motion, and, for D sufficiently small it can be used as a basis for perturbative calculations.

The path-integral formulation, which stresses not the *state* of the system, but the probabilities of different *processes*, can be useful if the thermodynamics of the process is considered. Let us consider the motion of a particle in a time-dependent potential force field, so that $v = \mu f(x,t) = -\mu[\partial U(x,t)/\partial x]$, where $U(x,t)$ is a potential of the force. If during the time interval Δt the particle has moved from x to $x + \Delta x$, its energy has changed from $U(x,t)$ to $U(x + \Delta x, t + \Delta t) \approx$

$U(x,t) - f\Delta x + (\partial U(x,t)/\partial t) \Delta t$. If the potential is fixed, our system is stationary and no work of the whatsoever external forces is performed inside the system. If the particle moves from x to $x + \Delta x$ and its energy changes, the necessary energy comes from (or is given back to) the heat bath. Thus, the heat received from the bath is $\delta Q = f\Delta x = F\frac{dx}{dt}\Delta t$. The second term is the work of the external source necessary for changing the potential. The work of the external force during the particle's displacement from is equal to $\delta W = (\partial U(x,t)/\partial t)\Delta t$. Thus, since the weight of each possible realization is known, the heat or work are given by the path integrals

$$Q = A \int_0^{t_{tot}} dt \int \mathcal{D}x(t) f(x,t)\dot{x}(t) \exp\left[-\frac{1}{2D}\int_0^t (\dot{x}(t') - \mu f(x,t'))^2 dt'\right] \quad (7.72)$$

and

$$W = A \int_0^{t_{tot}} dt \int \mathcal{D}x(t) \frac{\partial U(x,t)}{\partial t} \exp\left[-\frac{1}{2D}\int_0^t (\dot{x}(t) - \mu f(x,t'))^2 dt'\right]. \quad (7.73)$$

For $D \to 0$ when the overall motion tend to a deterministic limit, the heat production tends to $Q = \int_0^{t_{tot}} fv dt = \int_0^{t_{tot}} \mu f^2 dt$ which is the Joule's heat.

7.7 A non-Markovian Langevin equation

Let us shortly discuss how do non-Markovian Langevin equations typically appear. Let us remember that the Langevin equation describes the interaction of a test particle (typically taken to be considerably heavier than the particles of the medium) with the heat bath (in the case of Brownian motion, fluid medium). Therefore such an equation can be, in principle, derived in a way the canonical ensembles are typically treated: Let us consider the system at given temperature (in our case consisting of one probe particle) as a part of a larger microcanonical (Hamiltonian) one. The rest of the system is then considered as heat bath. The friction and the random force (noise) in the right-hand side of Eq. (7.5) describe the interaction between the particle and the bath, and depend on the bath's properties. In order to avoid difficulties connected with the discussion of real gases or fluids and with the necessity of at least Boltzmann-level description of the system, let us consider a trivial model of a heat bath consisting of harmonic oscillators, exactly the model Einstein and Debye were thinking about when putting down the thermodynamical theory of solids. Such a model is often called a Kac–Zwanzig heat bath, see Ford, Kac and Mazur, 1965, Zwanzig, 1973, and Zwanzig, 1980. Our discussion here follows the one of Zwanzig (2002) and of Kupferman (2003).

Let us consider a probe particle whose motion is described by the Hamiltonian

$$H_S = \frac{p^2}{2m} + U(x) \quad (7.74)$$

and the heat bath of N harmonic oscillators with the Hamiltonian

$$H_B = \sum_{j=1}^{N} \left(\frac{p_j^2}{2} + \frac{1}{2}\omega_j^2 q_j^2 \right) \qquad (7.75)$$

(we note that the momenta and the coordinates corresponding to the bath have not necessary to be the ones of the non-interacting particles in the bath, but might be the ones of normal modes of a complex system; the distribution of ω_j gives us then the spectrum of such modes; this is why we refrained from putting the mass of the "oscillator" particle in the Eq. (7.75); the corresponding momenta have the dimension of the square root of energy, and the coordinates have a dimension of square root of mass times length). We now switch on the coupling between the system and the bath

$$H = H_S + H_B + H_{\text{int}} \qquad (7.76)$$

by introducing a bilinear interaction term

$$H_{\text{int}} = \sum_{j} \gamma_j q_j x, \qquad (7.77)$$

which corresponds to the effectively harmonic coupling whose strength is given by the parameter γ_j. We will assume that the coupling strength depends only on the frequency of the corresponding mode so that $\gamma_j = \gamma(\omega_j)$ with some prescribed function $\gamma(\omega)$. The equations of motion for such system read:

$$m\ddot{x} = \dot{p} = -U'(x) - \sum_{j} \gamma_j q_j \qquad (7.78)$$

and

$$\ddot{q}_j = \dot{p}_j = -\omega_j^2 q_j - \gamma_j x. \qquad (7.79)$$

The corresponding system of equations can be solved analytically. Let us start from the second equation, and assume that $x(t)$ is some known function of t. The general solution of this equation then is:

$$q_j(t) = q_j(0) \cos \omega_j t + p_j(0) \frac{\sin \omega_j t}{\omega_j} - \gamma_j \int_0^t dt' x(t') \frac{\sin \omega_j (t - t')}{\omega_j}. \qquad (7.80)$$

The integral can be rewritten by integration by parts:

$$\int_0^t dt' x(t') \frac{\sin \omega_j (t - t')}{\omega_j} = -x(t') \frac{\cos \omega_j (t - t')}{\omega_j^2} \bigg|_0^t + \int_0^t dt' \dot{x}(t') \frac{\cos \omega_j (t - t')}{\omega_j^2}. \qquad (7.81)$$

We now introduce this solution into Eq. (7.78) and obtain:

$$m\ddot{x} = -U'(x) - \sum_j \gamma_j^2 \int_0^t dt' \dot{x}(t') \frac{\cos\omega_j(t-t')}{\omega_j^2}$$

$$- \sum_j \gamma_j \left[q_j(0) \cos\omega_j t + p_j(0) \frac{\sin\omega_j t}{\omega_j} \right]. \tag{7.82}$$

This equation has the structure of non-Markovian (time-delay) Langevin equation

$$m\ddot{x} = -U'(x) - \int_0^t dt' \dot{x}(t') K(t-t') + F(t) \tag{7.83}$$

where $K(t)$ is the friction memory-kernel,

$$K(t) = \sum_j \gamma_j^2 \frac{\cos\omega_j(t)}{\omega_j^2} \tag{7.84}$$

and $F(t)$ is the "noise" force,

$$F(t) = -\sum_j \gamma_j \left[q_j(0) \cos\omega_j t + p_j(0) \frac{\sin\omega_j t}{\omega_j} \right]. \tag{7.85}$$

Imagine that before switching on the interaction between the heavy particle and the heat bath the last one was in the state of thermodynamical equilibrium at temperature T. Due to symmetry, $\langle q_j(0) \rangle = 0$ and $\langle p_j(0) \rangle = 0$ (the averages here are the ensemble averages), so that $\langle F(t) \rangle = 0$ at all times, and moreover $\langle q_j(0) p_i(0) \rangle = 0$ (different oscillators are anyhow independent; for the same oscillator, the motion in either direction is possible at the same coordinate). Furthermore, due to energy equipartition $\langle \omega_j^2 q_j^2(0)/2 \rangle = \langle p_j^2(0)/2 \rangle = k_B T/2$. This gives us

$$\langle F(t) F(t') \rangle = \sum_j \gamma_j^2 \left[\langle q_j^2(0) \rangle \cos\omega_j t \cos\omega_j t' + \langle p_j^2(0) \rangle \frac{\sin\omega_j t \sin\omega_j t'}{\omega_j^2} \right]$$

$$= \sum_j \gamma_j^2 \frac{k_B T}{\omega_j^2} [\cos\omega_j t \cos\omega_j t' + \sin\omega_j t \sin\omega_j t']$$

$$= \sum_j \gamma_j^2 \frac{k_B T}{\omega_j^2} \cos\omega_j (t-t') = k_B T K(t-t'). \tag{7.86}$$

This is a corresponding fluctuation-dissipation theorem for a non-Markovian Langevin equation, which, being put into the spectral representation, is equivalent to the Nyquist theorem, $S(\omega) = k_B T K(\omega)$, where $S(\omega) = \int \langle F(t) F(t+\tau) \rangle e^{i\omega\tau} d\tau$ is the spectral density of the noise. Note that the factor 4, typically appearing in the Nyquist theorem, has to do with the introducing another, "causal" kernel $M(t)$, whose properties are discussed below.

If the number of oscillators N gets large (which we, of course, assume whenever their ensemble may be considered as a heat bath), we can pass from sums to integrals over ω so that

$$K(t) = \int_0^\infty g(\omega)\gamma^2(\omega)\frac{\cos\omega t}{\omega^2}d\omega, \qquad (7.87)$$

where $g(\omega)$ is the density of states of the thermostat's oscillators. The functions $K(t)$ and $F(t)$ are now the functionals of this density of states.

Note that the limits of integration in Eq. (7.83) make the response of the system to the external force causal, i.e. depending only on the values of x at previous instants of time. One can rewrite the equation in the form

$$m\ddot{x} = -U'(x) - \int_0^t dt'\dot{x}(t-t')M(t') + F(t), \qquad (7.88)$$

with $M(t) = K(t)\Theta(T)$, $\Theta(t)$ being a Heaviside Θ-function.

Let us now assume that there is no external force acting on the particle, so that $U(x) = 0$. In this case our equation is a closed linear integral equation for $v(t) = \dot{x}(t)$. Let us assume that the process has started long ago, so that the lower integration limit might be taken to tend to $-\infty$ (of course, we assume that the memory kernel M is integrable). The equation for the velocity can then be solved by applying the Fourier-transform:

$$i\omega v(\omega) = -M(\omega)v(\omega) + F(\omega) \qquad (7.89)$$

which gives us the velocity power-spectrum as

$$S_v(\omega) = \langle |v(\omega)|^2 \rangle = \frac{\langle |F(\omega)|^2 \rangle}{|M(\omega)|^2 + \omega^2}. \qquad (7.90)$$

Now, the value of $\langle |F(\omega)|^2 \rangle$ in the enumerator is exactly the spectral density of the random force, i.e. the Fourier-transform of the force-force correlation function $\langle F(t)F(t+t') \rangle$ in t', i.e. is proportional to the Fourier-transform of $K(t)$. The Fourier-transform of $M(t)$ is exactly $K(\omega)/2$, as it may be checked by direct inspection. Thus,

$$S_v(\omega) = \frac{4K(\omega)}{|K(\omega)|^2 + 4\omega^2}. \qquad (7.91)$$

Note that according to the Taylor–Kubo formula

$$D = \int_0^\infty \langle v(0)v(t)\rangle dt = \frac{1}{2}\int_{-\infty}^\infty \langle v(0)v(t)\rangle dt = \frac{1}{2}S_v(0), \qquad (7.92)$$

so that

$$D = \frac{2}{K(\omega=0)}. \qquad (7.93)$$

If the corresponding limiting value exists (i.e. $K(\omega)$ stays finite at small frequencies), the situation corresponds asymptotically to the normal diffusion. If it vanishes, then the diffusion coefficient diverges, and we have to do with the behavior which is superdiffusive (i.e. goes faster than diffusion, say, leads to $\langle x^2(t)\rangle \propto t^\alpha$ with $\alpha > 1$). If it diverges, the diffusion coefficient vanishes, and we have to do with the behavior which is subdiffusive (i.e. goes slower than diffusion, say, leads to $\langle x^2(t)\rangle \propto t^\alpha$ with $\alpha < 1$), giving rise to an interesting dynamical behavior (Bao and Zhuo, 2003). Note that $K(\omega) = g(|\omega|)\gamma^2(|\omega|)/\omega^2$, and thus using a heat bath with the corresponding density of states will lead to the anomalous diffusion properties. Assuming, for example, $K(\omega) \propto \omega^{\alpha-1}$ for ω small will lead exactly to the power-law behavior described above.

Although the model discussed above is of a formal nature, it has some physical implications: The density of vibrational states of a heat bath determines its thermodynamic properties, such as specific heat. Note that in the normal three dimensional continuous medium the spectrum of oscillations is dominated by the acoustic modes (phonons): This corresponds the Debye model of crystalline lattices. In this case $g(\omega) \propto \omega^2$, so that (under the assumption $\gamma(\omega) = const$) $K(\omega)$ stays finite at low frequencies and the normal diffusion takes place. This case corresponds to what one normally calls the "Ohmic" heat bath. The Einstein's model of a solid with $g(\omega) \propto \delta(\omega - \omega_0)$ also leads to $K(\omega \to 0) \to const$ and therefore to normal diffusion. On the other hand, in the Debye model in two dimensions the density of states would be $g(\omega) \propto \omega$, so that $K(\omega) \propto \omega^{-1}$, it diverges for $\omega \to 0$ and leads to strong subdiffusion ($\alpha = 0$). The physical reason for such anomalous diffusion is the high density of the low-frequency modes, giving rise to strond correlations in the particle's velocities at different time. This correlations decay too slow to warrant for decoupling at whatever delay time τ, and therefore lead to the violation of the Einstein's second postulate. The situation considered here is to some extent similar to the case of a particle in a two-dimensional fluid; the main difference between the cases corresponds to the type of the coupling (which for a Brownian particle is not a bilinear coupling via coordinates). Other non-Ohmic situations are also possible.

We note that in the absence of the external force on the particle, the position of a particle (starting at $t = 0$ at $x = 0$) at time t will be distributed accoring to a Gaussian Law: The situation here corresponds to the case of summation of random oscillations, the one considered by Rayleigh more than 120 years ago. This Gaussian nature of the displacement distribution is typical for many models of anomalous diffusion based on the picture of modes, and is absent in other situations (where the memory is introduced, say, through the sojourn times), which lead to strongly non-Gaussian distributions. Such situations will be discussed in some detail in Chapter 11.

Chapter 8

Fokker–Planck and Master Equations

8.1 Equations for the probability density

We note that the initial approach of Einstein was based on the discussion of the deterministic equations for the probability densities, while the Langevin's one emphasized the stochastic nature of single realizations. The generalization of the Einstein's approach leads to what is now known as a Fokker–Planck equation. Phenomenologically, the equation can be derived along the lines of the Fick's approach, by combining the linear response assumption and the continuity equation (Fick, 1855). The diffusion-like equation for the overdamped motion with drift was first proposed by A. Fokker in his dissertation in 1914, and discussed by M. Planck in 1918. The work of Planck starts from the note that Fokker published only the equation itself, without any derivation. Such derivation was promised but never published. The same equation was proposed by Smoluchowski in the work "*Über Brownsche Molekularbewegung unter Einwirkung äußerer Kräfte und deren Zusammenhang mit der verallgemeinerten Diffusionsgleichung*" (Smoluchowski, 1915).

Thus, the particle's density $n(x,t)$ or the probability to find a particle $p(x,t)$ fulfills the equation

$$\frac{\partial p}{\partial t} p = -\mathrm{div} \mathbf{J} \tag{8.1}$$

where $\mathbf{J}$ is the probability current. On the other hand, this current may appear due to the two factors: due to the concentration gradient and due to the external force $\mathbf{f}$:

$$\mathbf{J} = -D\nabla p + \mu \mathbf{f} p \tag{8.2}$$

where μ is the particle's mobility. Combining the equations we get the diffusion equation with drift:

$$\frac{\partial p}{\partial t} = \nabla(-\mu \mathbf{f} p + D\nabla p). \tag{8.3}$$

The standard derivation of the Fokker–Planck equation due to Smoluchowski, Kolmogorov et al. follows the lines of the Einstein's discussion and starts from the general Markovian assumption. One considers the process characterized by the transition probabilities $p(x,t|x',t')$, the probability for a particle to be at x at time t provided it started at x' at time $t' < t$. We note that x here has not necessarily to be considered as a scalar, but for the simplicity we use the scalar notation.

Our initial considerations are the ones already discussed in the end of Chapter 2. The Markovian property leads us to the assumption that probability to be at x at time t haven started at x_0 at time t_0 can be expressed through the integral

$$p(x,t|x_0,t_0) = \int dx' p(x,t|x',t') p(x',t'|x_0,t_0) \quad (8.4)$$

which essentially doesn't mean anything else than the statement of the fact that at time t' the particle has to be found at some place it really could get to. Depending on the community, this equation is termed as a Smoluchowski, or as a Chapman–Kolmogorov equation. We now consider the time t' as being close enough to t: subtracting $p(x',t'|x_0,t_0)$ from the both parts of the integral equation we get

$$p(x,t|x_0,t_0) - p(x',t'|x_0,t_0) = \int dx' p(x,t|x',t') p(x',t'|x_0,t_0) - p(x',t'|x_0,t_0). \quad (8.5)$$

We now discuss the situation when the difference $\Delta t = t - t'$ can be considered as small. The left-hand side can be approximated through $\frac{\partial}{\partial t'} p(x,t'|x_0,t_0) \Delta t$. We shall assume that typical changes in the coordinates during this short time are also to some extent small. Let us denote $r = x - x'$ and introduce the transition probabilities $w(y,r;t,\Delta t) = p(y+r,t+\Delta t|y,t)$. Now we rewrite Eq. (8.5) in the form

$$\frac{\partial}{\partial t} p(x,t|x_0,t_0) \Delta t = \int dr w(x-r,r;t,\Delta t) p(x-r,t|x_0,t_0) - p(x,t|x_0,t_0) \quad (8.6)$$

and start expanding the integrand in powers of r. We namely use the fact that

$$w(x-r,r;t,\Delta t) p(x-r,t|x_0,t_0) = w(x,r;t,\Delta t) p(x,t|x_0,t_0)$$

$$- r \frac{\partial}{\partial x} w(x,r;t,\Delta t) p(x,t|x_0,t_0)$$

$$+ \frac{1}{2} r^2 \frac{\partial^2}{\partial x^2} w(x,r;t,\Delta t) p(x,t|x_0,t_0) + \cdots \quad (8.7)$$

which lets us rewrite the first integral in the form

$$\frac{\partial}{\partial t} p(x,t|x_0,t_0) \Delta t = p(x,t|x_0,t_0) \int dr w(x,r;t,\Delta t)$$

$$- \frac{\partial}{\partial x} p(x,t|x_0,t_0) \int dr r w(x,r;t,\Delta t)$$

$$+ \frac{\partial^2}{\partial x^2} p(x,t|x_0,t_0) \int dr r^2 w(x,r;t,\Delta t) + \cdots$$

$$- p(x,t|x_0,t_0). \tag{8.8}$$

Noting that due to the normalization $\int drw(y,r;t,\Delta t) = 1$ we see that the first and the last terms of the expression in the right-hand side cancel. Performing the limiting transition $\Delta t \to 0$ one has

$$\frac{\partial}{\partial t} p(x,t|x_0,t_0) = \sum_{n=1}^{\infty} \frac{(-1)^n}{n!} \frac{\partial^n}{\partial x^n} K_n(x,t) p(x,t|x_0,t_0) \tag{8.9}$$

where the transition moments $K_n(x,t)$ are

$$K_n(x,t) = \lim_{\Delta t \to 0} \frac{1}{\Delta t} \int dr\, r^n w(x,r;t,\Delta t). \tag{8.10}$$

The expression Eq. (8.9) corresponds to the Kramers–Moyal expansion. For the case of the Brownian motion and other diffusive processes one finds that only 2 first transition moments are different from zero: there are

$$A(x,t) = \lim_{\Delta t \to 0} \frac{1}{\Delta t} \int dr\, r\, w(x,r;t,\Delta t) \tag{8.11}$$

and

$$B(x,t) = \lim_{\Delta t \to 0} \frac{1}{\Delta t} \int dr\, r^2 w(x,r;t,\Delta t). \tag{8.12}$$

All higher moments vanish under the limiting transition $\Delta t \to 0$ since

$$\int dr\, r^n w(x,r;t,\Delta t) = O(\Delta t^2) \text{ for } n \geq 3. \tag{8.13}$$

In this case the Kramers–Moyal expansion of the Chapman–Kolmogorov equation leads to the Fokker–Planck equation for $p(x,t|x_0,t_0)$

$$\frac{\partial}{\partial t} p(x,t|x_0,t_0) = -\frac{\partial}{\partial x} A(x,t) p(x,t|x_0,t_0) + \frac{1}{2} \frac{\partial^2}{\partial x^2} B(x,t) p(x,t|x_0,t_0). \tag{8.14}$$

Comparing this general equation with our phenomenological equation, Eq. (8.3), we see that it also has a form of a continuity equation,

$$\frac{\partial}{\partial t} p = \frac{\partial}{\partial x}\left[-Ap + \frac{1}{2}\frac{\partial}{\partial x} Bp\right]. \tag{8.15}$$

Note that the transition probability $p(x, t|y, s)$ is essentially a function of two spatial variables, x and y, the initial and the final particle's positions, and two times t and s. It is often desirable to have the equation which defines the function $p(x, t|y, s)$ with respect to its initial position and to the time $s = t_0$. This can be easily done by considering the increment $p(x, t|y, s) - p(x, t|y', s')$ and repeating the steps leaving to the Fokker–Planck equation. Under the same assumptions one gets

$$-\frac{\partial}{\partial s}p(x,t|y,s) = A(y,s)\frac{\partial}{\partial y}p(x,t|y,s) + \frac{1}{2}B(y,s)\frac{\partial^2}{\partial y^2}p(x,t|y,s). \qquad (8.16)$$

Equation (8.16) is called the backward Kolmogorov equation, while the Fokker–Planck equation in this context is called the forward one. The differential operator $\mathbf{L}^+$ acting on $p(x, t|y, s)$ in the right-hand side of this equation is adjoint to the differential operator in the right-hand side of the forward (Fokker–Planck) equation $\mathbf{L}$ with respect to the scalar product of two function $(f, g) = \int f(x)g(x)dx$ in the sense that $(f, \mathbf{L}g) = (\mathbf{L}^+ f, g)$.

We also note that the Pawula's theorem states that either the first two moments (having exactly the behavior described by Eqs. (8.11) and (8.12)) are enough for the full description, or the whole infinite series has to be used. Truncating the series after any other term than the second one leads to equations which do not guarantee the non-negativity of their solutions, which therefore cannot be interpreted as probability density functions. In this case the integral equation representations cannot be reduced to anything considerably more simple.

In the case of the overdamped motion the coefficients A and B can be easily obtained from the discussion of the overdamped Langevin equation, Eq. (7.51). From Eq. (7.66) we get the transition probability density $w(x, r; t, \Delta t)$ to be Gaussian

$$w(x, r; t, \Delta t) = \frac{1}{\sqrt{4\pi D \Delta t}}\exp\left(-\frac{(r - \mu f \Delta t)^2}{4D\Delta t}\right) \qquad (8.17)$$

from which we get that $A(t) = \mu f$ and $B = \Delta t^{-1}\left[(\mu f \Delta t)^2 + 2D\Delta t\right] \to 2D$ for $\Delta t \to 0$, and that the higher transition moments (being the combinations of the two lower ones) are of a higher order in Δt. Hence, the resulting Fokker–Planck equation reads

$$\frac{\partial}{\partial t}p = \frac{\partial}{\partial x}\left[-\mu f p + D\frac{\partial}{\partial x}p\right] \qquad (8.18)$$

(with D and μ connected by the fluctuation-dissipation relation $D = kT\mu$). Equation (8.18) is exactly the one following from our initial phenomenological considerations.

We note that taking the limits in Eqs. (8.11) and (8.12) corresponds essentially to the derivatives of the corresponding transition moments. The first one can be then interpreted as the mean velocity at point x, $\langle v(x) \rangle$, and the second one as the time derivative of the mean squared displacement from x, connected to a local diffusion coefficient.

The form of the Fokker–Planck equation stays the same if one considers the variable x as a vector in the coordinate or in the phase space: The only important assumptions are the Markovian nature of transitions and the existence of transition moments. The vector form of the Fokker–Planck equation reads (compare Chapter 2):

$$\frac{\partial}{\partial t}p = \sum_i \frac{\partial}{\partial x_i}\left[-A_i p + \frac{1}{2}\sum_j D_{ij}\frac{\partial}{\partial x_j}p\right]. \qquad (8.19)$$

8.2 Special stochastic processes

8.2.1 Example 1: The Ornstein–Uhlenbeck process revisited

Let us return to the Ornstein–Uhlenbeck process describing the behavior of an overdamped particle in the harmonic potential under the influence of white noise. The Smoluchowski equation is

$$\frac{\partial}{\partial t}p = \frac{\partial}{\partial x}\left[\frac{\kappa}{\gamma}xp + D\frac{\partial}{\partial x}p\right]. \qquad (8.20)$$

There are several methods to solve Smoluchowski equations; many of them are discussed in detail in Risken's book (Risken, 1988, 1994).

Note that a Smoluchowski equation (with coordinate-independent D) in a field of a potential force $f = -\nabla U$ can be reduced to a Schödinger equation by taking $p(x,t) = \sqrt{p_{eq}(x)}g(x,t)$, where $p_{eq}(x)$ is the equilibrium solution, $p_{eq}(x) = \exp\left[-U(x)/kT\right]$. The equation for $g(x,t)$ then reads

$$-\frac{\partial}{\partial t}g = D\left[-\frac{\partial^2}{\partial x^2} + U_{\text{eff}}(x)\right]g, \qquad (8.21)$$

with the effective potential

$$U_{\text{eff}}(x) = \left(\frac{1}{2kT}\frac{\partial U}{\partial x}\right)^2 - \frac{1}{2kT}\frac{\partial^2 U}{\partial x^2}. \qquad (8.22)$$

The advantage of putting the Smoluchowski equation in this form is the fact that the operator in the right-hand side of Eq. (8.21) is now Hermitian, and the equation itself allows for a much simpler analytical treatment. For example, it is a great convenience, that the right and left eigenfunctions of this operator coincide, so that the simple eigenfunction decomposition approach, known from the quantum mechanics, works:

$$g(x,t) = \sum_n \phi_n(x)\Phi_n(t) \qquad (8.23)$$

where the temporal functions are governed by ordinary differential equations of the type

$$\frac{d\Phi_n(t)}{dt} = -\lambda_n \Phi_n(t); \quad (8.24)$$

λ_n are the eigenvalues of the Schrödinger operator on the right-hand side. Moreover, the potential $U_{\text{eff}}(x)$ has an additional symmetry (it corresponds to a so-called supersymmetric quantum mechanics, Gendenstein and Krive, 1985) so that it immediately follows that one of the eigenvalues is zero, which corresponds to a steady (equilibrium) state, and that this state is non-degenerate.

For the quadratic potential corresponding to the Ornstein–Uhlenbeck process, the effective potential given by Eq. (8.22) is quadratic again. Thus, the Fokker–Planck equation describing the Ornstein–Uhlenbeck process is reduced to the Schrödinger equation for a harmonic oscillator. No wonder that all properties of the solutions are known. As a solution of the initial-value problem we get of course an already known result,

$$p(x,t|x_0,t_0) = \sqrt{\frac{1}{2\pi D\tau(1-e^{-2(t-t_0)/\tau})}} \exp\left(-\frac{(x-e^{-(t-t_0)/\tau}x_0)^2}{2D\tau(1-e^{-2(t-t_0)/\tau})}\right). \quad (8.25)$$

It is interesting to note, that the Ornstein–Uhlenbeck process (essentially the process with the exponentially decaying covariation) is the only diffusive process (with continuous trajectories) that is Gaussian and Markovian at the same time.

Let us consider a homogeneous Gaussian process, corresponding to the system at equilibrium. The joint probability density $p(x,t,x_0,t_0) = p(x,x_0,t-t_0)$ is fully defined by the dispersion σ and by the covariance $g(t-t') = \langle x(t)x(t')\rangle/\sigma^2$ and reads

$$p(x,t,x_0,t_0) = \frac{1}{2\pi\sigma^2\sqrt{(1-g^2)}} \exp\left(-\frac{(x^2 - 2gxx_0 + x_0^2)}{2\sigma^2(1-g^2)}\right). \quad (8.26)$$

We note that since $\lim_{dt \to 0} \langle x(t+dt)x(t)\rangle = \langle x(t)x(t)\rangle = \sigma^2$, one has $g(0) = 1$. To obtain the transition probability, which is the conditional probability for a particle to be at x at time t provided it was in x_0 at time t_0 we use a Bayes theorem, according to which $p(x,t,x_0,t_0) = p(x,t|x_0,t_0)p(x_0,t_0)$, and take $p(x_0,t_0)$ to be a one-point, equilibrium probability density,

$$p(x_0,t_0) = p(x_0) = \sqrt{\frac{1}{2\pi\sigma^2}} \exp\left(-\frac{x_0^2}{2\sigma^2}\right). \quad (8.27)$$

Combining Eqs. (8.26) and (8.27) we get:

$$p(x,t|x_0,t_0) = \sqrt{\frac{1}{2\pi\sigma^2(1-g^2)}} \exp\left(-\frac{(x-gx_0)^2}{2\sigma^2(1-g^2)}\right). \quad (8.28)$$

Inserting this transition probability into a Chapman–Kolmogorov equation

$$p(x,t|x_0,0) = \int dx' p(x,t|x',t') p(x',t'|x_0,0) \qquad (8.29)$$

and performing the integrations we get that

$$\sqrt{\frac{1}{2\pi\sigma^2(1-g(t)^2)}} \exp\left(-\frac{(x-g(t)x_0)^2}{2\sigma^2(1-g(t)^2)}\right)$$

$$= \sqrt{\frac{1}{2\pi\sigma^2(1-g(t-t')^2 g(t')^2)}} \exp\left(-\frac{(x-g(t-t')g(t')x_0)^2}{2\sigma^2(1-g(t-t')^2 g(t')^2)}\right) \qquad (8.30)$$

from which for any $t' \in [0,t]$ it follows that

$$g(t) = g(t-t')g(t'). \qquad (8.31)$$

The only differentiable solution of this functional equation is an exponential: taking $dt = t - t'$ to be small we get:

$$g(t) = g(t') + \frac{dg(t')}{dt'} dt = g(dt)g(t'). \qquad (8.32)$$

We thus get:

$$\frac{1}{g(t')} \frac{dg(t')}{dt'} dt = g(dt) - 1. \qquad (8.33)$$

Noting that $g(0) = 1$ and denoting $\tau^{-1} = -g'(0)$ we get

$$\frac{1}{g(t')} \frac{dg(t')}{dt} dt = -\tau^{-1} dt \qquad (8.34)$$

so that

$$\frac{d}{dt} \log g(t) = -\tau^{-1} \qquad (8.35)$$

and thus $g(t) = \exp(-t/\tau)$. The corresponding derivation can be generalized to multidimensional case as well.

8.2.2 Example 2: The Klein–Kramers equation

As an example of multivariate Fokker–Planck equation let us consider an underdamped case, corresponding to the genuine Langevin treatment. The mechanical state of the system is now characterized by a pair (v,x) of phase variables so that the probability density p depends on these two variables and on the time: $p(v,x,t)$. The overall equations for v and x read:

$$m\dot{v} = f(x) - \gamma v + \xi(t),$$
$$\dot{x} = v. \qquad (8.36)$$

The coefficients A and B are no more scalar: Thus, A is now a vector and B is a 2×2 matrix. The transition probabilities for the vector

$$\mathbf{r} = \begin{pmatrix} \Delta v \\ \Delta x \end{pmatrix} = \begin{pmatrix} v(t + \Delta t) - v(t) \\ x(t + \Delta t) - x(t) \end{pmatrix} \tag{8.37}$$

can be found by noting that $\Delta v = \frac{1}{m} \int_t^{t+\Delta t} [f(x) - \gamma v + \xi(t)] dt \simeq \frac{1}{m}[f(x) - \gamma v]\Delta t + O(\Delta t^2) + \frac{1}{m} \int_t^{t+\Delta t} \xi(t) dt$. The first two terms are deterministic, distribution of the last term is given by Eq. (7.26). The development of Δx is also deterministic, so that in the lowest order we have $\Delta x = v\Delta t$. Thus,

$$w(\Delta v, \Delta x, x, v, t, \Delta t)$$
$$= \frac{1}{\sqrt{2\pi kT\gamma \Delta t}} \exp\left\{-\frac{\left[\Delta v - \frac{1}{m}\left(f(x) - \gamma v\right)\Delta t\right]^2}{2kT\gamma \Delta t}\right\} \delta(\Delta x - v\Delta t). \tag{8.38}$$

The transition moments are:

$$A(x) = \frac{1}{\Delta t}\begin{pmatrix} \langle \Delta v \rangle \\ \langle \Delta x \rangle \end{pmatrix} = \begin{pmatrix} \frac{1}{m}f(x) - \frac{\gamma}{m}v \\ v(x) \end{pmatrix}, \tag{8.39}$$

and

$$B = \frac{1}{\Delta t}\begin{pmatrix} \langle \Delta v^2 \rangle & \langle \Delta x \Delta v \rangle \\ \langle \Delta x \Delta v \rangle & \langle \Delta x^2 \rangle \end{pmatrix} \to \begin{pmatrix} kT\gamma & 0 \\ 0 & 0 \end{pmatrix} \tag{8.40}$$

since the averages $\langle \Delta x \Delta v \rangle$ and $\langle \Delta x^2 \rangle$ are both proportional to Δt^2 and moreover $\langle \Delta v^2 \rangle = \langle \Delta v \rangle^2 + \sigma^2(\Delta t) = m^{-1}(f(x) - \gamma v)^2 \Delta t^2 + 4kT\gamma\Delta t \to 4kT\gamma\Delta t$. The overall Fokker–Planck equation for the probability density $p(v, x, t)$ now reads:

$$\frac{\partial}{\partial t}p = -\frac{\partial}{\partial x}vp - \frac{\partial}{\partial v}\frac{1}{m}(f - \gamma v)p + kT\gamma \frac{\partial^2}{\partial v^2}p. \tag{8.41}$$

This equation is often referred to as a Klein–Kramers equation (Klein, 1922, Kramers, 1940). Assuming that the force $f(x)$ is a potential one, $f(x) = -\frac{\partial}{\partial x}U(x)$, it is not hard to show that for the case when $p(v, x)$ is given by

$$p(v, x) = \frac{1}{Z}\exp\left(-\frac{mv^2/2 + U(x)}{kT}\right) \tag{8.42}$$

with

$$Z = \iint \exp\left(-\frac{mv^2/2 + U(x)}{kT}\right) dv dx, \tag{8.43}$$

the right-hand side Eq. (8.41) vanishes independently on the value of γ. Of course, Eq. (8.42) is a canonical distribution, and the absence of γ in this equation underlines the fact that the equilibrium properties of a system do not depend on the assumptions on its kinetic behavior. Note that the equilibrium distribution $p(v, x)$

factorizes into a product of the Maxwell distribution of the velocities and of the "barometric" distribution,

$$\frac{1}{\sqrt{2\pi kT}} \exp\left(-\frac{mv^2}{2kT}\right) \cdot \frac{1}{Q} \exp\left(-\frac{U(x)}{kT}\right) \qquad (8.44)$$

with $Q = \int \exp\left(-\frac{U(x)}{kT}\right) dx$. This may not be the case for steady states other than the equilibrium. We note that Fokker–Planck equation, Eq. (8.41) can be presented in its usual form of the continuity of the probability current

$$\frac{\partial}{\partial t} p(x,t) = -\frac{\partial}{\partial x} J_x - \frac{\partial}{\partial v} J_v \qquad (8.45)$$

with the probability current

$$\mathbf{J} = \begin{pmatrix} J_x \\ J_v \end{pmatrix} = \begin{pmatrix} vp \\ \frac{1}{m}(f - \gamma v)p + kT\gamma \frac{\partial}{\partial v} p(v,t) \end{pmatrix} \qquad (8.46)$$

Note that the probability current (which is the current in the phase space, corresponding to the fact that the phase coordinates of the system continuously change under its Hamiltonian evolution) does not vanish in the equilibrium, while the particles' current $I(x) = \langle J_x(x) \rangle = \int vp(x,v)dv$ definitely does.

We note that Eq. (8.41) for the joint probability distribution of the coordinate and the velocity can be considered as a combination of the continuity equation for the coordinate,

$$\frac{\partial}{\partial t} p(x,t) = -\frac{\partial}{\partial x} v(x,t) p(x,t) \qquad (8.47)$$

and the ordinary Fokker–Planck equation for the velocity

$$\frac{\partial}{\partial t} p(v,t) = -\frac{\partial}{\partial v} \frac{1}{m}(f - \gamma v) p(v,t) + kT\gamma \frac{\partial^2}{\partial v^2} p(v,t). \qquad (8.48)$$

which are coupled by the fact that the force f appearing in the last one may be x-dependent.

8.3 The Fokker–Planck equation and the Liouville equation

The phenomenological derivation of the diffusion equation by Fick (and our phenomenological derivation of the Fokker–Planck equation as a diffusion equation with drift) was based on the local continuity equation for the probability density (i.e. on the Liouville equation in the broader sense). The relation between the Fokker–Planck equation and the Liouville equation may be made clearer by the discussion of the immediate derivation of the first from the last. In our discussion we closely follow the book Zwanzig, 2000. We use here a one-dimensional, scalar notation, making more clear the main steps. The trivial generalization to the multivariate case will be discussed at the end of the paragraph.

Let us consider the random process $x(t)$ governed by the equation of motion

$$\frac{dx}{dt} = v(x) + \xi(t) \tag{8.49}$$

where $\xi(t)$ is a Gaussian random variable (noise), so that $\langle \xi_i(t) \rangle = 0$ and $\langle \xi(t)\xi(t') \rangle = 2D\delta(t-t')$. We are looking for the probability distribution function of the values of x_i at time t. This probability density $P(x,t)$ may be obtained from the one for a special realization of the noise $\xi(t)$, i.e. from the solution of the Liouville (continuity) equation. Assuming the local conservation law (the continuity equation) for P we can write that

$$\frac{\partial P}{\partial t} + \sum_i \frac{\partial}{\partial x}\left(\frac{\partial x}{\partial t} P\right) = 0. \tag{8.50}$$

Replacing the time derivative of x on the right-hand side of this equation by the expression given by Eq. (8.49) we get:

$$\frac{\partial P}{\partial t} = -\frac{\partial}{\partial x}[(v(x) + \xi(t))P]. \tag{8.51}$$

This is a stochastic differential equation giving $P(x,t)$ in a particular realization of the process.

Equation (8.51) is a linear differential equation, whose formal solution can be obtained as follows. Let us introduce the linear operator L corresponding to the deterministic part of the behavior:

$$\hat{L}P = \frac{\partial}{\partial x}(v(x)P). \tag{8.52}$$

Eq. (8.51) then reads:

$$\frac{\partial P}{\partial t} = -\hat{L}P - \frac{\partial}{\partial x}[\xi(t)P]. \tag{8.53}$$

The formal solution of this system is a generalization of the solution of a linear differential equation (say, Eq. (7.14)) and reads:

$$P(x,t) = e^{-\hat{L}t}P(x,0) - \int_0^t ds\, e^{-\hat{L}(t-s)} \frac{\partial}{\partial x}[\xi(s)P(x,s)]. \tag{8.54}$$

This solution shows explicitly that $P(x,t)$ depends on the noise at all times $s < t$. We now substitute this result into the last term in the right-hand side of Eq. (8.53) and get:

$$\frac{\partial}{\partial t}P(x,t) = -\hat{L}P(x,t) - \frac{\partial}{\partial x}\xi(t)e^{-\hat{L}t}P(x,0) + \frac{\partial}{\partial x}\xi(t)\int_0^t ds\, e^{-\hat{L}(t-s)}\frac{\partial}{\partial x}[\xi(s)P(x,s)]. \tag{8.55}$$

Now, the ensemble average of this probability density, $p(x,0) = \langle P(x,0) \rangle$ follows. The first term in the right-hand side, containing the $P(x,t)$ only is averaged easily. The second one, containing only the initial condition (independent on the realization of the noise) and $\xi(t)$ vanishes under the averaging, so that only the third one, with the paired noise factors, needs to be averaged explicitly:

$$\frac{\partial}{\partial t}p(x,t) = -\hat{L}p(x,t) + \nabla \int_0^t ds\, e^{-\hat{L}(t-s)} \nabla [\langle \xi(t)\xi(s)P(x,s)\rangle]. \tag{8.56}$$

Now, for $s \neq t$ $\xi(t)$ is uncorrelated with $\xi(s)$ for all $s < t$ and thus with the state of the system $P(x,s)$ (given by the values of the noise at earlier times). Due to the symmetry of the noise, the integrand is zero for all $s \neq t$, so that $\langle \xi(t)\xi(s)P(x,s)\rangle = 0$ for $s \neq t$. On the other hand, $P(x,s)$ given by an integral Eq. (8.54) over the times $s' < s = t$, is dominated by the values of the noise at the previous times, and thus can be considered as independent from $\xi(t)$ *exactly* at time t, so that $\langle \xi(t)\xi(s)P(x,s)\rangle = \langle \xi(t)\xi(s)\rangle \langle P(x,s)\rangle \equiv 2D\delta(t-s)p(x,t)$.

Thus,

$$\frac{\partial}{\partial t}p(x,t) = -\hat{L}p(x,t) + \frac{\partial}{\partial x}\int_0^t ds\, e^{-\hat{L}(t-s)} \frac{\partial}{\partial x} 2D\delta(t-s)p(x,s)$$

$$= \frac{\partial}{\partial x}(f(x)p) + \frac{\partial^2}{\partial x^2}Dp \tag{8.57}$$

which is a Fokker–Planck equation we looked for. We note that this is a typical derivation on the physical degree of the accuracy: we didn't bother too much about the properties of $e^{-\hat{L}(t-s)}$, the mathematician should do this.

The generalizations to the multidimensional case is trivial: we consider the multidimensional random process $\{x_i(t)\}$ governed by the equations of motion

$$\frac{dx_i}{dt} = v_i(x_1,\ldots,x_n) + \xi_i(t) \tag{8.58}$$

where $\xi(t) = (\xi_1(t),\ldots,\xi_n(t))$ is the vector of Gaussian random variables (noises) whose correlation properties are $\langle \xi_i(t)\rangle = 0$ and

$$\langle \xi(t)\xi(t')\rangle = 2\mathbf{B}\delta(t-t'). \tag{8.59}$$

We are looking for the probability distribution function of the values of x_i at time t. The local conservation law now reads

$$\frac{\partial P}{\partial t} + \sum_i \frac{\partial}{\partial x_i}\left(\frac{\partial x_i}{\partial t}P\right) = 0. \tag{8.60}$$

Replacing the time derivative of x_i on the right-hand side of this equation by the expression given by Eq. (8.49) we get:

$$\frac{\partial P}{\partial t} = -\sum_i \frac{\partial}{\partial x_i}[(v_i(x_1,\ldots,x_n) + \xi_i(t))P]. \tag{8.61}$$

This is a stochastic differential equation giving $p(\mathbf{x},t)$ in a particular realization of the process.

The linear operator L corresponding to the deterministic part of the behavior reads

$$\hat{L}P = \sum_i \frac{\partial}{\partial x_i}(v_i(x_1,\ldots,x_n)P), \tag{8.62}$$

so that performing the same steps as before we get

$$\frac{\partial p}{\partial t} = -\nabla \mathbf{v}(\mathbf{x})p + \nabla \mathbf{B}(\mathbf{x})\nabla p. \tag{8.63}$$

8.4 Transition rates and master equations

The Kramers–Moyal expansion used in Section 8.1 is a formal trick which allows for mathematically rigorous derivation of a Fokker–Planck equation; this trick however is physically not quite transparent. Therefore we discuss here another variant of the approach, which has immediate thermodynamical implications and connects our discussion of Brownian motion with the canonical formalism of statistical physics.

Let us assume the following structure for the transition probability during the time Δt, $w(x, r; t, \Delta t) = p(x + r, t + \Delta t | x, t)$:

$$w(x, r; t, \Delta t) = \left[1 - \Delta t \int dx' w(x'|x+r)\right] \delta(r) + w(x+r|x)\Delta t + O(\Delta t^2). \tag{8.64}$$

The meaning of the assumption is as follows: Imagine that at time t the system was in state x; during the time Δt it made a transition to a state $x + r$. The probability of such a transition is now expressed through the transition rate $w(x + r|x)$; the function $w(x|y)$ is the probability of transition from y to x per unit time. The first term follows from the normalization condition $\int dr w(x, r; t, \Delta t) = 1$ which must be fulfilled for any Δt. Its interpretation is that at time t the system could already have been in state $x + r$; this probability of not changing the state is assumed to decay as Δt grows.

Assuming the form, Eq. (8.64) we get instead of Eq. (8.6) the following equation:

$$\frac{\partial}{\partial t}p(x,t|x_0,t_0)\Delta t = \left[-\int dx' w(x'|x)p(x,t|x_0,t_0)\right.$$
$$\left.+ \int dr w(x|x-r)p(x-r,t|x_0,t_0)\right]\Delta t + O(\Delta t^2). \tag{8.65}$$

Taking the limit $\Delta t \to 0$, we get after evident transformations, the following equation for $p(x, t|x_0, 0)$:

$$\frac{\partial}{\partial t}p(x,t|x_0,0) = \int dx' \, w(x|x')p(x',t|x_0,0) - \int dx' \, w(x'|x)p(x,t|x_0,0). \tag{8.66}$$

Equation (8.66) is called a master equation, and has a transparent physical meaning: The change in the probability $p(x,t|x_0,0)$ per unit time is due to the two competing processes: to the inflow due to the fact that the system may changed its state from some other state x' to the state x and to the outflow, due to the fact that it may have already been in state x and may have left it for another state. Of course, we can obtain further a Fokker–Planck equation from the master equation, assuming that the transitions are only very-short-ranged, so that at least the two first moments of Δx are finite. Using Eq. (8.64) we get

$$A(x) = \int dr\, r w(x+r|x) \tag{8.67}$$

and

$$B(x) = \int dr\, r^2 w(x+r|x). \tag{8.68}$$

If these two moments are finite, the Fokker–Planck equation follows from the master equation in a usual way, and reads

$$\frac{\partial}{\partial t} p = \frac{\partial}{\partial x}\left[-vp + D\frac{\partial}{\partial x}p\right] \tag{8.69}$$

with $D = B/2$ and $v = A - \frac{1}{2}\frac{\partial}{\partial x}B$. If these lower two moments diverge, we have to do with a genuine jump-process, whose description lies outside of the range of applicability of standard Fokker–Planck equations, see Chapter 11, but still can be treated within the master equation formalism.

For a discrete system, where the states are numbered by whole numbers n instead of a continuous variable x the situation gets even more transparent:

$$\frac{\partial}{\partial t}p(n,t|n_0,0) = \sum_{n'} w_{n'\to n} p(n',t|n_0,0) - \sum_{n'} w_{n\to n'} p(n,t|n_0,0) \tag{8.70}$$

where $w_{n'\to n}$ is the transition rate from the state n' to the state n. The coefficients $w_{n'\to n}$ form a matrix W. Such discrete equations are in many cases more reasonable instruments of description than the continuous ones, especially in the cases when n can be interpreted not as a coordinate, but as a number of particles (birth-death processes, chemical reactions, etc.).

8.5 Energy diffusion and detailed balance

Master equations are often introduced on a quite formal basis, and are intimately connected with the thermodynamical formalism. Let us assume a closed system in a contact with the heat bath at temperature T. According to the Zeroth Law, whatever nonequilibrium the initial state of the system is, in course of the time the system will tend to an equilibrium one, if all external perturbations are switched off. Let ν enumerate the states of this system. Let moreover E_ν be the energy of the state ν. Then, the final, equilibrium state of the system will be characterized by the Boltzmann distribution

$$p_{eq}(\nu) = \frac{1}{Z} \exp\left(-\frac{E_\nu}{kT}\right). \tag{8.71}$$

We now consider the relaxation of our system from a nonequilibrium state characterized by the some distribution function $p(\nu)$ to the equilibrium state. We note that during this evolution the states themselves (which depend on the external conditions which are assumed not to change) do not change; the relaxation is fully due to the change of the probabilities for a system to be in a state ν. Let us assume these probabilities to follow the master equation with constant transition rates. This assumption follows from the one of the time-independent nature of the states and of the heat bath.

We note that the assumption of the time-independent states fully characterized by their energy may have quite different physical implications. The simplest situation is that the states E_ν are the localized quantum states (say of an electron at different impurities in a doped semiconductor), and the incoherent transitions between them are caused by the interaction with the heat bath. Just as simple are the Ising-like systems without any internal dynamics. Here the only possible changes are the flips of single spins or their clusters. Each such discrete event may (or may not) change the energy of the system. A similar situation takes place at a coarse-grained level of description in a classical system corresponding to a particle in a rugged energy landscape (see Ebeling et al., 1984; Engel and Ebeling, 1987, Haus and Kehr, 1987). An underdamped situation corresponds to a much more complex situation: Here, even at a constant energy, a complex dynamics corresponding to the "microcanonical" (i.e. purely Hamiltonian) evolution of the system takes place. In order to be able to assume that the state is fully characterized by the value of its energy, we have to consider a "state" as a microcanonical ensemble of systems, equally populating the energy surface $H(x,p) = E_\nu$. The interaction of the system with the heat bath introduces the transitions between different energy surfaces, the rates of such transitions are $w_{\nu' \to \nu}$.

The equilibrium state is a stationary (time-independent) solution of the master equation:

$$0 = \sum_{\nu'} w_{\nu' \to \nu} p_{eq}(\nu') - \sum_{\nu'} w_{\nu \to \nu'} p_{eq}(\nu). \tag{8.72}$$

Thus, the Zeroth Law requires that

$$\sum_{\nu'} [w_{\nu' \to \nu} p_{eq}(\nu') - w_{\nu \to \nu'} p_{eq}(\nu)] = 0. \tag{8.73}$$

This requirement is tightened by the Second Law: one namely has to assume that not only the sum, but each term separately vanishes, so that

$$w_{\nu' \to \nu} p_{eq}(\nu') - w_{\nu \to \nu'} p_{eq}(\nu) = 0. \tag{8.74}$$

It follows that

$$\frac{w_{\nu \to \nu'}}{w_{\nu' \to \nu}} = \frac{p_{eq}(\nu')}{p_{eq}(\nu)} \equiv \exp\left(-\frac{E_{\nu'} - E_{\nu}}{kT}\right). \tag{8.75}$$

This assumption is called the *principle of detailed balance*. If the external conditions are time-dependent, the detailed balance principle has to hold at each time for time-dependent rates, as long as Markovian dynamics holds. The same assumptions are of course valid in the continuous case.

The meaning of the principle is as follows: Imagine that the transition rates between the states 1 and 2 of the system do not follow Eq. (8.74), so that the number of the transitions from state 1 to the state 2 is not balanced by the backwards transitions. For example, let us assume that in equilibrium more transitions take place immediately from 1 to 2 than back from 2 to 1: the back flow from 2 to 1 follows through some intermediate state(s) 3, so that a perpetual current $1 \to 2 \to 3 \to 1$ flows (The perpetual superconductive currents or the probability currents discussed above are not the currents between the states, but currents within a state; they are not forbidden by the following consideration). If the energy of the state 2 is lower than the energy of state 1, then at first step the energy is dissipated to the heat bath, and the second step, $2 \to 3 \to 1$, is thermally activated: the energy is taken from the bath. The equilibrium state is stable, so that small external perturbation wouldn't change the state considerably. Thus, if an ingenious gadgeteer would be able to use the unbalanced $1 \to 2$ current for producing work against small external force, this work will be produced on the cost of cooling the only one heat reservoir, which is explicitly forbidden by the second law. We refrain here from the discussion of possible constructions of such perpetual mobile. The discussion of the thermodynamical implications of detailed balance was given in Bridgman, 1928.

Now let us return to our master equation in the case when it can be reduced to a Fokker–Planck one and show that the Einstein's relation follows in general from the detailed-balance principle (essentially the initial discussion by Perrin about the exact equilibration of the two currents was a kind of use of it!). Note that in the case when the master equation can be reduced to a Fokker–Planck one, i.e. in the case when $w(x+r|x)$ decays sufficiently fast as a function of r, the detailed balance requires the connection between the transition moments: In the overdamped case one can assume $E(x) = U(x) = -\int f(x)dx$, so that the energy difference between the state $x+r$ and the state x is $E(x+r) - E(x) = -\int f(x)dx$, and the transition rates between these states are connected by

$$w(x+r|x) = w(x|x+r) \exp\left(\int_x^{x+r} \frac{f(x)}{kT} dx\right). \tag{8.76}$$

Let us now assume that our system is homogeneous, so that $w(x+r|x) = w(x|x-r)$. (The inhomogeneous situation will be considered in some detail later on). Combining the homogeneity with Eq. (8.76) we get:

$$w(x-r|x) = w(x+r|x) \exp\left(\int_x^{x+r} \frac{f(x)}{kT} dx\right). \quad (8.77)$$

Furthermore, we assume that also for the case $f = 0$ the corresponding moments exist, so that the situation is characterized by a constant diffusion coefficient and by a constant mobility. We moreover consider the force as so weak, that the lowest-order (linear) approximation in $f(x)$ is sufficient. The integrals for A and B then read

$$A(x) = \int_{-\infty}^{\infty} dr\, rw(x+r|x) = \int_0^{\infty} r\left[w(x+r|x) - w(x-r|r)\right] dr$$

$$= \int_0^{\infty} rw(x+r|x)\left[1 - \exp\left(\int_x^{x+r} \frac{f(x)}{kT} dx\right)\right]$$

$$\approx \int_0^{\infty} w(x+r|x)r\left[1 - 1 - r\frac{f(x)}{kT}\right]$$

$$= -\frac{f(x)}{kT}\int_0^{\infty} w(x+r|x)r^2 dr. \quad (8.78)$$

Note that the term of the zeroth order vanishes. The lowest-order approximation for the coefficient B is zeroth order in f and reads

$$B = \int_{-\infty}^{\infty} dr\, r^2 w(x+r|x) = \int_0^{\infty} r\left[w(x+r|x) - w(x-r|r)\right] dr$$

$$= \int_0^{\infty} w(x+r|x)r^2\left[1 + \exp\left(\int_x^{x+r} \frac{f(x)}{kT} dx\right)\right]$$

$$\approx 2\int_0^{\infty} w(x+r|x)r^2 dr. \quad (8.79)$$

Thus, one readily infers that $A(x) = -\frac{f(x)}{2kT}B$. Remembering that $B/2$ is exactly the diffusion coefficient D we get

$$\frac{\partial}{\partial t}p = \frac{\partial}{\partial x}\left[-\frac{D}{kT}f(x)p + D\frac{\partial}{\partial x}p\right], \quad (8.80)$$

i.e. the particles' mobility μ is connected to D via $D = kT\mu$.

8.6 System in contact with several heat baths

The detailed balance principle guarantees that in equilibrium the distribution over the energy states in the isotherimc system is the Boltzmann one. It thus connects the forwards and the backwards rates of the transitions taking place in presence of the heat bath. However, in many cases, at least not too far from the equilibrium, the rates of different transitions can be considered as independent from each other, and can be taken to depend only on the local temperature at which the transition takes place. This allows us for discussion systems in a contact with several heat baths. A very simple toy model of such a system is considered in what follows.

As an example we consider here a simple system with three energy levels 1, 2, and 3, see Fig. 8.1. The results discussed here can easily be generalized to more complex situations or to continuous systems. The discrete system considered here is a very simple one and allows for full mathematical description. The potential differences between states 1 and 2, 3 and 2, and 1 and 3 are U_{12} and U_{32}, respectively, both of them positive (the energy difference between 3 and 1 is then $U_{31} = U_{12} - U_{32}$). If the system is kept at a constant temperature, the currents through whatever bond vanish. Moreover, the probabilities to find a particle on the site i does not depend on whether the bond 31 is present or absent. This is the consequence of the transitivity of thermodynamical equilibrium.

Let us now consider the system in contact with two heat baths, so that the bond 1-2 is kept at temperature T_1, and the transition 2–3 at temperature T_2. In what follows we assume $T_1 > T_2$ and $U_{12} > U_{32}$, so that the difference $U_{31} = U_{12} - U_{32}$ is positive. Physically, it can be considered as a minimalistic model of a thermocouple (state 1: "electron in the conductor 1", state 2 – "electron in the conductor 2", state 3 – "electron in the conductor 3") or as a model for the water circulation in the atmosphere (state 1: "water molecule on the surface of the ocean", state 2 – "water molecule in the rain cloud", state 3 – "water molecule in the mountain lake"). However in all these cases the potential energies are essentially the chemical potentials, which themselves depend on temperature and other parameters, so that the whole thermodynamics of the system gets extremely involved. Here we dispense from thermodynamics and fully concentrate on the dynamics of the system. For simplicity, we take all downhill transition rates to be equal to $w_{ij}^- = w$ (the so-called Metropolis prescription, often used in numerical Monte-Carlo simulations of thermodynamical system), the uphill rates will then be $w_{ij}^+ = w_- \exp(-U_{ij}/k_B T)$.

Then, the probabilities (or the particle densities) in states 1, 2, and 3 are given by the stationary solution of the master equation

$$\frac{d}{dt}p_1 = wp_2 + wp_3 - w\left(e^{-\frac{U_{12}}{k_B T_1}} + e^{-\frac{U_{13}}{k_B T_2}}\right)p_1$$

$$\frac{d}{dt}p_2 = we^{-\frac{U_{12}}{k_B T_1}}p_1 + we^{-\frac{U_{32}}{k_B T_2}}p_3 - 2wp_2 \qquad (8.81)$$

$$\frac{d}{dt}p_3 = we^{-\frac{U_{13}}{k_B T_1}}p_1 + wp_2 - w\left(1 + e^{-\frac{U_{32}}{k_B T_2}}\right)p_3.$$

where p_i is the probability to find the particle in the state i. Note that the sum of the probabilities $p_1 + p_2 + p_3 = 1$ is the integral of the motion, so that the three equations are not independent. Let us consider the stationary situation. The

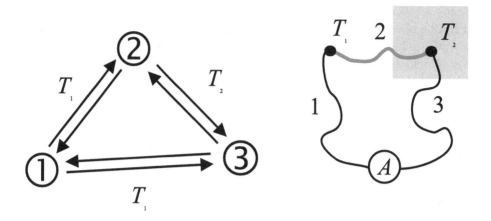

Fig. 8.1 Left: A simple discrete model of a three-level system considered in the text. Right: One of possible continuous realizations of the model: A thermocouple, in which the two contacts of different metals (shown in black and in gray) are kept at different temperatures. For the circuit with an ideal amperemeter A with zero resistivity the potential difference U_{31} is always zero.

corresponding solutions for p_1 and p_2 then read

$$p_1 = \frac{1}{1+e^{-\frac{U_{12}}{k_B T_1}} + e^{-\frac{U_{13}}{k_B T_2}}}, \qquad (8.82)$$

$$p_2 = \frac{e^{-\frac{U_{12}}{k_B T_1}} + e^{-\frac{U_{12}}{k_B T_1}} e^{-\frac{U_{32}}{k_B T_2}} + e^{-\frac{U_{32}}{k_B T_2}} e^{-\frac{U_{13}}{k_B T_1}}}{\left(1+e^{-\frac{U_{12}}{k_B T_1}} + e^{-\frac{U_{13}}{k_B T_1}}\right)\left(2+e^{-\frac{U_{32}}{k_B T_2}}\right)}, \qquad (8.83)$$

and

$$p_3 = \frac{e^{-\frac{U_{12}}{k_B T_1}} + 2e^{-\frac{U_{13}}{k_B T_2}}}{\left(1+e^{-\frac{U_{12}}{k_B T_1}} + e^{-\frac{U_{13}}{k_B T_1}}\right)\left(2+e^{-\frac{U_{32}}{k_B T_2}}\right)}. \qquad (8.84)$$

Knowing p_i we can also calculate the currents through the bonds, $I_{ij} = w_{ij} p_i - w_{ji} p_j$. For example,

$$I_{12} = \frac{e^{-\frac{U_{13}}{k_B T_1}} e^{-\frac{U_{32}}{k_B T_2}} - e^{-\frac{U_{12}}{k_B T_1}}}{\left(1+e^{-\frac{U_{12}}{k_B T_1}} + e^{-\frac{U_{13}}{k_B T_1}}\right)\left(2+e^{-\frac{U_{32}}{k_B T_2}}\right)}. \qquad (8.85)$$

Of course, the stationary currents through all three bonds are the same. We see that due to the fact that $U_{13} + U_{32} = U_{12}$ the current vanishes exactly for $T_1 = T_2$ and is nonzero if the temperatures of the transitions are different. In our discrete model this is exactly the thermocurrent. Now, we can in principle tap the 31 bond to win energy from this current. However, the efficiency of our system will depend on the particular mechanism of how do we win the energy from the current.

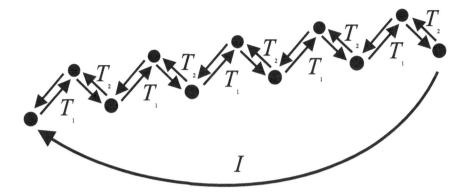

Fig. 8.2 A cascade of N three-level systems with transitions kept at different temperatures as a prototype of continuously working heat engine.

In order to avoid complications let us consider a system shown in Fig. 8.2. It is a cascade of N such systems switched in series, so that now the site 3 of the previous system is exactly the site 1 of the next one. The existence of the overall current means now that particles injected in the left-hand side of the system are transported uphill (the corresponding model is simply a discrete variant of one discussed by Buttiker, 1987). When reaching the top, the particles are used to produce useful work, say, to charge the battery or to rotate the wheel. Since the energy difference between the upper right and the lower left site is very large, the spontaneous backward transitions may be neglected, so that we really don't need to think about how do the thermal fluctuations of the battery's voltage really look like. (The impossibility of backward transitions in the working medium of the system is a rather typical case, since, say, the way from the ocean to the lake is only possible through evaporation, but not through swimming up the river. The mechanism forbidding this is, however, different, and has to do with the properties of collective motion, an aspect which is absent in our simple model). Under stationary conditions, due to the periodicity, we have, as an equation for p_1,

$$0 = 2wp_2 - w\left(e^{-\frac{U_{12}}{k_B T_1}} + e^{-\frac{U_{32}}{k_B T_2}}\right)p_1 \qquad (8.86)$$

which is complemented by the fact that $p_1 + p_2 = 1$. We thus have

$$p_1 = \frac{2}{2 + e^{-\frac{U_{12}}{k_B T_1}} + e^{-\frac{U_{32}}{k_B T_2}}} \qquad (8.87)$$

and

$$p_2 = \frac{e^{-\frac{U_{12}}{k_B T_1}} + e^{-\frac{U_{32}}{k_B T_2}}}{2 + e^{-\frac{U_{12}}{k_B T_1}} + e^{-\frac{U_{32}}{k_B T_2}}}. \qquad (8.88)$$

We can also calculate the stationary current through the system (as a current through the 12-bond),

$$I = p_1 e^{-\frac{U_{12}}{k_B T_1}} - p_2 = \frac{e^{-\frac{U_{12}}{k_B T_1}} - e^{-\frac{U_{32}}{k_B T_2}}}{2 + e^{-\frac{U_{12}}{k_B T_1}} + e^{-\frac{U_{32}}{k_B T_2}}}. \qquad (8.89)$$

We see that for the case $U_{12} > U_{32}$ and $e^{-\frac{U_{12}}{k_B T_1}} - e^{-\frac{U_{32}}{k_B T_2}} > 0$ i.e. for

$$\frac{T_2}{T_1} < \frac{U_{32}}{U_{21}} \qquad (8.90)$$

the system indeed pumps particles uphill, against the overall potential difference.

Now, imagine that the current between 3 and 1 produces a useful work. The useful power then is $P = \dot{A} = INU_{31}$. We also can calculate the heat taken from the hot reservoir (at temperature T_1). Since the energy of the particle at sites 1 and 2 differ by exactly U_{12}, this energy is the one that has to be dissipated to the bath when a downhill transition from 2 to 1 takes place, and the one which has to be gained from the bath under the uphill transition. The corresponding heat power then reads

$$\dot{Q} = NU_{12}I, \qquad (8.91)$$

and the efficiency η of the system reads

$$\eta = \frac{\dot{A}}{\dot{Q}} = \frac{U_{31}}{U_{12}} = \frac{U_{12} - U_{32}}{U_{12}} = 1 - \frac{U_{32}}{U_{12}}. \qquad (8.92)$$

The efficiency of our system is thus independent on temperature. According to Eq. (8.90) we thus have

$$\eta \leq 1 - \frac{T_2}{T_1} \qquad (8.93)$$

and moreover we see that the Carnot value is achieved when $\frac{T_2}{T_1} = \frac{U_{32}}{U_{21}}$, so that the current vanishes (stalling conditions, work at zero power) which exactly corresponds to the quasistatic situation in usual, cyclically working heat engines. It is also interesting to calculate the efficiency at largest power. Let us now fix T_1 and T_2 and U_{12} and tune U_{32} until the maximum power

$$P = N(U_{12} - U_{32}) \frac{e^{-\frac{U_{12}}{k_B T_1}} - e^{-\frac{U_{32}}{k_B T_2}}}{2 + e^{-\frac{U_{12}}{k_B T_1}} + e^{-\frac{U_{32}}{k_B T_2}}} = \max. \qquad (8.94)$$

is achieved. We note that in this case the maximal efficiency is not only the function of the quotient between the temperatures of the heat baths, but depend on their relation to potentials. If the temperatures are low, the overall efficiency stays close to the Carnot one since the currents are small. However, when both temperatures, T_1 and T_2 are high compared to the energetic barriers, another result appears, and the maximal efficiency tends to exactly one half of the Carnot value. We note that no general expression for the efficiency at maximal power exists, which would be valid for all systems. The well-known expressions like the the celebrated Novikov–Curzon–Ahlborn formula $\eta = 1 - \sqrt{T_2/T_1}$ (Novikov, 1958; Curzon and Ahlborn, 1975) are pertinent to rather special situations. In application to cyclically working machines these problems are the issue of the so-called "finite time thermodynamics" (see Andersen, 1984).

Several other systems with different heat baths were considered in the literature, mostly inspired by the Feynman's ratchet-and-pawl device or by an earlier "thermal fluctuation rectifier" with a semiconductor diode discussed by L. Brillouin (Brillouin, 1950). The diode thermal fluctuation rectifier is only a slightly more complicated machine than our simple discrete model (Sokolov, 1998; Sokolov, 1999). It unveils, however, the important property of all realistic devices of such kind, namely the possibility of heat transport through the fluctuations of "mechanical" degrees of freedom, which makes them intrinsically irreversible and lowers the efficiency even under stalling conditions in comparison with the Carnot's value. The genuine Feynman's device (consisting of a vane at temperature T_1 and of a ratchet-and-pawl mechanism at temperature T_2 connected by a solid axle) is a much more intricate system. Feynman himself assumed its efficiency to be a Carnot one; however Parrondo and Espanol (1996) have shown that it is not the case.

The more general issue of thermodynamics of the ratchet-like devices under different types of forcing has found much attention in the last decade, since their way of functioning is closely related to the one of many biological molecular motors", see Astumian and Hänggi, 2002. Rather exhaustive review of different aspects of such motors can be found in Reiman (2002), Frey (2002), Parrondo and De Cisneros (2002).

Chapter 9
Escape and First Passage Problems

The possibility to calculate the probability density function or its moments, like the mean square displacement, does not exhaust the whole class of problems of stochastic theory. One of the most important classes of other problem settings are the first passage problems for a stochastic process. Mathematically, the task is to calculate the probability density $\phi(t)$ (as a function of time) for a process $\mathbf{x}(t)$ to reach for the first time a point x, or a set of points (a typical one-dimensional problem position), or to cross a boundary of a spatial domain in spatial dimension more than one. Physical problems leading to this mathematical formulation are abundant. The genuine first-passage problems are often pertinent to reaction kinetics, where the two particles interact if they approach each other at distance a. These situations correspond to crossing the boundary, so to say, from the outside. An opposite situation of the crossing a boundary from inside, emerges when describing a decay of a bounded state in some given potential (a problem emerging by the description of, say, dissociation of a complex molecule, Ebeling, Schimansky-Geier and Romanivsky, 2002). This situation is often termed as an escape problem. In ecology, the first passage through some prescribed value, might mean the extinction of the population, or the start of the epidemic outbreak. The literature discussing the first passage problems both from the mathematical, and from the physical point of view, is abundant. A simple introduction into the topic is given in the book Redner, 2001. In this Chapter we consider only some simple situations. Understanding these situations is, however, necessary for understanding fluctuation effects in chemical reactions, Chapter 10, or the emergence of anomalous diffusion in complex potential landscapes, Chapter 11.

9.1 General considerations

In this chapter we first concentrate on one-dimensional problems. Let us first look at the realizations of the corresponding random process and try to find a mathematical formalization of the problem. The physical situation here can be formulated as follows: At time $t = 0$ a particle is introduced into a system at $x = 0$. As soon as particle reaches the place x_0 the realization is stopped, the time necessary to reach the boundary (the first passage time) is recorded. Then, we can obtain the distribution of these times, and use it for further calculations. It is important to note that we are interested in the frequencies of the realizations of a random process, i.e. in the trajectories (paths) of the particle. In order to get the distribution of the first passage times, we have to know, in how many realizations the path of the particle crossed the point $x = x_0$ at times $t' < t$. In order to count them, it is enough to assume an absorbing boundary condition at $x = x_0$: Particles having touched the point disappear and their trajectories are disregarded at future times. We assume the overdamped regime of motion, where the probability density to find a particle's at point x is governed by Smoluchowski equation

$$\frac{\partial p(x,t)}{\partial t} = -\mu f \frac{\partial p(x,t)}{\partial x} + D \frac{\partial^2 p(x,t)}{\partial x^2}. \quad (9.1)$$

Here the force $f = -dU/dx$. In cases corresponding to the systems considered in thermodynamics (i.e. ones possessing true thermodynamical equilibrium) the mobility μ and the diffusion coefficient D are connected to each other through the Einstein's relation $D = \mu k_B T$. The equation, Eq. (9.1) has to be solved under the boundary condition $p(x_0, t) = 0$ and for the initial condition $p(x, 0) = \delta(x)$, so that the corresponding solutions are essentially the Green's functions of the Fokker–Planck equation, $p(x, t) = G(x, t|0, 0)$. It is important to stress here that the equivalence of our assumption that all particles touching x_0 disappear and the actual condition that the trajectories crossing x_0 have to be disregarded at larger times assumes the continuity of the trajectories, which is the case for Brownian motion, but might be violated for some processes described by generalizations of a Fokker–Planck equation (Chechkin et al., 2003), so that care has to be taken when generalizing the results discussed here to the processes other than Fickian diffusion.

After solving the equation for all times, we can calculate the overall probability for the particle to stay within the interval, which is

$$P(t) = \int_{-\infty}^{x_0} p(x,t) dt \quad (9.2)$$

and note that the change in $P(t)$ between the times t and $t + dt$ is exactly the probability to leave the interval during dt. This is,

$$\psi(t) = -\frac{dP(t)}{dt}. \quad (9.3)$$

By noting that

$$\frac{dP(t)}{dt} = \int_{-\infty}^{x_0} \frac{\partial p(x,t)}{\partial t} dx \quad (9.4)$$

and using Eq. (9.1) we get

$$\psi(t) = \int_{-\infty}^{x_0} \left[-\mu f \frac{\partial p(x,t)}{\partial x} + D \frac{\partial^2 p(x,t)}{\partial x^2} \right] dx. \tag{9.5}$$

Applying partial integration, and using the natural boundary condition $p(x,t) \to 0$ and for $x \to -\infty$ we get:

$$\psi(t) = -D \left. \frac{\partial p(x,t)}{\partial x} \right|_{x=x_0}, \tag{9.6}$$

i.e. that the first passage time probability density is equal to the diffusion current through the absorbing boundary. If we are interested in the case of two absorbing boundaries at the ends of an interval the corresponding currents give us the probability per unit time to leave the interval through the corresponding end.

This approach based on the solution of the forward equation assumes the knowledge of its time-dependent solution, and is the typical "physicist's" approach to the problem, which is not necessarily the simplest or the most elegant one. However, this is the one which works and, moreover, the one which works reliably even when the coefficients in the Fokker–Planck equation are time-dependent.

9.2 The renewal approach

The renewal approach to the first passage problem uses explicitly the continuity of sample paths and the Markovian nature of the problem. It reduces the solution of a problem with absorbing boundary condition to one for the free problem, the one with "natural" boundary conditions, which is sometimes much simpler. The deficiency of the approach is however that it is only effective in the one-dimensional case.

Let us start from the simple case of a single boundary situated at $x = x_0 > 0$. We consider the process starting at $t = 0$ at $x = 0$. Imagine that at some time $t_f > 0$ the particle is found to the right of the boundary, at some point $x_f > x_0$. The probability of this event is given by the transition probability density function of the "free" process $p(x_f, t_f | 0, 0)$.

Due to the continuity of trajectories, the process has to have passed the point x_0 before reaching x at some time $t < t_f$ (and might have passed the point x_0 one or several times after this). Thus the realizations of the process leading from x to x_f may be uniquely classified with respect to the time they first crossed x_0. Let us denote $\psi(t, x_0)$ the first passage time distribution through point x_0, and concentrate on all realizations of the process contributing to $\psi(t, x_0)$.

Here the Markovian nature of the process comes in play. The probability that the particle reaches $x = x_f$ at $t = t_f$ provided it was at x_0 at time t depends only on x_0 and t but not on the history of the process at $t < t_0$ and is given by the transition probability density function of the free process $p(x_f, t_f | x_0, t_0)$. This means that the probability density to cross the point x_0 at time t and then to reach x_f at time t_f is simply a product $p(x_f, t_f|x_0, t)\psi(t, x_0)$. Summing over all possible times $t < t_f$ one gets

$$p(x_f, t_f|0, 0) = \int_0^{t_f} p(x_f, t_f|x_0, t)\psi(t, x_0)dt. \quad (9.7)$$

This is an exact equation determining the first passage time for a Markovian process.

A next step can be done if the process $x(t)$ in a "free" motion is stationary (i.e. whenever the coefficients in our Fokker–Planck equation are time-independent). In this case the transition probability density depends only on the difference between its time arguments: $p(x_2, t_2|x_1, t_1) = G(x_2, x_1, t_2 - t_1)$ and is equal to the Green's function solution of the Fokker–Planck equation with time-independent coefficients. The integral in the r.h.s. of Eq. (9.7) has now a form of a convolution. Another simplification stems from the fact that for diffusion processes one can take a limit $x_f \to x_0$ (since the corresponding Green's function is nonsingular in this limit), so that

$$G(x_0, 0, t_f) = \int_0^{t_f} G(x_0, x_0, t_f - t)\psi(t, x_0)dt. \quad (9.8)$$

This is a Volterra integral equation, which easily can be solved numerically. The simplest way to its analytical solution is to take Laplace transforms of the both sides of the equation, so that $\tilde{G}(x_2, x_1, u) = \int_0^\infty G(x_1, x_2, t)e^{-ut}d\tau$. Applying this transform we get $\tilde{G}(x_0, 0, u) = \tilde{G}(x_0, x_0, u)\tilde{\psi}(u, x_0)$, i.e.

$$\tilde{\psi}(u, x_0) = \frac{\tilde{G}(x_0, 0, u)}{\tilde{G}(x_0, x_0, u)}. \quad (9.9)$$

Of course, in many situations, the inverse Laplace transform has to be performed numerically. However, the expression for the mean first passage $\tau = \int_0^\infty t\psi(t, x_0)dt = \frac{d}{du}\tilde{\psi}(u, x_0)\big|_{u=0}$ follows easily.

The renewal approach can also be applied for calculating splitting probabilities. Let us consider a particle moving between the two absorbing boundaries. Here the two first passage probabilities, $\psi(t, x_L)$ for first crossing the left boundary situated at x_L and $\psi(t, x_R)$ of first crossing the right boundary situated at x_R. Parallel to our previous consideration one can say that if the particle which started at $x = 0$ at $t = 0$ is found at time t to the right from the right boundary (at some $x_f > x_R$) it might either have first crossed the right boundary without touching the left one at some time $t' < t$ and then made the way from x_R to x_f or crossed first the left boundary and then made the way from x_L to x_f. The probability to touch the left boundary for the first time without previously touching the right one between times t and $t + dt$ is given by the function $\psi(t, x_L)dt$ and the probability to first touch the right boundary (without touching the left one before) is given by $\psi(t, x_R)dt$. Note that $\psi(t, x_L)$ and $\psi(t, x_R)$ are not proper probability density functions: both integrals

$P_L = \int_0^\infty \psi(t, x_L) dt$ and $P_R = \int_0^\infty \psi(t, x_R) dt$ (representing the probabilities to leave the interval through its left or through its right boundary, the so-called splitting probabilities) are in general less than unity. The normalization condition for the splitting probabilities is given by $P_L + P_R = 1$. Following the same scheme as before we obtain

$$G(x_f, t|0, 0) = \int_0^t G(x_f, t|x_R, t') \psi(t', x_R) dt' + \int_0^t G(x_f, t|x_L, t') \psi(t', x_L) dt'. \quad (9.10)$$

This is an equation determining both $\psi(t, x_L)$ and $\psi(t, x_R)$ since it has to be valid for *any* x_f. To obtain the explicit equations we take another x_f to lie to the left from the left boundary, and then make the limiting transition, taking the corresponding x_f's to tend to the boundaries of the interval from the outside. Since for stationary case the integrals in both such equations are of the convolution type we can make the Laplace-transform of the both and get

$$G_{RR}\psi_R + G_{RL}\psi_L = G_{R0},$$
$$G_{LR}\psi_R + G_{LL}\psi_L = G_{L0}, \quad (9.11)$$

where the following shorthand notation is introduced for the Green's functions: $G_{RR} = \tilde{G}(x_R, x_R; u)$, $G_{LL} = \tilde{G}(x_L, x_L; u)$, $G_{RL} = \tilde{G}(x_R, x_L; u)$, $G_{LR} = \tilde{G}(x_L, x_R; u)$, $G_{R0} = \tilde{G}(x_R, 0; u)$, and $G_{L0} = \tilde{G}(x_L, 0; u)$. Moreover, $\psi_L = \tilde{\psi}(u, x_L)$ and $\psi_R = \tilde{\psi}(u, x_R)$. The solution of the system of equations (9.11) is

$$\psi_L = \frac{G_{R0} G_{LL} - G_{L0} G_{RL}}{G_{RR} G_{LL} - G_{LR} G_{RL}} \quad (9.12)$$

and

$$\psi_R = \frac{G_{RR} G_{L0} - G_{LR} G_{R0}}{G_{RR} G_{LL} - G_{LR} G_{RL}} \quad (9.13)$$

so that the probabilities can be found explicitly. The splitting probabilities P_L and P_R are simply given by the limiting values of the corresponding functions at $u \to 0$.

For non-Markovian processes Eq. (9.8) might or might not hold, depending on the exact nature of the process. Thus it still holds for "semi-Markovian" situations like continuous-time random walk models discussed in Chapter 11. In this case the renewal approach is effective and preferable because the probability density for free motion might be obtained by the alternative methods, without explicitly solving corresponding (non-Markovian, fractional) Fokker–Planck equations. Equation (9.8) does not in general hold for the processes described by non-Markovian Langevin equations or by moving averages. However, for a stationary non-Markovian process Eq. (9.8) might still be a reasonable approximation: it simply assumes that having arrived to a position of the absorbing boundary, the particle has practically forgotten initial conditions, so that its further behavior can be described by a *new* initial condition problem. Of course it does not imply that the corresponding solution of the initial-value problem resembles to any extent the Green's function of any Markovian process. In this case the renewal approximation based on Eq. (9.8) is equivalent to the so-called Wilemski–Fixman approximation (Wilemski and Fixman, 1974, Sokolov, 2003) being widely used for description of reactions involving polymers.

9.3 Example: Free diffusion in presence of boundaries

As an example let us consider the first passage time distribution for free diffusion (Brownian motion), where $G(x_2, x_1, t) = (4\pi D t)^{-1/2} \exp\left[-(x_2 - x_1)^2/4Dt\right]$, so that

$$\tilde{G}(x_2, x_1, u) = \frac{1}{\sqrt{4Du}} \exp\left(|x_2 - x_1|\sqrt{\frac{u}{D}}\right). \tag{9.14}$$

This delivers $\tilde{\psi}(u, x_0) = \exp(|x_0|\sqrt{u/D})$ from which it follows that

$$\psi(t, x_0) = \frac{|x_0|}{\sqrt{4\pi D} t^{3/2}} \exp\left(\frac{x_0^2}{4Dt}\right). \tag{9.15}$$

This distribution of the first passage time to a boundary is called Smirnov, Lévy–Smirnov or sometimes "inverse Gaussian" distribution, and decays for long times as $\psi(t, x_0) \propto t^{-3/2}$ so that it does not have the first moment: the mean first passage time diverges. Note that the function $\psi(t, x_0)$ is a proper probability density function, so that $\int_0^\infty \psi(t, x_0) dt = 1$. This means that the particle in 1d is eventually captured at the boundary. This fact has to do with the recurrence of the one-dimensional Wiener process, which visits any point on the line with probability 1 at longer times.

We note that our consideration here are based on the fact that the trajectories of the process are continuous. The result that at long times $\psi(t, x_0) \propto t^{-3/2}$ is, however, valid also for a large class of jump processes, i.e. the ones with discontinuous trajectories. However in the case $\psi(t, x_0)$ has to be considered not as the distribution of first passage times, but as the distribution of times at which the particle is for the first time found on the other side of x_0 than its initial position $x = 0$ was, see Chechkin et al., 2003. This statement is one of the important consequences of the Sparre–Andersen theorem from the theory of random walks, see Feller, 1991.

Let us now consider splitting probabilities for a particle starting at $X = 0$ on the interval with absorbing boundaries at $x_L < 0 < x_R$. Using Eqs. (9.12) and (9.13) and our Laplace-transformed Green's function, Eq. (9.14) we get after expanding the exponential up to the first order in their arguments:

$$P_R = 1 - P_L = \frac{1}{2} + \frac{|x_L| - |x_R|}{2|x_R - x_L|}. \tag{9.16}$$

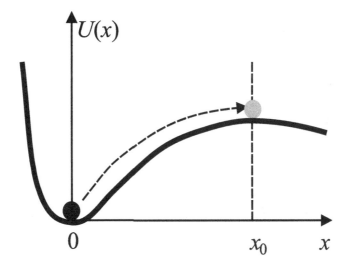

Fig. 9.1 The typical potential discussed in the escape problem: a potential well with a simple quadratic minimum and flat maximum.

9.4 Mean life time in a potential well

One of the most important results following from the theory based on the Fokker–Planck equation is the typical life-time in a potential well, i.e. the time necessary to overcome a potential barrier. Since this result will be repeatedly used in what follows, we shall discuss it here in some detail.

The situation considered here is depicted in Fig. 9.1. Imagine, a particle starts at the minimum of the potential, at point $x = 0$. We say that the particle overcame a barrier if it arrived for the first time at a maximum of the potential curve at $x = x_0$. The value of the potential energy of the particle there is $U(x_0)$ while the minimum of the potential energy at $x = 0$ is taken for the reference point, $U(0) = 0$.

9.4.1 *The flow-over-population approach*

In order to find the *mean* first passage time, we don't need to solve the whole time-dependent problem.

Let us return to the physical formulation of the problem, and again discuss the experiment with putting the particles into a system. However, in order to obtain the result, it is not necessary to put the particles one by one, and wait until the corresponding realization terminates. Let us imagine that at the point $x = 0$ a constant current of strength I is flowing into the system, see Fig. 9.2. At the beginning, after switching on such current, the concentration of particles in the system will grow, and the output current, leaving the system at point x_0, will be smaller than the input current I. Eventually, the steady state is reached, the input and the output currents equilibrate, and the steady-state concentration

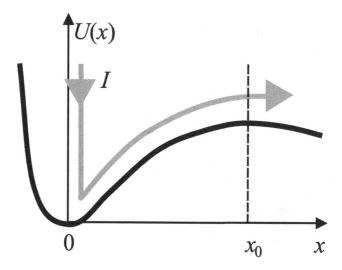

Fig. 9.2 The flow over population approach: We assume the particles are introduced at $x = 0$ and leave the system through the absorbing boundary at $x = x_0$. This corresponds to the constant current of particles (or probability) through the system.

profile establishes itself. Now, if the mean first passage time, i.e. the time a particle on the average spends within the system, is τ, the mean number of particles within the system will be exactly $I\tau$.

This mean number is nothing else than the integral over the steady-state concentration of the particles over the whole system (in our case over the semi-infinite interval $-\infty < x < x_0$). Thus, we have:

$$\tau = I^{-1} \int_{-\infty}^{x_0} p(x)dx \qquad (9.17)$$

where $p(x)$ is the steady-state solution of the Fokker–Planck equation (9.1). Note that the Fokker–Planck equation is essentially the continuity equation for the probability current, and that

$$I = \mu(x)f(x)p(x,t) - D(x)\frac{d}{dx}p(x,t). \qquad (9.18)$$

Here we remind that $f = -dU/dx$. This is an ordinary linear differential equation, which can be rewritten in the form:

$$\frac{dp(x)}{dx} + p(x)\frac{\mu(x)}{D}\frac{d}{dx}U(x) = -\frac{I}{D(x)}. \qquad (9.19)$$

Equation (9.19) is a linear differential equation, whose formal solution reads:

$$p(x) = \exp(-V(x))\left[p(0) - \int_0^x \frac{I}{D(x)}\exp(V(x'))dx'\right] \qquad (9.20)$$

with $V(x) = \mu(x)U(x)/D(x) = U(x)/k_BT$. The boundary condition $p(x_0) = 0$ leads to the expression for $p(0)$:

$$p(0) = \exp(-V(0)) \int_0^{x_0} \frac{I}{D(x)} \exp(V(x'))dx'. \tag{9.21}$$

From here on we assume for simplicity that the mobility and the temperature are constant throughout the whole system. According to our choice of the reference energy in Fig. 9.2 we have $\exp(-V(0)) = 1$, which simplifies the expressions. Note that the probability current to the left from point $x = 0$ vanishes, so that to the left of this point the steady-state solution coincides with the equilibrium solution $p(x) = p(0)\exp(-V(x))$. We thus have:

$$N(x) = \begin{cases} \dfrac{I}{D}\exp(-V(x)) \int_0^{x_0} \exp(V(x'))dx' & \text{for } x \leq 0 \\ \dfrac{I}{D}\exp(-V(x)) \int_x^{x_0} \exp(V(x'))dx' & \text{for } 0 < x < x_0 \end{cases} \tag{9.22}$$

We thus have the following expression for the mean first passage time in the system:

$$\tau = \frac{1}{D}\int_{-\infty}^0 \exp(-V(x'))dx' \int_0^{x_0}\exp(V(x'))dx'$$
$$+ \frac{1}{D}\int_0^{x_0}\left[\exp(-V(x''))\int_{x''}^{x_0}\exp(V(x'))dx'\right]dx'', \tag{9.23}$$

the expression that can be rewritten in a form

$$\tau = \frac{1}{D}\int_0^{x_0} dy' \int_{-\infty}^{y'} dy'' \exp[V(y') - V(y'')]. \tag{9.24}$$

Let us now concentrate on the case of a deep well or low temperatures. In this case we can make simple estimates for our integrals which now depend only on the properties of the potential close to points $x = 0$ and $x = x_0$.

9.4.2 The Arrhenius law

Let us assume that close to the maximum at point $x = x_0$ one has $V(x) \simeq V(x_0) - (2k_BT)^{-1}U''(x_0)(x_0 - x)^2$ and evaluate Eq. (9.23). In this case the integral $\int_0^{x_0}\exp(V(x'))dx'$ in the first line of the last expression and the integral $\int_{x''}^{x_0}\exp(V(x'))dx'$ (which, for x'' not too close to x_0, is practically a constant) are approximately equal to each other and can be evaluated using the Laplace method:

$$\int_0^{x_0}\exp(V(x'))dx' \simeq \frac{\sqrt{\pi k_BT}}{\sqrt{2|U''(x_0)|}}\exp[V(x_0)]. \tag{9.25}$$

If we assume that the potential has a simple quadratic minimum close to $x = 0$, i.e. that $V(x) \simeq (2k_BT)^{-1}U''(0)x^2$, the same type of an approximation for

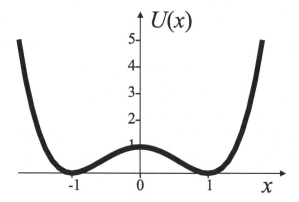

Fig. 9.3 A symmetric double-well potential $U(x) = x^4 - 2x^2 + 1$.

$\int_{-\infty}^{0} \exp(-V(x'))dx'$ and for $\int_{0}^{x_0} \exp(-V(x'))dx'$ again shows that the integrals are approximately equal to each other and can be approximated by

$$\int_{0}^{x_0} \exp(-V(x'))dx' \simeq \frac{\sqrt{\pi k_B T}}{\sqrt{2U''(0)}} \exp[-V(0)] \qquad (9.26)$$

(in our case we have $V(0) = 0$). The overall expression for τ the reads

$$\tau \simeq \frac{\pi k_B T}{D\sqrt{U''(x_0)U''(0)}} \exp\left(\frac{U(x_0)}{k_B T}\right) : \qquad (9.27)$$

the mean first passage time depends exponentially on the height of the barrier relative to the thermal energy of particles. This exponential growth is often termed as the Arrhenius law.

9.4.3 Diffusion in a double-well

As the next example let us consider the behavior of a particle in a double-well potential consisting of two wells of depths U_1 and U_2 measured with respect to the top of potential barrier. Then the mean first passage time from one well to another one is determined by the corresponding well's depths. Figure 9.3 shows a symmetric potential well $U(x) = x^4 - 2x^2 + 1$ (having two minima at $x = \pm 1$ separated by a barrier of height 1 with the maximum at $x = 1$. A trajectory of the diffusive process $x(t)$ for $\mu = 1$ and for $T = 0.15$ is shown in Fig. 9.4. Note that the behavior of the trajectory shown in Fig. 9.4 suggests its possible description as a sequence transitions between the two well-defined states, in which the particle is localized either in the left well (around $x = -1$) or in the right one (around $x = 1$). Let us now identify the position of the particle to the left from the top of the barrier with the L-state of the system, and the position to the right from the top with the R-state. The average life-time in either state is 2τ (since after reaching the top of the potential barrier the particle can either change the well or return to the initial

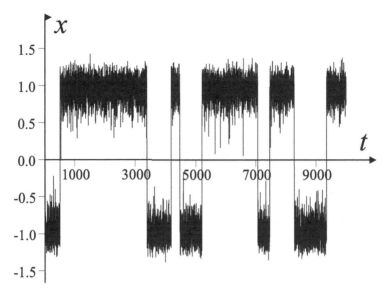

Fig. 9.4 A realization of diffusion in a double-well potential shown in Fig. 9.3 for $\mu = 1$ and for $T = 0.15$.

one, with equal probability), which defines the transition rate between the wells, k, which corresponds to the mean number of transitions in either direction per unit time. For the wells of different depth, two different rates, k_+ for the transition in positive direction and k_- for the transition in negative direction; each of them being equal to a half of the inverse of the corresponding mean first passage time to the top of the barrier.

The transition between the L- and the R states of the system can be considered as a simplest reversible isomerization reaction,

$$L \rightleftharpoons R. \tag{9.28}$$

and the probabilities P_L and P_R to find a system in either state are given by the system of ordinary differential equations,

$$\frac{dP_L}{dt} = -k_- P_L + k_- P_R,$$
$$\frac{dP_R}{dt} = -k_+ P_R + k_+ P_L. \tag{9.29}$$

This coarse-grained approximation corresponds to the approach known as formal kinetics. If we consider the ensemble of such two-level systems, than the number of systems being in states R and L are proportional to the corresponding probabilities, and their sum equals to the overall number of systems N.

The equilibrium probabilities following from Eq. (9.29) are $P_L = k_-/(k_+ + k_-)$ and $P_L = k_+/(k_+ + k_-)$ so that

$$\frac{P_L}{P_R} = \frac{k_-}{k_+}. \tag{9.30}$$

Expressing the transition rates through the first passage times and using the expression for the first passage time derived above we get

$$\frac{P_L}{P_R} = \sqrt{\frac{U''(x_L)}{U''(x_R)}} \exp\left(\frac{U(x_R) - U(x_L)}{k_B T}\right). \tag{9.31}$$

Note that this is exactly the result which one would obtain if one simply uses the equilibrium Boltzmann distribution $p(x) = A\exp(-U(x)/k_B T)$ with A being the normalization constant and define $P_L = \int_{-\infty}^{0} p(x)dx$ and $P_R = \int_{0}^{\infty} p(x)dx$. The corresponding integrals are exactly of the type of Eq. (9.26) so that

$$P_L = A\sqrt{\pi k_B T/2U''(x_L)} \exp[-U(x_L)/k_B T] \tag{9.32}$$

and

$$P_R = A\sqrt{\pi k_B T/2U''(x_R)} \exp[-U(x_R)/k_B T], \tag{9.33}$$

and their quotient is exactly given by Eq. (9.31). This connection between the rates and the equilibrium populations is rather universal and holds whenever rates exist. It is closely related to the mass action law of equilibrium thermodynamics, see Chapter 10.

9.5 Moments of the first passage time

In the previous section we have seen how much information is contained even in the first moment of the first passage time distribution, i.e. in the mean first passage time. Sometimes one also needs higher moments of the distribution. The trivial way of obtaining $\psi(t, x_0)$ and then integrating it to obtain

$$\langle t^n \rangle = \int_0^\infty t^n \psi(t, x_0) dt \tag{9.34}$$

might be too complicated, so that it is reasonable to derive equations which give us these moments immediately. Here we first return to the potential shown in Fig. 9.1 where the particle leaves the potential well through the point x_0. Let us consider now a general initial condition, i.e. let the particle start at some point y within the well and obtain the first passage time probability density $\phi(t, y) = \psi(t, x_0|y)$ as a function of y. We have

$$\phi(t, y) = -\frac{\partial}{\partial t} \int_{-\infty}^{x_0} p(x, t|y, 0) dx. \tag{9.35}$$

We now can interchange the sequence of temporal differentiation and integration over x. Moreover, if the coefficients in the Fokker–Planck equation are time-independent, the probability density $p(x,t|y,t')$ depends only on the difference $t-t'$ and thus as a function of y is given by the backward Kolmogorov equation

$$\frac{\partial}{\partial t}p(x,t|y,0) = \left[\mu f(y)\frac{\partial}{\partial y} + D\frac{\partial^2}{\partial y^2}\right] p(x,t|y,0), \qquad (9.36)$$

so that

$$\phi(t,y) = -\left[\mu f(y)\frac{\partial}{\partial y} + D(x)\frac{\partial^2}{\partial y^2}\right] \int_{-\infty}^{x_0} G(x,t|y,0)dx. \qquad (9.37)$$

where we have interchanged the sequence of integration over x and differentiation over y. Taking the temporal derivative from both parts of this equation and using the definition of $\phi(t,y)$, Eq. (9.35) we get the closed equation for $\phi(t,y)$ in form of a backward equation

$$\frac{\partial}{\partial t}\phi(t,y) = -\left[\mu f(y)\frac{\partial}{\partial y} + D\frac{\partial^2}{\partial y^2}\right] \phi(t,y). \qquad (9.38)$$

The corresponding initial and boundary conditions follow from those for the backward Kolmogorov equation, but can be easily put down also from the purely physical considerations: $\phi(0,y) = 0$ for $y \neq x_0$ (the particle needs finite time to reach the boundary from any point inside the well) and $\phi(t,x_0) = \delta(t)$, since the particle needs zero time to rich the boundary in the case it was there from the very beginning. Now, we use the definition of the moments, i.e. multiply both sides of Eq. (9.38) by t^n and perform the integration over t to get

$$\left[\mu(y)f(y)\frac{d}{dy} + D\frac{d^2}{dy^2}\right]\langle t^n(y)\rangle = -n\langle t^{n-1}(y)\rangle \qquad (9.39)$$

giving the recursive hierarchy of equations for moments. The boundary condition $\langle t^n(x_0)\rangle = 0$ ($n < 0$) follows from those for the δ-functional form of $\phi(t,x_0)$. Note that $\langle t^0 \rangle = 1$ due to normalization. For the mean first-passage time we thus get

$$\left[\mu(y)f(y)\frac{d}{dy} + D(y)\frac{d^2}{dy^2}\right]\langle t(y)\rangle = -1. \qquad (9.40)$$

This equation was first derived by Pontryagin, Andronov and Witt (1933). The splitting probabilities can be, of course, discussed by introducing the corresponding boundary conditions at two ends of the interval.

Note that Eq. (9.40) can be easily integrated (by introducing first $z(y) = \frac{d\langle t(y)\rangle}{dy}$ and assuming that this function vanishes for $y \to -\infty$) and leads to the same results as the flow over population method: The general solution of the equation reads

$$\tau = \langle t(y)\rangle = \int_y^{x_0} dy' \int_{-\infty}^{y'} dy'' \frac{exp[V(y') - V(y'')]}{D(y'')} \qquad (9.41)$$

where

$$V(x) = -\int_{-\infty}^{x} \frac{\mu(z)f(z)}{D(z)} dz = \frac{U(x)}{k_B T} \qquad (9.42)$$

is the same as in section 9.4.1. The overall scheme however allows for obtaining the higher moments as well.

The corresponding results may also be obtained from the immediate discussion of the trajectories of the process, see van Kampen, 1992, which approach is a bit less formal, but needs starting again from the general Master equation for the whole process.

9.6 An underdamped situation

This is only the overdamped motion in 1d (described by the Smoluchowski equation) for which many beautiful exact solutions are readily available. For more general cases, say the ones described by the Klein–Kramers equation, exact expressions are not known (in the sense that you have first to solve equation to get the result). Here a lot of nice numerical work is done, and several approximations based on special properties of the process are known, see Risken, 1988; Hänggi, Talkner and Borkovec, 1990.

However, the limiting case of extremely low friction is also amenable for theoretical investigation. The treatment here might be based on the idea of energy diffusion discussed in Chapter 8. For $\mu \to \infty$ (small friction) the Klein–Kramers equation is reduced to the Liouville equation: We have now to do with the microcanonical ensemble with conserved energy $E = mv^2/2 + U(x)$, defining the states of the system. The external noise causes the transitions between the different states. In this case we can approximately reduce the two-dimensional Klein–Kramers equation to a one-dimensional one. From formal reasons it is better to characterize the states not by the energy itself but by the action $I(E)$ being the function of energy.

The relation between the energy and the action

$$I = \oint p dx = 2 \int_{x_{\min}}^{x_{\max}} \sqrt{2m[E - U(x)]} dx \qquad (9.43)$$

follows from the Hamiltonian dynamics of the system; $x_{\min}$ and $x_{\max}$ are the turning points of the corresponding trajectory.

The instantaneous probability of transitions between the two states depends on where the phase point is situated during the transition. The next step is then averaging over all such transitions to get the corresponding second moment giving the diffusion coefficient in the corresponding Smoluchowski equation. This procedure, introduced by Kramers (see Kramers, 1940 and Zwanzig, 1959) leads to the following expression for the probability density in I:

$$\frac{\partial p(I,t)}{\partial t} = \gamma \frac{\partial}{\partial I} I \left[1 + \frac{2\pi k_B T}{\omega(I)} \frac{\partial}{\partial I} \right] p(I,t) \qquad (9.44)$$

where $\omega(I)$ is the frequency corresponding to the action I:

$$\omega(I) = 2\pi \frac{\partial E}{\partial I}. \qquad (9.45)$$

Equation (9.44) is a one-dimensional equation for which the discussions of our previous section hold. Using the method based on the backward equation, Eq. (9.40), we can get an explicit result for the mean life time in a well.

The main difference between the over- and the underdamped situation can, however, be grasped without really solving Eq. (9.44). Let us consider, for example, the motion in a harmonic potential with a cutoff at $E = E_{\max}$. For this case $\omega(I) = \omega_0 = \sqrt{k/m}$, where k is the elasticity constant of the spring, and $I(E) = 2\pi E/\omega_0$. Eq. (9.44) now reads*:

$$\frac{\partial p(I,t)}{\partial t} = \gamma \frac{2\pi}{\omega_0} \frac{\partial}{\partial I} \left[\frac{\omega_0}{2\pi} I + k_B T \frac{\partial}{\partial I} \right] p(I,t) \qquad (9.46)$$

Comparing this equation with a typical overdamped form

$$\frac{\partial}{\partial t} p(x,t) = \mu \frac{\partial}{\partial x} \left[-f(x) + k_B T \frac{\partial}{\partial x} \right] p(x,t) \qquad (9.47)$$

we see that the situation is equivalent to the overdamped case with μ changed for $2\pi\gamma/\omega_0$ and the force being $-\omega_0 I/2\pi$. Therefore we can immediately say that the corresponding rate (proportional to μ) will linearly grow with friction coefficient γ, a situation opposite to the overdamped case, where it decays as γ^{-1}. Thus, the overall situation corresponds to a nonmonotonous γ-dependence: For initial energies below $E_{\max}$ the rate, as a function of γ first grows and then starts to decay. The crossover between these two types of behavior at moderate damping was discussed almost half a century later, in the works by Melnikov and Meshkov, 1986, and by Pollak, Grabert and Hänggi, 1989, see Hänggi, Talkner and Borkovec, 1990 for a review.

---

*The simplification consisting the assumption that the maximum of the potential corresponds to a cusp at $E_{\max}$, is not always realistic. Typically, the escape escape takes place through a saddle point of the potential (quadratic maximum). The frequency on the trajectory passing through this saddle point vanishes, so that the diffusion coefficient $2\pi k_B T \gamma I/\omega(I)$ formally diverges. Thus, in the underdamped problem the behavior close to the separatrix is much less important than for the overdamped case. Estimates show that our simple result is not too far from the reality, see van Kampen, 1988.

Chapter 10

Reaction Kinetics

One of the typical and important cases pertinent to the domain of nonequilibrium thermodynamics or statistical physics corresponds to the reaction kinetics. One has to stress that what we call the "reactions" has a much broader meaning than a real *chemical* reaction in a gas or liquid phase, but includes many "physical" phenomena such as recombination of ions and electrons in plasma, the recombination of electrons and holes in semiconductors, different classes of processes leading to luminescence in solids and liquids, to creation and healing of the radiation defects, etc. Some special examples of these reactions will be discussed in the following chapters. Here we start from what is called the classical kinetic theory, the approach based on the assumption that there exist well defined reaction rates, i.e. the probabilities of reaction per particle per unit time. Before going into details of kinetics, let us say a few words about the thermodynamics of reactions.

10.1 The mass action law

The typical chemical reaction

$$adducts \rightleftharpoons products \qquad (10.1)$$

eventually leads to establishing equilibrium between the adducts and the products of the reaction; their equilibrium concentrations depend on the thermodynamical properties of adducts and products and on the reaction heat, and is given by the mass action law, being a law of equilibrium thermodynamics. For example, if our reaction is a simple bimolecular one,

$$A + B \rightleftharpoons C \qquad (10.2)$$

(say, simple dissociation $H + H \rightleftharpoons H_2$, or ionization in plasma $H^+ + e^- \rightleftharpoons H$ or in electrolytes, or electron-hole recombination on a semiconductor, or recombination of Frenkel defects in an irradiated solid) the equilibrium concentration of the product C and its dependence on temperature and pressure are prescribed by thermodynamics, see (Landau and Lifshitz, 1990). The concentration of the reactant A (B, C) will throughout this chapter be denoted by the italic A (B, C). Each reaction equation of the type above can be put down in a form

$$n_1 R_1 + n_2 R_2 + \cdots + n_n R_n \rightleftharpoons n_{n+1} R_{n+1} + \cdots + n_M R_M \qquad (10.3)$$

where R_i with $i = 1, \ldots, n$ are the adducts and R_i with $i > n$ are the products of the reaction. In our example above we had $\nu_1 = \nu_2 = \nu_3 = 1$. Equation (10.3) can be rewritten as

$$\nu_1 R_1 + \nu_2 R_2 + \cdots + \nu_n R_n + \nu_{n+1} R_{n+1} + \cdots + \nu_M R_M = 0, \qquad (10.4)$$

where the numbers ν_i are the stoichiometric coefficients. Here we have $\nu_i = n_i$ for the adducts and $\nu_i = -n_i$ for the products of the reaction. Let us concentrate on the case when only one reaction is possible in the system. The reaction equation dictates then that the changes in the numbers of corresponding species are not independent: the reaction equation, Eq. (10.4) represents a conservation law for a given combination of the particle numbers $N_1, N_2, \ldots, N_M$. Thus, if the number of particles of the sort R_1 changed by δN_1, then the number of particles R_2 has to change by $\delta N_2 = (\nu_2/\nu_1)\delta N_1$, etc. The changes in the numbers of the corresponding particles $N_1, N_2, \ldots, N_M$ are connected via

$$\frac{\delta N_1}{\nu_1} = \frac{\delta N_1}{\nu_1} = \cdots = \frac{\delta N_M}{\nu_M} = \delta \alpha. \qquad (10.5)$$

The equilibrium state of the system is then characterized by the minimum of Gibbs potential (free enthalpy) G being the function of $N_1, N_2, \ldots, N_M$, of the pressure p, and of the temperature T. The change in this potential is

$$\delta G = \sum_{i=1}^{M} \frac{\partial G}{\partial N_i} \delta N_i = \delta \alpha \sum_{i=1}^{M} \nu_i \frac{\partial G}{\partial N_i}. \qquad (10.6)$$

In equilibrium, the Gibbs potential is minimal, so that

$$\sum_{i=1}^{M} \nu_i \frac{\partial G}{\partial N_i} = 0. \qquad (10.7)$$

Let us now turn to the simplest situation which emerges when the reaction takes place in well-mixed gaseous phase or in dilute solutions. In this case the overall value of G can be put into the form

$$G = \sum_{i=1}^{M} \left(N_i \mu_i - k_B T N_i \ln \frac{N_i}{N} \right), \qquad (10.8)$$

where μ_i is the chemical potential of the pure component, and the second term corresponds to the mixing entropy. This last one is taken in a form it has in ideal gases or in dilute solutions. Here N denotes the overall number of particles $N = N_1 + N_2 + \cdots + N_M$. Using Eqs. (10.7) and (10.8) we get:

$$\sum_{i=1}^{M} \nu_i \left[\mu_i - k_B T \left(\ln \frac{N_i}{N} + 1 \right) \right] = 0, \qquad (10.9)$$

which equation can be rewritten in the form

$$\sum_{i=1}^{M} \nu_i \ln \frac{N_i}{N} = \sum_{i=1}^{M} \nu_i \left(\frac{\mu_i}{k_B T} - 1 \right). \qquad (10.10)$$

Introducing the relative concentrations $c_i = N_i/N$ we get

$$c_1^{\nu_1} \cdots \cdots c_M^{\nu_M} = K_c \qquad (10.11)$$

where K_c is the reaction equilibrium constant,

$$K_c = \exp \left[\sum_{i=1}^{M} \nu_i \left(\frac{\mu_i}{k_B T} - 1 \right) \right]. \qquad (10.12)$$

Note that K_c is the function of the pressure p and the temperature T being the natural variables of chemical potentials. This is the mass action law, formulated by Guldenberg and Waage in 1867. This law connects the relative equilibrium concentrations of the particles with chemical potentials of reacting components. The dependence of the equilibrium constant on temperature and pressure was examined by van't Hoff, whose investigations brought him the (very first!) Nobel prize in chemistry in 1901. The volume concentrations $R_i = c_i N/V = c_i p/(k_B T)$ of the products satisfy the similar equation,

$$R_1^{\nu_1} \cdots \cdots R_M^{\nu_M} = K_c(T, p) \left(\frac{p}{k_B T} \right)^{\nu} \qquad (10.13)$$

with $\nu = \nu_1 + \cdots + \nu_M$. In our simple bimolecular reaction taken as an example the combination in the left hand side would correspond to AB/C.

The mass action law and the corresponding van't Hoff's equations are the consequences of thermodynamics, and do not depend on the exact dynamical mechanism of the reaction. Note however, that here the condition of "well-mixedness" is explicitly assumed. The Gibbs potential (the free enthalpy) of the system depends, through the entropic contribution, on the correlations between the particle's positions, which might appear either through interactions or dynamically, through the reaction itself. Thus, the mass action law might be violated, which doesn't mean the violation of any thermodynamical principles. Note that some correlations might be destroyed by effective mixing: this is how mixing procedures may influence chemical equilibria.

10.2 Classical kinetics

Let us now turn to the *kinetics* of the reaction. In many cases both the forward and the backward reaction can be described by the rates k_+ and k_-:

$$A + B \underset{k_-}{\overset{k_+}{\rightleftharpoons}} C \qquad (10.14)$$

so that the kinetic equations for the corresponding concentrations (denoted by the same symbols, now in italic) follows a system of ordinary differential equations:

$$\frac{d}{dt}A = \frac{d}{dt}B = -k_+ AB + k_- C,$$
$$\frac{d}{dt}C = k_+ AB - k_- C. \qquad (10.15)$$

Here we assume that the two adduct molecules A and B have a well-defined probability to meet and to react per unit time, and that the product molecule C has a well-defined probability to dissociate per unit time. The first probability is assumed to be proportional to the product of concentrations A and B (which means that for a given one A-molecule the number of reactive encounters with B is proportional to the actual B-concentration in the system; this probability per unit time is exactly k_+); the second probability is assumed to be constant and is exactly k_-.

The system of equations of classical reaction kinetics can be easily solved, and shows a relaxation to an equilibrium state given by the situation when the time-derivatives in the left hand sides vanish. In this case

$$\frac{AB}{C} = \frac{k_-}{k_+}. \qquad (10.16)$$

We note that $\frac{AB}{C}$ is exactly the combination whose value is prescribed by the mass action law. Thus, if the reaction rate coefficients k_+ and k_- exist, they are not independent, but connected to each other via the constant of the mass action law. However, as we hasten to note, the existence of well-defined, time-independent reaction rates is essentially an additional hypothesis which is often violated, so that the real kinetics of the reaction hardly resembles the one prescribed by the system of first-order ordinary equations of the formal kinetics, or even the reaction-diffusion equations.

Van Kampen summarizes typical conditions under which the reaction rate approach applies (see van Kapmen, 1992):

(1) The mixture must be homogeneous, so that the density at each point is the same. If the reaction is sufficiently slow, such homogeneity might be achieved by stirring.

(2) The non-reactive elastic collisions must be sufficiently frequent to ensure the Maxwell velocity distribution at temperature T. Otherwise the collision frequency could not be proportional to the product of densities. This assumption is typically satisfied in the presence of a solvent or inert gas. It is, of course, always the case if the transport process can be approximated by diffusion.
(3) The internal degrees of freedom of the molecules are supposed to be in thermal equilibrium, at the same temperature T.
(4) The temperature must be constant is space and time so that one may treat the reaction rates as constants even though they depend strongly on temperature.

In many cases, the homogeneity of the system as a whole might be violated. However, just in spirit of non-equilibrium thermodynamics, in many cases one can still assume local equilibrium, so that the reaction can be described by concentrations, temperatures, etc., which are coordinate-dependent. This approach assumes that the characteristic size of inhomogeneities is large compared with the interparticle distance and with their mean free path. Moreover, the approach assumes that some "microscopic" reaction rate constants exist, which govern the local course of the reaction, so that all inhomogeneities may be considered on the "mesoscopic" level, i.e. assuming that the reaction takes place in a locally homogeneous medium, however, the local concentrations or temperatures may differ from place to place. This assumption leads to the reaction-diffusion or reaction-diffusion-advection equations, being mesoscopic approximations. The corresponding continuous-medium approximations (e.g. reaction-diffusion equations) describe a large variety of effects. However, these approaches fail whenever the clear-cut scale separation gets impossible.

10.3 The Smoluchowski approximation

It is important to stress that real kinetics of chemical reactions depends both on the properties of the elementary act of the reaction (through the probability of forwards/backwards reactions) and on the transport process, bringing the reacting particles together. The last one depends strongly on the properties of the reaction medium, and may be different for example for the cases when this medium is stirred or not. Note that the chemical equilibrium, depending essentially on the thermodynamic (i.e. static) properties of products and adducts is typically determined solely by the nature of the products and adducts, but not by transport, whenever the overall homogeneity is guaranteed.

The first theory taking into account the properties of the transport process stems again from Smoluchowski, whose model was the coagulation of diffusing particles in absence and in presence of flows (Smoluchowski, 1917).

The approach of Smoluchowski was rather simple but effective. Let us consider a bimolecular reaction of particles A and B in a quiescent medium. Let at the beginning imagine that only particles A diffuse, and the B-particles are immobile (fixed). The physical picture exactly corresponding to this situation will be discussed in the next paragraph.

Let us first consider the reaction taking place with probability 1 on contact between A and B particles. The reaction rate k of the forward reaction can then be interpreted as a current (Berg 1994, Rice, 1985) of A-particles onto the surface of a B one: This is a number of encounters of A's and a chosen B pro unit time. Imagine, the motion of A's is pure diffusion, and the reaction takes place when the A particle approaches the B one at a distance a, called the reaction radius. The overall situation is assumed to be spherically symmetric. The reaction rate of such purely *diffusion-controlled* reaction can be obtained through the solution of the diffusion equation for the concentration n of the A-particles surrounding the chosen B. To get the number of such encounters per unit time it is the enough to calculate the overall current of A's onto the surface of the chosen B. This is a diffusion current, which in three dimensions is given by

$$J_{diff} = -4\pi a^2 D \nabla n(r \to a_+, t); \qquad (10.17)$$

the reaction rate is nothing else as $k = J_{diff}/A$. Here the concentration $n(r,t)$ is a solution of the diffusion equation

$$\frac{\partial}{\partial t} n(r,t) = D \Delta n(r,t). \qquad (10.18)$$

There are no A-particle within the reaction sphere, so that the boundary condition $n(a,t) = 0$ applies. The initial condition corresponds to a well-premixed system, in which the A-concentration is constant, so that $n(r,0) = A\theta(r-a)$.

This equation is not too hard to solve. Applying the Laplace-transform in time (changing to $n(r,u) = \int_0^\infty n(r,t)e^{-ut}dt$) and using the spherical symmetry of the problem we get:

$$un(r,u) - A\theta(r-a) = D\frac{1}{r^2}\frac{\partial}{\partial r}\left(r^2 \frac{\partial n(r,u)}{\partial r}\right) \qquad (10.19)$$

with boundary conditions $n(a,u) = 0$, $n(r,u) \to A$ for $r \to \infty$. The change of variables $n(r,u) = \psi(r,u)/r$ reduces this equation to one with constant coefficients, which is readily solved. The inverse Laplace transform then gives

$$n(r,t) = A\left[1 - \frac{a}{r}\text{erfc}\left(\frac{r-a}{2\sqrt{Dt}}\right)\right]. \qquad (10.20)$$

Note that for $t \to \infty$ the solution tends to a stationary one,

$$n(r,u) = A\left(1 - \frac{a}{r}\right), \qquad (10.21)$$

i.e. to the solution of the three-dimensional Laplace equation fulfilling the boundary conditions. Equation (10.20) gives us the current onto the reaction surface

$$J_{diff} = A4\pi Da \left(1 + \frac{a}{2\sqrt{Dt}}\right) \tag{10.22}$$

which after a transient tends to a constant giving the famous result

$$k_{diff} = 4\pi Da \tag{10.23}$$

for the reaction rate constant (Smoluchowski, 1917). If both reactants are mobile, the diffusion coefficient D is changed for the mutual diffusion coefficient $\tilde{D} = D_A + D_B$ being the sum of the diffusion coefficients of reactants. The same approach was used by Debye (Debye, 1942) to get the effective rates of recombination in electrolytes, see also Falkenhagen, 1971, Rice, 1985.

The fact that the reaction takes place not with the probability 1 on encounter can be taken into account by using a partially reflecting (von Neuman) boundary condition and leads to a simple result

$$k^{-1} = (k_{diff} + k_r)^{-1} \tag{10.24}$$

where k_r is the reaction-controlled contribution, connected with the reaction probability (see Rice, 1985, Ovchinnikov, Timashev and Belyi, 1986). Since many real chemical reactions in fluid or solid phases are not "fast" (i.e. purely diffusion-controlled) so that the probabilities of the reactions on a contact are small, the effective rate is mostly governed by the reaction probability (Kramers rate), and therefore the deficiencies of the Smoluchowski's approach are not relevant. However, in the cases with slow diffusion and rather fast reaction these deficiencies get evident. Such cases are often pertinent to luminescent energy transport or to interactions of radiation defects in solids. Interestingly enough, here the theorists were first to discuss the situation; the results were then experimentally proved. The reason why the effects were not paid much attention are typically as follows: The systems are anyhow complex. The exact reaction schemes may include intermediate stages about which nothing is known. Including such stages allows to fit whatever experimental curve. On the other hand, theoretical approaches starting from simple models have clearly shown that the deviations might be strong and that they might persist asymptotically.

Considering the physical meaning of $n(r,t)$ in Eq. (10.18), it is easy to understand that it is proportional to $c_{AB}(r,t)$, the two-point correlation function of the positions of A and B-particles, so that the equation of formal kinetics,

$$\frac{\partial}{\partial t} c_A = -k c_A c_B \tag{10.25}$$

with

$$k = -4\pi a^2 D \nabla c_{AB}(r,t) \tag{10.26}$$

for the one-point concentrations $c_A = A$ and $c_B = B$ and the Smoluchowski's diffusion equation

$$\frac{\partial}{\partial t} c_{AB}(r,t) = D \Delta c_{AB}(r,t) \tag{10.27}$$

with boundary condition $c_{AB}(a,t) = 0$, and initial condition $c_{AB}(r,0) = \theta(r-a)$ are essentially the first and the second equation in the BBGKY-hierarchy, with a truncation of the second one. (The correct second equation should also contain a functional of the three-point correlation function in its r.h.s., Kuzovkov and Kotomin, 1988; 1996.) This means, that the Smoluchowski equation is clearly a two-particle approximation, fully disregarding higher correlations. So are also many of the reaction-diffusion schemes in use. The relative importance of many-particles effects in nonequilibrium dynamics of such reactions may be assessed by means of more accurate theories, or by numerical simulations. Note that although the characteristic scale of the change in A-concentration, which might be considered as the effective range of the "interaction" caused by chemical reaction, given by Eq. (10.21) is the reaction radius a, i.e. is microscopic, the "interaction" itself is long-range. This indicates that the situations might occur in which the two-particle picture by Smoluchowski, as well as other two-particle approximations lose their validity. By now, several classes of reactions are known, in which the approximations dramatically fail to predict kinetics; three ones will be considered in what follows.

We note that, as it is often the case, the BBGKY-hierarchy itself, although being formally the universal theoretical instrument, does not provide a comfortable base for practical calculations, so that only few effects (like Ovchinnikov–Zeldovich slowing-down, *vide infra*) can be reasonably described within these schemes; moreover even here the description relies on uncontrolled approximations like Kirkwood decomposition or equivalent. On the other hand, several special cases can be discussed with high degree of rigor using alternative (probabilistic or field-theoretical) approaches, see e.g. Mattis and Glasser, 1998. We would like to stress that for the quantitative description of such cases the random-walks approaches based on the picture put forward by Elliot Montroll and Georges Weiss were proved superior to Langevin schemes or continuous approximations. However, it is not our aim here to discuss the modern theory of diffusion-controlled reactions in full depth: For us it is only valuable as an example of how surprising non-equilibrium effects emerge from in the situations, in which the equilibrium behavior is clear and boring. The interested reader finds a lot of material in a book by Kuzovkov and Kotomin, 1996.

These are not only the multiparticle effects that may lead to failure of the reaction rate approximation. The notion of the constant reaction rates is intrinsically inappropriate in lower dimensions (capillaries, surfaces, involved fractal geometries of porous media, see Kopelman 1988, Blumen, Klafter and Zumofen, 1986).

Let us consider as an example a purely one-dimensional situation (the coordinate is now denoted by x, the B-particle is placed in the origin). Once again the concentration of A is given by Eqs. (10.25)–(10.27). In one dimension, however,

Eq. (10.27) does not possess any stationary solution, to which the actual solution of the initial-value problem would tend at longer times. The solution of Eq. (10.27) in one dimension is given by

$$c_{AB}(x,t) = \text{erfc}\left(-\frac{|x|-a}{2\sqrt{Dt}}\right) \qquad (10.28)$$

and is not a stationary, but a self-similar one. The corresponding reaction rate $k(t)$ is a function of time and continuously decays, indicating the ineffectiveness of the one-dimensional transport:

$$k(t) = -2D\frac{d}{dx}c_{AB}(x,t)|_{x=a} = \frac{2\sqrt{D}}{\sqrt{\pi t}} \qquad (10.29)$$

(note that the dimension of the reaction rate in d dimensions is $[L^d/T]$, so that in 1d it has the dimension of velocity).

In two dimensions the reaction rate also decays, and goes as $1/\ln t$. The behavior is 1d and 2d is intrinsically connected with the recurrence of random walks in these dimensions, which leads to a compact exploration of the reaction volume. We return to this discussion in Chapter 11, devoted to the random walk approaches to nonequilibrium processes.

10.4 Fluctuation effects in chemical kinetics

In many cases considered in thermodynamics fluctuations play a subordinated role, and are typically neglected. One of such cases is pertinent to the theory of chemical reactions, where one often uses classical reaction schemes as if they were immediately following from thermodynamics. However, one has to be aware that such schemes may be very week and crude approximations, which are valid only for restricted values of parameters (say, homeopathic concentrations or high temperatures) or under special additional conditions (say, vigorous mixing).

Let us consider a few simple examples illustrating the deficiencies of the classical concept of reaction rates. We consider a few classes of reactions which are especially "popular" due to their relevance for physics. Thus we consider

$$A + B \to B$$

with B immobile ("trapping"), showing strongly non-classical kinetics, as compared to

$$A + B \to B$$

with A immobile ("scavenging"), whose kinetics follow a classical pattern,

$$A + B \to 0$$

("annihilation", nonclassical) as compared to

$$A + A \to 0$$

("quenching", classical, at least at small concentrations), as well as

$$A + B \to 2B,$$

being the simplest autocatalytic reaction. Some effects considered here are discussed in much more detail in in the review article (Mikhailov, 1989) and in the book by Kotomin and Kuzovkov (1996).

10.4.1 The Vaks–Balagurov kinetics in trapping

The Smoluchowski's result was derived for the case when the A-particle is immobile and the B-particles diffuse. This situation is often termed as scavenging. The further discussion supposed that only the relative diffusion coefficient matters, so that the trapping reaction, in which A is mobile, and the positions of B's are fixed (trapping) has essentially the same kinetics. The physical interest to the trapping reaction comes from the luminescence of mixed molecular crystals. A dipole-inactive excitation (say, a triplet exciton) has an extremely long life-time and may jump from one host molecule to another. However, if the excitation comes close to a defect or to an impurity (guest) molecule, leading to a change in the symmetry of the corresponding molecular orbital, the excitation acquires a dipole moment and decays rapidly emitting the light: The reaction equation has essentially the form $A + B \to B + h\nu$. The photon, however, immediately leaves the system, making the relaxation after the irradiation of the crystal by a short but intensive laser pulse practically irreversible. The impurity atom or the defect B acting as a catalyst stays where it was. This variant of the reaction is called *trapping* in physical literature: this corresponds to catching a moving particle by an immobile trap.

Assuming classical kinetics, $A + B \xrightarrow{k} B$, one easily obtains the equation for the concentration A (the concentration B stays unchanged):

$$\frac{d}{dt}A = -kAB \tag{10.30}$$

whose solution is a simple exponential

$$A(t) = A(0)\exp(-kBt). \tag{10.31}$$

This solution, however, does not apply in many cases, when strong departure from a simple exponential decay are observed. The effect was first found in a theoretical work by Vaks and Balagurov, 1973, who adapted the Ilya Lifschitz's optimal fluctuation argumentation to study an irreversible process. Then it was independently rediscovered by Donsker and Varadhan, 1979, who used a rigorous but hard-to-follow probabilistic approach. This result was then translated into physical language

Fig. 10.1 The one-dimensional trapping reaction with perfect traps in one dimension. The traps (shown as black circles) act as absorbing boundaries for A-particles (white circles).

by Grassberger and Procaccia, 1982, leading to the argumentation rather close to the one of Vaks and Balagurov. However, before discussing general properties of trapping reactions, we consider here the one-dimensional case, which is exactly solvable, and which gives the possibility to easily grasp the physical nature of the effect considered.

10.4.1.1 The Vaks–Balagurov problem in one dimension

Although the problem of diffusion-controlled reactions one-dimensional systems is somewhat different from ones in high dimensions, it is reasonable to discuss here first such a one-dimensional problem, for which the deficiencies of the Smoluchowski theory get evident. We remind that, at variance with dimensions 3 or higher, the Smoluchowsky approximation in 1d would propose $k(t) \propto t^{-1/2}$ and therefore

$$A(t) \propto \exp(-\gamma t^{1/2}) \tag{10.32}$$

with γ being some constant, i.e. a slower decay than a simple exponential one which is predicted by classical kinetics. However, in reality the decay is even slower, due to fluctuation effects. Let us now discuss the problem in more detail.

In one dimension immobile B-particles divide the line (see Fig. 10.1) into independent compartments filled by A-particles. These A particles do not interact with each other and perform independent diffusive motion. Let us first concentrate on one interval and give its ends the coordinates 0 and L. Let us consider one particular A particle and calculate its survival probability as a function of time.

The probability $p(x,t)$ to find the particle at site x at time t is given by a diffusion equation

$$\frac{\partial p}{\partial t} = D \frac{\partial^2 p}{\partial x^2}. \tag{10.33}$$

The initial condition is $p(x,0) = 1/L$, since we assume that the A-particles are initially homogeneously and independently distributed within the whole system. The fact that an A-particle disappears, touching one the B-particles at the boundaries of each interval, is taken into account by imposing absorbing boundary conditions at the ends of the interval.

The solution of the diffusion equation for each interval is given by the well-known series following from the eigenfunction decomposition:

$$p(x,t) = \sum_k a_k T_k(t) X_k(x) \tag{10.34}$$

where a_k are number coefficients, and the functions $T_k(t)$ and $X_k(x)$ are the solutions of the equations

$$\frac{dT_k}{dt} = \lambda_k T_k \tag{10.35}$$

and

$$\frac{d^2 X_k}{dx^2} = D^{-1} \lambda_k X_k . \tag{10.36}$$

The solutions of the second equation are trigonometric functions, and the ones fulfilling the boundary conditions are given by

$$X_k = \sin\left(\frac{\pi k x}{L}\right) \tag{10.37}$$

so that

$$\lambda_k = -\frac{\pi^2 k^2 D}{L^2} . \tag{10.38}$$

Therefore the solutions of the first equation read

$$T_k = \exp\left(-\frac{\pi^2 k^2 D}{L^2} t\right) . \tag{10.39}$$

The prefactors a_k in Eq. (10.34) follow from the initial conditions:

$$a_k = \frac{1}{L}\frac{2}{L}\int_0^L \sin\left(\frac{\pi k x}{L}\right) dx = \begin{cases} 4A_0/\pi k L & k \text{ odd}, \\ 0 & k \text{ even}. \end{cases} \tag{10.40}$$

Thus, the concentration of A-particles on the interval is given by

$$p(x,t) = \sum_{m=0}^{\infty} \frac{4}{L\pi(2m+1)} \sin\left(\frac{\pi(2m+1)x}{L}\right) \exp\left(-\frac{\pi^2(2m+1)^2 D}{L^2} t\right) \tag{10.41}$$

and the overall survival probability of the particle on the interval of length L

$$P_L(t) = \int_0^L p(x,t) dx = \sum_{m=0}^{\infty} \frac{8}{\pi^2 (2m+1)^2} \exp\left(-\frac{\pi^2 (2m+1)^2 D}{L^2} t\right) . \tag{10.42}$$

Now, the probability density for a particle to be initially found in the interval of length L is proportional to $Lp(L)$, where $p(L)$ is the probability density to find a B-free interval of length L. To get this probability normalized, we have to take it equal to

$$\frac{Lp(L)}{\int_0^\infty Lp(L) dL} = \frac{Lp(L)}{\langle L \rangle} . \tag{10.43}$$

Assuming B particles to be distributed according to the Poisson law, we have

$$p(L) = B \exp(-BL) \tag{10.44}$$

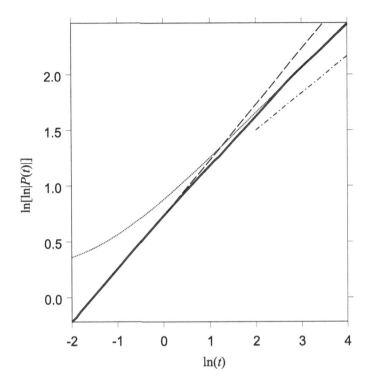

Fig. 10.2 The survival probability $P(t)$ in a trapping reaction in one dimension shown on the scales $\ln(|\ln(P(t))|)$ vs. $\ln(t)$. The exact survival probability as following from numerical evaluation of Eq. (10.46) with $D = 1$ and $B = 1$ is shown as a full line, and the approximate result, Eq. (10.48) as a dotted line. Note that at short times $P(t)$ follows $P(t) \propto \exp\left(-const \cdot t^{1/2}\right)$ (dashed line), in agreement with the prediction of the Smoluchowski theory. This behavior appears as as a straight line with slope 1/2 when on the scales of the plot. At longer times the reaction slows down, deviating from the dashed line. The dashed-dotted line has a slope 1/3, which will eventually be reached at very long times.

where B is the concentration of B-particles and $\langle L \rangle = B^{-1}$. Therefore, the survival probability for an A-particle chosen at random at $t = 0$ is given by

$$P(t) = B^2 \int_0^\infty P_L(t) L p(L) dL$$

$$= B^2 \sum_{m=0}^\infty \frac{8}{\pi^2(2m+1)^2} \int_0^\infty \exp\left(-\frac{\pi^2(2m+1)^2 D}{L^2}t\right) L \exp(-BL) dL . \quad (10.45)$$

Although the overall expression is somewhat awkward, it is easy to obtain the long-time asymptotic behavior of $P(t)$, since for t large only the first term in the sum, the one with $m = 0$ plays the role. The corresponding integral can be easily evaluated using the Laplace method: according to this we have to find the maximum of the exponent,

$$f(L) = -\frac{\pi^2 D}{L^2}t - BL, \qquad (10.46)$$

$f'(L_0) = 0$ giving $L_0 = (2\pi^2)^{1/3}(Dt/B)^{1/3}$ and approximate the exponential close to this point by

$$\exp(-f(L)) \approx \exp(-f(L_0))\exp\left(\frac{f''(L_0)}{2}(L-L_0)^2\right) \qquad (10.47)$$

(with $f''(L_0) < 0$). The integration over L (when moving the lower limit of this integration to $-\infty$) then gets trivial and one obtains

$$\begin{aligned} P(t) &\simeq B^2 \frac{8}{\pi^2}\sqrt{\frac{\pi}{2f''(L_0)}}L_0 \exp(-f(L_0)) \\ &= \frac{16}{\sqrt{3\pi}}B\sqrt{Dt}\exp\left[-\frac{3\pi^{2/3}}{2^{2/3}}B^{2/3}(Dt)^{1/3}\right]. \end{aligned} \qquad (10.48)$$

The concentration of A-particles is given by multiplying this result by their initial concentration A_0. The asymptotic behavior at long times is governed by the exponential term which decays slower than the one which follows from the Smoluchowski approximation.

10.4.1.2 *General asymptotic behavior in trapping*

Let us now discuss the situation in higher dimensions, $d \geq 2$. We will follow practically the same line of argumentation. In higher dimensions this approach only gives us the lower bound of the concentration and does not allow for obtaining correct preexponentials, however it provides correct long-time asymptotic behavior.

We again concentrate on the behavior of one A-particle, as shown in Fig. 10.3. The reaction rate k depends on the reaction probability of between A and B, however, whatever this reaction probability is, the reaction cannot take place before the particles A and B meet. Our further considerations will aim at getting a crude approximation for the probability of such encounter from above.

Let us first concentrate on the three-dimensional situation and consider a sphere with the center at the initial position of A and touching the closest of the B particles. The A-particle cannot react before it reached the surface of the sphere; however, even if it did, the reaction only takes place, if it hits the B. Thus, the assumption, the the A disappears whenever it touches the surface of the sphere, gives us a very rough estimate of the reaction probability from above.

Calculating the survival probability of a particle starting in a center of an absorbing sphere is a simple exercise in using the diffusion equation

$$\frac{\partial p}{\partial t} = D\Delta p \qquad (10.49)$$

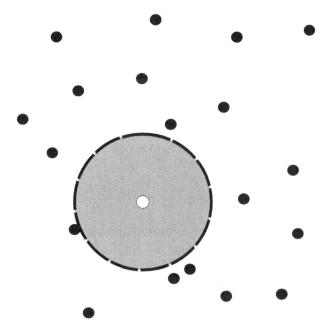

Fig. 10.3 The geometrical construction used in our argumentation. The interior of the circle shown in gray is free of traps (shown as black circles); the A-particle (white circle) starts in the middle of the circle and definitely survives until it crosses its boundary.

with the boundary condition $p(R) = 0$, where R is the radius of the sphere. Since the problem possesses spherical symmetry, we can change to spherical coordinates and put down

$$\frac{\partial p}{\partial t} = D \frac{1}{r^2} \frac{\partial}{\partial r} \left(r^2 \frac{\partial p}{\partial r} \right). \tag{10.50}$$

The solution is readily obtained using the eigenfunction decomposition: as in the one-dimensional case we assume that it has a form

$$p(r, t) = \sum_n a_n T_n(t) S_n(r); \tag{10.51}$$

with T_i fulfilling the equation

$$\frac{\partial T_n}{\partial t} = \lambda_n T_n \tag{10.52}$$

and S_i being the solution of the equation

$$\frac{1}{r^2} \frac{\partial}{\partial r} \left(r^2 \frac{\partial S_i}{\partial r} \right) = D^{-1} \lambda_n S_n \tag{10.53}$$

with the boundary condition $S_i(R) = 0$ and regular at $r = 0$. The solutions of Eq. (10.53) are, evidently, spherical Bessel functions of the first kind: $S_n(r) =$

$j_0\left(\sqrt{-D\lambda_n}r\right)$ so that $\lambda_i < 0$ follows from the equation

$$j_0\left(\sqrt{-\lambda_n/D}R\right) = \left(\sin\sqrt{-\lambda_n/D}R\right)/\left(\sqrt{-\lambda_n/D}R\right) = 0. \tag{10.54}$$

From this we get $\lambda_n = -\tau_n^{-1} = -\pi^2 n^2 D/R^2$, $n \geq 1$. Thus,

$$p(r,t) = \sum_n a_n e^{-t/\tau_n} j_0\left(r/\sqrt{D\tau_n}\right). \tag{10.55}$$

Note that τ_n vary quadratically with n, so that at longer time only the first term of the expansion is important, so that

$$p(r,t) \simeq a\exp(-t/\tau_1)\frac{\sin(\pi r/R)}{\pi r/R}. \tag{10.56}$$

The non-vanishing prefactor a can be easily found through the initial condition, $p(r,0) = \delta(r)$, but is of no importance for the further discussion. The overall survival probability within the sphere is then

$$P(t) = 4\pi \int_0^R p(r,t)r^2 dr \propto const \cdot \exp(-t/\tau_1(R)) \tag{10.57}$$

where we explicitly state that the characteristic time τ_1 is R-dependent. Thus, at longer times, the survival probability of a particle in a sphere decays exponentially, with the characteristic time proportional the square of the sphere's radius. The same is valid also in dimensions d other than $d = 3$: Although the exact forms of spatial eigenfunctions differ, the overall decay form is always

$$P(t) \simeq C(d)\exp\left(-c(d)\frac{Dt}{R^2}\right), \tag{10.58}$$

where $C(d)$ and $c(d)$ are the constants depending on the spatial dimension d.

Our next step will be to average the survival probability over many A-particles starting at different points in our system which within the adopted approximation corresponds simply to averaging over the radii of spheres. The distribution of such radii is given by the Poisson distribution of the B-particles. Since the probability not to find a B-particle within the volume V is exactly $\exp(-BV)$, the corresponding distribution is given by $p(R) = \exp(-BV(R))\frac{dV}{dR}$. In $d = 3$ we have

$$p(R) = 4\pi R^2 \exp\left(-\frac{4}{3}\pi BR^3\right), \tag{10.59}$$

and in d dimensions in general $p(R) = dk(d)R^{d-1}\exp\left(-k(d)BR^d\right)$ with $k(d) = 2d^{-1}\pi^{d/2}/\Gamma(d/2)$.

The averaged survival probability of a particle starting at an arbitrary point in a system in d dimensions is given by:

$$\langle P(t)\rangle \simeq C(d)dk(d)\int_0^\infty R^{d-1}\exp\left(-c(d)D\frac{t}{R^2} - k(d)BR^d\right)dR. \tag{10.60}$$

This integral at large t can be easily evaluated by the Laplace method: since the extremum (maximum) of the exponent is reached at $R_0 = [2C(d)Dt/(dk(d)B)]^{1/(d+2)}$ we get that

$$\langle P(t) \rangle \simeq \exp\left(-\kappa(d) B^{\frac{2}{d+2}} D^{\frac{d}{d+2}} t^{\frac{d}{d+2}}\right), \qquad (10.61)$$

where $\kappa(d)$ is a number constant, depending on d, k, and C. Due to the fact that the survival probability is larger than $\langle P(t) \rangle$ (since in reality our imaginary sphere is not impenetrable) we have:

$$A(t) \geq A(0) \exp\left(-\kappa(d) B^{\frac{2}{d+2}} D^{\frac{d}{d+2}} t^{\frac{d}{d+2}}\right). \qquad (10.62)$$

Note that the overall decay pattern is not a simple exponential, but a much slower stretched exponential. Note moreover that this effect persists in any dimension. In $d = 3$ we have $A(t) \propto \exp(-const \cdot t^{-3/5})$ instead of $\exp(-t)$, and that the result in $d = 1$ is $A(t) \propto \exp(-const \cdot t^{-1/3})$ instead of $A(t) \propto \exp(-const \cdot t^{-1/2})$ proposed by using the time-dependent rate, Eq. (10.29). Let us stress that the slowing-down of reaction found by Vaks and Balagurov is a multiparticle effect (since it does not emerge within the two-particle approximation) and that it is a fluctuation effect. It would be absent for example in the case if B-particles would be ordered in a superlattice: in this case the radii of all spheres in Fig. 10.3 would be the same.

If both types of particles diffuse, the exponential character of the decay of A-concentration typical for scavenging is restored at longer times (Szabo, Zwanzig, and Agmon, 1988), however, in this case the overall kinetics of reaction depends on the details of the system (here, on both diffusion coefficients) in a nontrivial manner. The same can be said about applying mixing procedures (processes, other than diffusion, leading to changes in positions of B-particles) (Palszegi, Sokolov and Kauffmann 1998). Except for pure scavenging, there is hardly a situation in which the effective reaction rate is given by the Smoluchowski approximation.

10.4.2 Fluctuation-dominated kinetics in A + B → 0 Reaction

The A + B → 0 irreversible strictly bimolecular reaction is the one in which the fluctuation effects can (to some extent) be adequately described by the diffusion reaction scheme. This is due to the fact that these effects lead to the development of a large-scale structure (the word "development" here describes a process which is really similar to the development of latent image in a photographic process). Before we turn to the description within the reaction-diffusion scheme, we present a simple qualitative explanation of the effect within the scaling theory stemming from Kang and Redner (1984).

Let us consider the stoichiometric case, in which the initial concentrations of A and B are equal, $A = B = c(0)$, and the reactants are well-premixed, so that equilibrium Poissonian fluctuations of concentrations of both reactants establish.

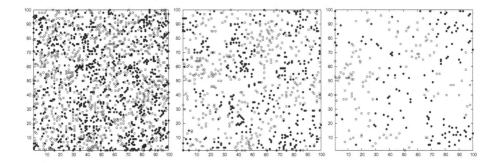

Fig. 10.4 Results of a simulation of an irreversible A + B → 0 reaction on a 100 × 100 lattice with periodic boundary conditions in both directions. Initially each site of the lattice is occupied other by A or by B; at $t = 0$ there are exactly 5000 particles of each sort. Particles perform random walks on the lattice (see Chapter 11) and react on contact. Multiple occupation of a site by particles of the same sort is allowed. The three panels of the figure correspond to times $t = 5$, $t = 20$ and $t = 100$ in units of a random walk's timestep. The A-particles are shown as empty circles, the B ones as filled circles. The formation of larger and larger domains occupied only by particles of the same sort gets evident in the course of time.

Let us start from the initial particle configuration and follow the destiny of A-particles within some given volume. During the time t the initial configuration gets mixed due to diffusion, the particles might look for their reaction partners within the volume L^d defined by the typical diffusion length $L \simeq \sqrt{Dt}$. However, due to Poissonian fluctuations the numbers of reaction partners within such volumes fluctuate: The numbers of A and B within the volume $V \simeq (Dt)^{d/2}$ might differ by $\Delta N \simeq N^{1/2} = (c(0)V)^{1/2} = c(0)^{1/2}(Dt)^{d/4}$. The particles which are not able to find reaction partners definitely survive. Their concentration will be $n = \Delta N/V = c(0)^{1/2}(Dt)^{-d/4}$. Since the particles' concentration at time t is definitely larger than n (because the particles which might have found a partner within the mixing volume must not necessarily have reacted) one has

$$c(t) \geq c(0)^{1/2}(Dt)^{-d/4}. \tag{10.63}$$

On fractal structures the dimension d in Eq. (10.63) has to be understood as a *spectral dimension* of the structure, see e.g. (Sokolov, 1986). Note that the slowing-down goes hand in hand with geometrical segregation of particles of different types. The slowing down has to do with the formation of large A- and B-rich domains (clusters), so that the reaction takes place only within thin reaction (mixing) zones on the boundaries of different clusters, as it is evident from the simulation results shown in Fig. 10.4. The slowing-down of the reaction in a stoichiometric AB-mixture was first theoretically predicted by Ovchinnikov and Zeldovich (1978), who also considered a reversible and a non-stoichiometric situation in three dimensions, and started from the standard chemical situation with well-premixed reactants. In 1983 the effect was independently rediscovered by Toussaint and Wilzcsek (1983), who discussed the matter-antimatter distribution in the Universe and mainly concentrated on the properties of matter-antimatter segregation, trying to explain how this evident fact may be brought into accordance with the assumption of the intrinsic matter-antimatter symmetry.

Note that putting down classical kinetic equations (pertinent to spatial dimension $d > 2$)

$$\frac{d}{dt}A(t) = -kAB$$
$$\frac{d}{dt}B(t) = -kAB \qquad (10.64)$$

under the initial condition $A(0) = B(0) = c(0)$ we get

$$c(t) = \frac{1}{c(0)^{-1} + kt} \qquad (10.65)$$

so that at longer times $c(t) \propto t^{-1}$, which would be a faster decay than the one given by Eq. (10.63). In dimensions $d = 1$ and 2 (and on fractal structures with spectral dimension lower than 2) the reaction rate coefficient is intrinsically time-dependent, $k(t) \propto t^{-d/2}$, so that at longer time the classical kinetic result would read $c(t) \propto t^{-d/2}$, which is anyhow a faster decay than the one predicted by Eq. (10.63). Equation (10.63) gives the true asymptotic behavior of the process in all dimensions lower than 4 (Bramson and Lebowitz, 1988), which is the upper critical dimension for this effect. This fluctuation effect has an essentially multiparticle nature, and cannot be described within the two-particle (Smoluchowski or analogue) approximations, however it is reproduced by some truncation schemes, see (Kuzovkov and Kotomin, 1988). Mixing procedures faster than diffusion destroy the effect, and, depending on the quality of mixing procedure, either restore classical kinetics or introduce new kinetics, which depends on the properties of the mixing procedure.

To understand the effect on the semiquantitative level and to assess the possible role of mixing, let us consider it within the reaction-diffusion (or reaction-diffusion-advection) scheme, where, again, the initial fluctuations are taken into account through the initial conditions. In what follows we assume the diffusion coefficients for A and B to be the same; this simplifies our considerations strongly, the qualitative behavior is the same in all cases. One has then

$$\frac{\partial}{\partial t}A(t) = D\Delta A - kAB$$
$$\frac{\partial}{\partial t}B(t) = D\Delta B - kAB\,. \qquad (10.66)$$

The next step is to change to new variables, the sum $S(t) = A(t) + B(t)$ and the difference $Q(t) = A(t) - B(t)$ of local concentrations, and to get the equations for these variables:

$$\frac{\partial}{\partial t}Q(t) = D\Delta Q \qquad (10.67)$$

$$\frac{\partial}{\partial t}S(t) = D\Delta S - \frac{k}{4}(S^2 - Q^2). \tag{10.68}$$

It is important to note that the equation for Q does not contain S and is a linear partial differential equation of the diffusion type, and the equation for S contains Q as a parameter.

In the limit of fast reaction ($k \to \infty$) we simply have $S^2 = Q^2$, which assumes $S = |Q|$ so that $\langle S \rangle = \langle |Q| \rangle$, the overall concentration S simply follows Q. The limit of fast diffusion is rather unrealistic, however it gives us the exact lower bound on the concentration, since one assumes that there are only A-particles in the domains where A-particles are in excess and only B-particles there where B's are in excess, so that in reality $\langle S \rangle \geq \langle |Q| \rangle$. However, the picture of segregated A and B particle is quite realistic: As already stated, numerical simulations clearly show that the system separates into A- and B-clusters, and the reaction takes place only in thin reaction zones on the clusters' boundaries. The geometry of clusters and their boundaries within the reaction-diffusion scheme (based on Eq. (10.68)) was first considered by Sokolov (1986). Havlin et al. (1993) have shown that the one-dimensional situation is vastly different. The review of the results (within the field-theoretical approach) can be found in Mattis and Glasser, 1998.

Let us now turn to the Eq. (10.67) which is rather simple to solve: since it has the form of a diffusion equation, the formal solution of this equation is given through the Green's function of the diffusion:

$$Q(\mathbf{x}, t) = \int d\mathbf{x}' G(\mathbf{x} - \mathbf{x}', t) Q(\mathbf{x}', 0) \tag{10.69}$$

where $Q(\mathbf{x}, 0)$ corresponds to the initial condition. We also note that for the initial conditions that do not show long-range correlations in space, the distribution of $Q(\mathbf{x}, t)$ tends to be Gaussian at longer times since it is a weighted sum of independent random increments. Let us now calculate the second moment of $Q(\mathbf{x}, t)$. Just like in the well-known Taylor-trick, we get

$$Q(\mathbf{x}, t) Q(\mathbf{y}, t) = \iint d\mathbf{x}' d\mathbf{y}' G(\mathbf{x} - \mathbf{x}', t) G(\mathbf{y} - \mathbf{y}', t) Q(\mathbf{x}', 0) Q(\mathbf{y}', 0). \tag{10.70}$$

Taking $\mathbf{x} = \mathbf{y}$ and averaging over the realizations of initial conditions we get

$$\langle Q^2(\mathbf{x}, t) \rangle = \iint d\mathbf{x}' d\mathbf{y}' G(\mathbf{x} - \mathbf{x}', t) G(\mathbf{x} - \mathbf{y}', t) \langle Q(\mathbf{x}', 0) Q(\mathbf{y}', 0) \rangle, \tag{10.71}$$

where we use the fact that the Green's function of the diffusion equation $G(\mathbf{x}, t) = (4Dt)^{-d/2} \exp(-\mathbf{x}^2/4Dt)$ is a nonrandom function of its arguments. For the Poissonian initial distribution of reactants the values of Q at different points are uncorrelated, so that, as it is easy to show $\langle Q(\mathbf{x}', 0) Q(\mathbf{y}', 0) \rangle = 2c(0)\delta(\mathbf{x}' - \mathbf{y}')$ and thus

$$\langle Q^2(\mathbf{x}, t) \rangle = 2c(0) \int d\mathbf{x}' G^2(\mathbf{x}', t). \tag{10.72}$$

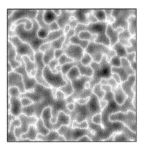  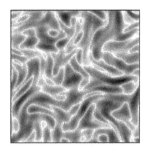

Fig. 10.5 Concentration of A or B particles (shown in greyscale) in the stoichiometrical reacting mixture, as described by Eq. (10.73). The structures evolved from homogeneous initial conditions with Poisson fluctuations of concentrations. Such a plot can be considered as a course-grained variant of the situation shown in Fig. 10.4. The left panel corresponds to the case of no mixing, the medium panel represents reaction under shear flow, the right panel is obtained under turbulent mixing procedure. Note the well-defined reaction zones at cluster boundaries (due to low reactant concentrations there they are seen as white bands).

The integration then gives $\langle Q^2(\mathbf{x}, t)\rangle = 2c(0) \int d\mathbf{x}' G^2(\mathbf{x}', t)$, i.e. $\langle Q^2(\mathbf{x}, t)\rangle = (8\pi)^{-d/2}(Dt)^{-d/2}$.

Note that in the case of a Gaussian distribution $\langle |Q|\rangle = \sqrt{2/\pi}\,\langle Q^2\rangle^{1/2}$, so that $c = S/2 \geq 2^{-d/2}(2\pi)^{-d/4}(Dt)^{-d/4}$.

The picture of clusters and reaction zones in a simple A + B → 0 reaction is given in Fig. 10.5, along with pictures obtained in numerical simulations of reaction-diffusion-advection scheme

$$\frac{\partial}{\partial t}A(t) - \mathbf{v}(\mathbf{r},t)\nabla A = D\Delta A - kAB$$
$$\frac{\partial}{\partial t}B(t) - \mathbf{v}(\mathbf{r},t)\nabla B = D\Delta B - kAB$$
(10.73)

with different mixing velocity fields, namely for a laminar shear flow and for a model turbulent-like random flow. The data for Fig. 10.5 are ones used in (Reigada et al., 1997).

Discussing the overall effect of mixing on the kinetics of A + B → 0 reaction we should note that (corresponding to the very effective mixing procedure, the one representing the continuous stage of the baker's transformation) is effective enough to restore classical kinetics at longer times, see (Sokolov and Blumen, 1991a, 1991b). The shear flow is effective in three dimensions but ineffective in 2d (Sokolov and Blumen, 1992).

Let us make a short note about the properties of mixing procedures and their effect on reaction kinetics. Reactions under mixing are very important topic in physical chemistry; however, the approximations used by chemists are typically very crude. The reason for this is that in many cases the quality of mixing does not really matter. The effective procedures based on dilatation flow like baker transformation (technically realized in so-called static mixers) or horseshoe transformation (exactly the one a smith performs forging iron) are applied only in the situations where the medium to be mixed is very viscous and the energetic aspect of performing mixing comes to the foreground. The work of smith consists in flattening a piece of iron, which let its surface oxidize in contact with the atmospheric air, and then repeatedly folding it to mix the oxygen up to small scales letting it oxidize the excess carbon. However, when the possibility appeared to increase temperature up to the melting point of pure iron, then the effective (but energy consuming) horseshoe mixing was changed for notoriously ineffective shear-flow procedure (puddle iron), for convective procedures or for blowing oxygen through the iron in the Bessemer process. More about properties of laminar mixing procedures can be found in the book by Ottino (1989). The turbulent procedures are effective on the viscous scales (where they correspond to chaotic laminar procedure), but have practically the same (low) efficiency as a shear flow on the inertial scale, where the persistent eddy structures of the flow prevent fast dissolution of long-range gradients.

If the reader wants to learn more about mixing, one may recommend to look in the geophysical literature: The structure and the compositions of the minerals in the earth crust strongly depends on mixing in the upper mantle of the earth, so that quantitative theories were pursued and model experiments have been performed in order to understand the properties of mixing processes taking place on geological time scales, see (Sokolov and Blumen, 1991c) for further references.

We note that fluctuation effects related to considered here are pertinent to the A + B → 0 reaction under other conditions than simple stoichiometric pre-mixed system under irreversible reaction. Thus, in the case of a reversible reaction A + B ⇌ C the equilibrium concentration C_{eq} stays unchanged, but approaching it gets very slow. Here, several approximate solutions have been obtained before it was rigorously shown that for a sufficiently long time t, the approach to the equilibrium takes the form of a power law $C - C_{eq} \simeq At^{-d/2}$, for any spatial dimension d; (Rey and Cardy, 1999, Gopich and Szabo, 2002). The electrostatic interaction of charges in plasmas prevents cluster formation, but if the charges are screened, the short-range interaction is not enough to fully neutralize the effect, see (Oshanin et al., 1996).

For the case of geminate creation of particles in close pairs (a situation typical for creation of radiation-induced defects in solids) the reaction shows memory effects: the kinetics of defect healing depends not only on defect concentration but on the time protocol of irradiation (Sokolov and Blumen, 1994). However, the more important property of such recombination reaction is the interaction of the defects through mechanical stress field. The clusterization of vacancies then may lead to formation of voids making the material porous and brittle.

The situations other than simple well-premixed case (layered systems) were considered in (Muzzio and Ottino 1989, 1990; Sokolov and Blumen, 1990; Sancho et al., 1996), more complex cases were discussed in (Monson and Kopelman, 2000), and experimentally realized by Monson and Kopelman (2004).

The possibility to describe the effects within the reaction-diffusion scheme in higher dimensions has to do with the fact that the structures formed by the reaction in this case are macroscopic, i.e. have the typical size which is at all times much larger than both the interparticle distance $\lambda(t) = c(t)^{-1/d}$ and the reaction radius a. Whenever this is not the case, the reaction-diffusion, or reaction diffusion-advection schemes fail, often only on the quantitative level, but sometimes also qualitatively. We also must stress that these schemes are intrinsically inappropriate in one-dimension and might be rather inaccurate in two dimensions.

10.4.3 Autocatalytic reactions and front propagation phenomena

Let us consider another very important reaction, the simple quadratic autocatalytic one of the type A+B → 2B which describes, say the infection propagation (Murray, 1990). Let us assume that the concentration of A is initially constant $A = c$; a B-particle (a source of infection) is introduced in the system at some instant of time. The particles are assumed to diffuse and to react on contact. The reaction (which is essentially renaming of A-particles into B-ones) corresponds to the propagation of the reaction front separating the B-region (infected zone) from the still-uninfected A-population. In what follows we discuss several facets of this reaction. The aim of our discussion is to show, that it is not only the mathematical analysis of the equations, but the physical analysis of the situation, which is necessary to obtain the correct solution, see (Panja, 2004) for more details.

The classical reaction-diffusion scheme for this reaction reads

$$\frac{\partial A}{\partial t} = D\Delta A - kAB,$$
$$\frac{\partial B}{\partial t} = D\Delta B + kAB. \tag{10.74}$$

Noting that $c = A + B$ is the integral of motion of this system of equations, one can reduce these two equations to only one equation for the concentration B:

$$\frac{\partial A}{\partial t} = D\Delta B + kB(c - B). \tag{10.75}$$

This equation was first put forward by Fisher (1937), who described the propagation of a favorable gene in a population, and mathematically investigated by Kolmogorov, Petrovski and Piskunov (1937), and thus is often abbreviated as an FKPP-equation. Considerable information about the properties of front solutions to this equation can be found in the book by Murray (1990).

The initial problem was one-dimensional (the coast-line), however, even in 2 and 3 dimensions the fronts from a point-source will grow in radius and eventually flatten, so that the discussion will concentrate on the propagation of a flat front. The mathematical treatment of the equation typically concentrates on the stability analysis of the propagating front solution of a constant form; indeed, the numerical solution of Eq. (10.75) shows that the initial concentrated (localized) perturbation of concentration eventually develops into a front propagating with a constant velocity

v, say, along the x-direction. The next step of standard discussion then consists in changing to a comoving frame, in which the front's form stays time-independent. This form, given by $a(t)$ then fulfills the equation

$$-vb' = Db'' + kb(c - b) \qquad (10.76)$$

where the primes denote the differentiation with respect to the spatial coordinate in the comoving frame. Whatever mathematical tools are applied, the following step will be to consider the asymptotic behavior of the front in the far region which is termed as its leading edge. The comprehensive mathematical information about the propagation of such "pulled fronts" can be found in (van Saarlos, 2003). Physicists prefer to speak about the fronts propagating into the unstable state (the A-phase is unstable in contact with the B-one).

In the leading edge of a front the concentration of B particles is low, so that one can linearize Eq. (10.76) which then reads

$$-vb' = Db'' + kbc. \qquad (10.77)$$

This linear equation has an exponential solution corresponding to the decay of B-concentration far into A-region: assuming the form of the solution $b = b_0 \exp(-\lambda x)$ we get for λ

$$v\lambda = D\lambda^2 + kc \qquad (10.78)$$

which equation has solutions

$$\lambda_{1,2} = \frac{v}{2D} \pm \sqrt{\frac{v^2}{4D} - \frac{kc}{D}}. \qquad (10.79)$$

The important point here however is that concentration has to be real and positive, which means that there exists the minimal propagation velocity of a stable front, namely $v_{\min} = 2\sqrt{kc/D}$. The value of $\lambda = v/2D$ defines the length scale $W = \sqrt{D/kc}$ which turns out to be the characteristic front's width. The aims of further analysis was to explain why this is exactly the velocity the typical front will assume, the fact known as the marginal stability, which is clearly demonstrated also by numerical solutions of Eq. (10.75). However, if one turns to the numerical simulations of the underlying process, one immediately observes, that there is nothing special about the velocity $v_{\min} = 2\sqrt{kc/D}$: the velocities which are both larger and smaller than this one can be observed in the simulations, $v_{\min}$ is essentially the velocity assumed under some limiting conditions, which will get clear in what follows.

In numerical simulations, the velocity of the front can be defined using different procedures. Three of them were discussed in the work (Mai et al., 1996):

1) One can fix the time t and calculate the average concentration profile of the front from several independent runs. The front's position $x(t)$ is then defined as the point where $A(x(t)) = C/2$ and the velocity is $v = x(t)/t$. This method does

not allow to define the front velocity in each realization separately. This is rendered possible by the following two definitions:

2) Another possibility is given through the total number $N_A(t)$ of A-particles at time t: We then set $x(t) = N_A(t)/c$.

3) One determines the position of the rightmost A-particle (RAP) in each realization and calculates v from $x(t) = x_{RAP}(t)$, again as $v_{RAP}(t) = x_{RAP}(t)/t$.

Different procedures imply different types of averaging: The velocities obtained by 1) are already averaged over all realizations, the result given by 2) is in some sense preaveraged; in method 3) v_{RAP} depends on the specific realization. The mean velocities obtained through these three procedures coincide, but their fluctuation properties differ. It is interesting to note that calculating the front's positions by methods 1 and 2 and then calculating the front's form with respect to the corresponding positions leads to the forms drastically different from the ones predicted by FKPP equation: They are strongly broadened by the fluctuation effects. Using the third method we obtain the fronts whose form resembles the predictions of FKPP in the middle part, however, it is clear that the front does not possess any "leading edge": there are no B-particles to the right of the rightmost B.

First, let us stress, that the purely one-dimensional situation (under assumption of reaction on contact with probability 1) is very specific (due to the absence of time-independent reaction rate k), so that the front's velocity in this case is proportional to c and not to $\sqrt{c}$, i.e. is much smaller for small concentrations (Mai, 1996). The qualitative explanation of this behavior is as follows: In 1d, under the assumption of immediate reaction on a contact, the front is essentially abrupt: If we concentrate on the position of the rightmost B-particle, we see that to the right of it only A-particles are present, while to the left of it only the B-ones are situated. Let us assume that the typical distance l between the rightmost B and the leftmost A-particle is of the order of c^{-1} (there is no other relevant length scale in our system, however, due to the fluctuation effects, this distance is not exactly c^{-1} (Warren et al., 2001)). The front moves over this distance during the time $\tau \propto l^2/D = 1/c^2 D$, which is necessary for A and B particles on the both sides of the gap to diffuse towards each other. The velocity of the front is thus of the order of $v \propto l/\tau = cD$. This explanation is somewhat oversimplified, but qualitatively correct. Assuming reaction taking place with probability p on contact one will see that the dependence on c changes and gets close to $v \propto \sqrt{c}$ for $p \to 0$ (Mai et al., 2000). This probability p plays essentially the role of the effective reaction rate. The introduction of such a probability describes a gradual transition to a three-dimensional situation, in which the front propagates in a capillary of a finite cross-section, where we now consider the projection of the particles' concentration onto the axis of the capillary.

For situations with relatively large p and in dilute situations (very small c) the number of particles within the front region is very small, there are no particles outside of this region. In this case the condition that the far edge of the front (which anyhow contains no particles) does not show the oscillatory behavior of the concentration does not give a right description of the propagation velocity.

On the other hand, for a given, fixed c decreasing p makes the front broader, and the number of particles in the front region larger, facts which let $v(p)$ approach $v_{\min}$. Now for large c one finds that $v(p)$ approaches $v_{\min}$ from above, see (Kessler, 1998, Brunet and Derrida, 1997, 2001). On the other hand when c is low, the propagation velocity $v(p)$ can be considerably less than $v_{\min}$; it only tends to $v_{\min}$ when p tends to zero approaching it from below.

The aim of our discussion was to show that continuous-medium schemes based on reaction-diffusion equations might lead to solutions which are far from reality. The corresponding schemes are based on the assumption of strong scale separation (the one that the typical scale of the concentration gradients in the system is much larger than both the reaction radius and the interparticle distance). If it is not the case, the best thing is to rely on explicit numerical simulations of reactions.

Chapter 11
Random Walk Approaches

The diffusion behavior governing fluctuations close to equilibrium and the time evolution of many nonequilibrium systems is ubiquitous, and the description of this diffusive behavior within the Langevin or Fokker–Planck schemes is a commonplace. However, as already said, these approaches are based on extrapolating mesoscopic type of behavior to microscopic scales, which is not always advantageous. It might be reasonable to start from a simple microscopic kinematic model (the one very close to the multiple scattering picture proposed by P. Drude in his electronic theory of metals), and to build the theory along slightly different lines. Although the approaches considered in this chapter are very close in idea to the ones proposed by Einstein and Smoluchowski, they are not always even mentioned in books on nonequilibrium thermodynamics. However, random walks approaches are one of the methods of the widest use at least in the discussion of simplified nonequilibrium models, and give clues to understanding of such complex phenomena as dispersive transport in amorphous solids or aging in glass-like materials. The versatility of random walk models and the possibility to adapt them to a variety of essentially non-Markovian situations make them an extremely valuable tool in the investigation of kinetic phenomena in the cases when the approaches based on the Markovian dynamics fail. The weakness of the approach in its classical form, put forwards by Elliot Montroll and George Weiss, is that it strongly relies on the homogeneity of the system: It is not always easy to adapt this approach to the case of motion in the external fields; however, this is sometimes possible and leads to beautiful mathematics (fractional Fokker–Planck equations). General information about the random walk approaches to different physical problems can be bound in the review articles by Haus and Kehr (1987), by Bouchaud and Georges (1990) and by Isichenko (1992).

11.1 The random walk approach to transport processes

Parallel to the Einstein's treatment of the Brownian motion, let us considers the motion of a particle as a sequence of independent steps. Each of these steps takes some time τ and brings a particle some distance s away from its initial position. The step's duration τ and length s may be correlated or uncorrelated and are taken from some probability distribution $\psi(\tau, s)$. At the beginning we assume the step lengths and times to be independent on the particle's position and on the actual time: As one can anticipate, the random walk approach reduces our transport problem to the mathematical problem of the distributions of the sums of independent, equally distributed increments. For example, the situation discussed in the Einstein's work on Brownian motion is really very close to a genuine random walk picture put forward by Bachelier, Pearson, and Rayleigh, see Chapter 1.

To stress the connection of random walks with our previous approaches let us return to the Einstein's discussion of the Brownian motion as following from the independence of the particles' motion and from the two additional postulates, namely the ones that the displacements of a particle during two subsequent intervals of the duration τ are independent and that the distribution of these displacements, $\phi(s)$, possesses the finite dispersion. The displacement of the particle can be considered as a result of many small, independent, equally distributed steps. The displacement of the particle after n such steps is:

$$X_n = s_1 + s_2 + \cdots + s_n. \tag{11.1}$$

In order to find the distribution of the sum of independent random variables, a useful instrument is provided by *characteristic functions*: A characteristic function of a probability distribution given by the probability density $p(x)$ is defined as the mathematical expectation of e^{ikx}, i.e. it is simply the Fourier-transform of $p(x)$. Note that the Fourier-transform can be inverted, so that the density of a probability distribution can be found through its characteristic function via the inverse Fourier-transform (under the well-known conditions and restrictions). Note also that since

$$\frac{d^n}{dk^n} f(k) = \int_{-\infty}^{\infty} (ix)^n e^{ikx} p(x) dx, \tag{11.2}$$

the characteristic function of a distribution is a generating function of its moments: if the n-the derivative of the characteristic function exists then

$$M_n = \int x^n p(x) dx = (-i)^n \left. \frac{d^n}{dk^n} f(k) \right|_{k=0}. \tag{11.3}$$

Now let us consider the characteristic function of the sum of n independent random variables. Let us start from $n = 2$. The probability $p(y)$ that the sum of two independent random variables x_1 and x_2 having the probability densities $p_1(x_1)$

and $p_2(x_2)$ is equal to y is given by

$$p(y) = \int dx_1 p_1(x_1) p_2(y - x_1) \tag{11.4}$$

(i.e. is a convolution of the probability distributions p_1 and p_2). To see this it is enough to note that for any value of x_1 the value of the sum $y = x_1 + x_2$ is attained for the well-prescribed value of $x_2 = y - x_1$, with the probability $p_2(y - x_1)dx_1$, which has to weighted over all possible values of x_1. A characteristic function of $p(y)$ is, of course, the Fourier-transform of this convolution, i.e. the product of the corresponding characteristic functions:

$$f(y) = \int e^{iky} p(y) dy = f_1(k) \cdot f_2(k). \tag{11.5}$$

For the sum of n independent, identically distributed variables we can proceed recursively and show that the characteristic function of the distribution of a random variable X given by Eq. (11.1) will be

$$f_X(k) = f_1(k) \cdot f_2(k) \cdot \ldots \cdot f_n(k). \tag{11.6}$$

If all $f_i(k) = f(k)$ are the same, we simply have

$$f_X(k) = [f(k)]^n. \tag{11.7}$$

Thus, the characteristic function of the distribution of the sum of n identically distributed, independent variables is the n-th power of the characteristic function of the distribution of each of them. We also note that if the corresponding variable s is nonnegative, the Laplace-transform of the the probability density, $\hat{L}\{p(X)\}$ will have the same property, namely,

$$\hat{L}\{p_X(X)\} = \left[\hat{L}\{p_s(s)\}\right]^n. \tag{11.8}$$

Let us remind that the Laplace transform is defined as

$$\hat{f}(u) \equiv \hat{L}\{f(y)\} \equiv \int_0^\infty e^{-uy} f(y) dy. \tag{11.9}$$

Let us first discuss the genuine Einstein's (or Pearson's) problem. The random walk approach allows us for finding the probability distribution of the particle's position $p(x)$ immediately through $\phi(s)$, without the intermediate step of putting down the corresponding differential equation for it. Let us assume that $\phi(s)$ possesses at least two finite moments. Then one has

$$\hat{\phi}(k) = 1 + M_1 k + \frac{1}{2} M_2 k^2 + o(k^2). \tag{11.10}$$

After n steps we then have

$$\hat{f}(k) = \hat{\phi}(k)^n = \left[1 + iM_1 k - \frac{1}{2} M_2 k^2 + o(k^2)\right]^n. \tag{11.11}$$

Let us now concentrate on the case $n \to \infty$ and use the well-known limiting relation
$$(1+x)^{1/x} \to e \tag{11.12}$$
for $x \to 0$. One thus has for k small:
$$f(k) \simeq \left\{[1+(\phi(k)-1)]^{1/(\phi(k)-1)}\right\}^{(\phi(k)-1)n} \to e^{n(\phi(k)-1)}. \tag{11.13}$$
Thus, for k small enough we have
$$f(k) \simeq \exp\left[n\left(iM_1 k - \frac{1}{2}M_2 k^2\right)\right] \tag{11.14}$$
which is a characteristic function of a Gaussian distribution
$$p(x) = \frac{1}{\sqrt{2\pi n\sigma}} \exp\left[-\frac{(x-nM_1)^2}{2n\sigma^2}\right] \tag{11.15}$$
with $\sigma = M_2 - M_1^2$. Since in the Einstein's treatment the distribution $\phi(s)$ was supposed to be symmetric, $M_1 = 0$ and
$$p(x) = \frac{1}{\sqrt{2\pi n\sigma}} \exp\left[-\frac{x^2}{2n\sigma^2}\right]. \tag{11.16}$$
Again, as in the Einstein's approach we can take $n = [t/\tau]$, where $[\ldots]$ denotes the whole part of the number. For t large enough we can simply take $n = t/\tau$. Inserting this expression in Eq. (11.16) we get
$$p(x,t) = \frac{1}{\sqrt{4\pi Dt}} \exp\left[-\frac{x^2}{4Dt}\right] \tag{11.17}$$
with $D = \sigma^2/2\tau$, as already discussed. Due to the assumed homogeneity in space and time we can also immediately put down the transition probability (the Green's function)
$$p(x,t|x_0,t_0) = \frac{1}{\sqrt{4\pi D(t-t_0)}} \exp\left[-\frac{(x-x_0)^2}{4D(t-t_0)}\right]. \tag{11.18}$$

Of course, what we did up to now, is nothing else than a hand waving derivation of the Central Limit Theorem first stated by Gauss and Laplace; its full mathematical formulation is due to Lévy and Cramér; the criteria of convergence and the possibility to drop out an $o(k^2)$-term in Eq. (11.11) are given by the Berry–Esseen theorem and its generalizations (see Feller, 1991).

11.1.1 Random walks on lattices

Let us discuss random walks on lattices, which are often considered as *the* random walk model. A vast mathematical literature is devoted to precisely this subject; the already mentioned books by Spitzer (1976), by Feller (1991) as well as one by Georges Weiss (1994) are the examples. However, also physically the model is quite reasonable for description of, say, energy transfer in crystals (say luminescent reactions in molecular crystals as exemplified by trapping, see (Montroll and Weiss 1965; Montroll 1969). The model follows by allowing the particle to jump only to next-neighboring lattice sites, i.e. to fixing the function $\phi(x) = C^{-1} \sum_j \delta(x - \mathbf{r}_j)$, where $j = 1, \ldots, C$ numbers the nearest neighbors of the corresponding site and C is the coordination number of the lattice.

Of course, using the general scheme of a Master equation one can immediately describe the particles' displacements either exactly or within the Fokker–Planck approximation. However, there are other approaches which might be more suitable for calculation of some special properties. We concentrate here first on random walks with discrete time (fixed time needed to perform a step). The continuous-time results follow then by the subordination procedure, as discussed in the next section.

Many useful results for random walks are obtained using the generating function formalism. The idea here is as follows. Let $P_N(\mathbf{r})$ be the probability for a particle starting at 0, to be at the point $\mathbf{r}$ after N steps. The values of $P_N(\mathbf{r})$ for given $\mathbf{r}$ form a number sequence. Let us represent this sequence as a sequence of Taylor-coefficients of a function

$$P(\mathbf{r}; z) = \sum_{N=0}^{\infty} P_N(\mathbf{r}) z^N \qquad (11.19)$$

which is said to be a generating function for the sequence $P_N(\mathbf{r})$. If the generation function exists, it contains the whole information about $P_N(\mathbf{r})$, and allows for obtaining many results in a closed, analytical form.

Let us first consider a general situation. Imagine that $f(z) = \sum_{N=0}^{\infty} f_N z^N$ is a generating function of a sequence f_N (here we assume that the series converge at least in some range of z). The z-transformation leading from the sequence f_N to its generating function is a discrete analogue of the Laplace-transform: taking $z = \exp(u)$, one recognizes in Eq. (11.19) an integral sum for the corresponding Laplace integral. The z-transformation shears with the Laplace transform some important properties: thus, it is linear, $\mathcal{Z}(\{f_n\} + \{g_n\}) = \mathcal{Z}\{f_n\} + \mathcal{Z}\{g_n\}$, and the $\mathcal{Z}$-transform of a discrete convolution of two sequences $\{f_n\}$ and $\{g_n\}$, i.e. of a sequence $\{h_n\} = \{f_n\} * \{g_n\}$ with the elements $h_n = \sum_{i=0}^{n} f_i g_{n-i}$ is a product of the corresponding generating functions: $h(z) = f(z)g(z)$. Moreover, under some regularity conditions the generating function can even be approximated by a Laplace transform.

Now let us say a few words about the back transformation. Of course we can use the tables, or approximate the inverse z-transform by an inverse Laplace transform. However, for a special class of functions we meet in what follows (i.e. those which asymptotically behave as power-laws), we will not need it. The back transform is essentially given by so-called Tauberian theorems, see (Feller 1991, Vol. 2,

Chap. XIII, Section 5).

Imagine $g(y) = \sum_{N=0}^{\infty} g_n e^{-yn}$ with $g_n > 0$ (note that $y = \ln z$). Let us consider the functions which behave essentially as power laws, so that for y small $g(y) = y^{-\gamma} L(1/y)$ and $L(x)$ is a *slowly changing function* of x, i.e. $\lim_{x\to\infty} \frac{L(Cx)}{L(x)} = 1$ for any positive constant C. An example of a slowly changing function are not only functions tending to a finite limit when $x \to \infty$, but also, say, $\log(x)$ or any powers of the logarithm. Let us introduce now a new function $\varphi(y)$ so that $g(y) = \varphi(1/y)$. Then the partial sum of the series can be approximated as

$$g_1 + g_2 + \cdots + g_n \simeq \frac{\varphi(n)}{\Gamma(\gamma+1)}. \tag{11.20}$$

If the sequence $\{g_n\}$ is monotonous,

$$g_n \simeq \frac{\varphi'(n)}{\Gamma(\gamma+1)}. \tag{11.21}$$

The quality of approximation is typically very good.

Having the instrument, we can start discussing some important results connected with the first passage probabilities and with the overall number of visited sites.

Let us discuss for example the probability $F_N(\mathbf{r})$ of visiting the site $\mathbf{r}$ for the first time at step N. We can use here the discrete analogue of the renewal approach based on the relation between $P_N(\mathbf{r})$ and $F_N(\mathbf{r})$. Let us consider a random walk starting at $\mathbf{0}$. The probability to be at step N at site $\mathbf{r}$ is then given by:

$$P_N(\mathbf{r}) = \delta_{N,0}\delta_{\mathbf{r},\mathbf{0}} + \sum_{j=0}^{N} P_{N-j}(\mathbf{0}) F_j(\mathbf{r}). \tag{11.22}$$

The particle starting from $\mathbf{0}$ and being at $\mathbf{r}$ after N steps might have first visited the site $\mathbf{r}$ for the first time at step j and then returned to it. The probability of such return after $N-j$ steps is $P_{N-j}(\mathbf{0})$ due to the homogeneity of the lattice. Applying the z-transformation to this equation we get: $P(\mathbf{r}, z) = \delta_{\mathbf{r},\mathbf{0}} + P(\mathbf{0}, z) F(\mathbf{r}, z)$, from which

$$F(\mathbf{r}; z) = \frac{P(\mathbf{r}; z) - \delta_{\mathbf{r},\mathbf{0}}}{P(\mathbf{0}; z)}. \tag{11.23}$$

The overall probability to visit the site $\mathbf{r}$ is thus $F(\mathbf{r}) = \sum_{n=0}^{\infty} F(\mathbf{r}, N)$. Let us calculate for example the overall return probability $F(\mathbf{0})$. Note that $F(\mathbf{r}) = F(\mathbf{r}; 1)$, and thus

$$F(\mathbf{0}) = 1 - \frac{1}{P(\mathbf{0}; 1)}. \tag{11.24}$$

Thus, the random walk returns to the origin with probability 1 if $P(\mathbf{0}; 1)$ diverges, and the return probability is finite if $P(\mathbf{0}; 1)$ is finite. Now, using the characteristic function of the random walk, $f_N(\mathbf{k}) = \lambda^n(\mathbf{k})$, $P_N(\mathbf{0})$ can be easily found by back Fourier-transform: $P_N(\mathbf{0}) = \left(\frac{1}{2\pi}\right)^d \int_\Omega f_N(\mathbf{k}) d\mathbf{k}$ (integration over the Wigner-Seitz cell of the lattice), so that $P(\mathbf{0}; z)$ may be obtained by integration the geometric series. Interchanging integration and summation we obtain $P(\mathbf{0}; z) = \left(\frac{1}{2\pi}\right)^d \int_\Omega \sum_{i=0}^\infty f_N(\mathbf{k}) d\mathbf{k} = \left(\frac{1}{2\pi}\right)^d \int_\Omega \sum_{i=0}^\infty [f(\mathbf{k})]^N d\mathbf{k}$, i.e. arrive at

$$P(\mathbf{0}; z) = \left(\frac{1}{2\pi}\right)^d \int_\Omega \frac{d\mathbf{k}}{1 - z f(\mathbf{k})}. \tag{11.25}$$

The integral for $P(\mathbf{0}; 1)$ can only diverge if for some $\mathbf{k}$-vectors one has $f(\mathbf{k}) = 1$. For example, for hypercubic lattices $f(\mathbf{k}) = \frac{1}{d} \sum_{j=1}^d \cos(k_j a)$ and divergence can take place only for $k = 0$. For small k, $f(\mathbf{k}) \approx 1 - \frac{1}{2} a^2 k^2 + \ldots$, so that

$$P(\mathbf{0}; 1) \simeq \left(\frac{1}{2\pi}\right)^d \int_\Omega \frac{k^2 dk}{1 - (1 - \frac{1}{2} a^2 k^2)} \simeq 2 \left(\frac{1}{2\pi}\right)^d a^{-2} \int_\Omega k^{d-3} dk. \tag{11.26}$$

The corresponding integral diverges for $d = 1$ and 2: the simple random walks in one and two dimensions are recurrent. In $d = 3$ the integral converges, so that the random walk does not necessarily return to the origin (is transient). Some known return probabilities are:

$$F(\mathbf{0}) = \begin{cases} 0.3405 \text{ for SC lattice} \\ 0.2822 \text{ for BCC lattice} \\ 0.2563 \text{ for FCC lattice} \end{cases} \tag{11.27}$$

(Here SC denotes the simple cubic lattice, and BCC and FCC the body-centered and the face-centered cubic lattices, respectively). A more important property of the lattice random walk is the number of different sites visited. Indeed, it is a property which is intimately connected with the rates of diffusion-controlled reactions on such lattices. Let $\langle S_N \rangle$ be the mean number of different sites visited by a particle A. Let moreover consider a scavenging reaction where single mobile A-particle removes immobile B's met on its way. The mean number of B removed during N steps is connected with $\langle S_N \rangle$ via $N_B = c_B \langle S_N \rangle$ where c_B is the concentration of B (a probability that a lattice site is occupied by B). Thus, the rate of scavenging reaction on a lattice is proportional to the increase in S_N per unit time. Of course, the corresponding picture can be translated to the continuous limit and gives rise to the visited-volume approach to chemical reactions (stemming from Montroll and Weiss) as opposed to the Smoluchowski approach.

Let us now calculate $\langle S_N \rangle$ following Dvoretzky and Erdös (1951). One first notes that

$$\langle S_N \rangle = 1 + \sum_{j=1}^N \Delta_j \tag{11.28}$$

where Δ_j is the mean number of the sites visited for the first time on step j: $\Delta_j = \sum_\mathbf{r} F(\mathbf{r}, j)$. For $\Delta(z)$, the generating function of Δ_j, we thus have

$$\Delta(z) = \frac{z}{(1-z) P(\mathbf{0}; z)} \tag{11.29}$$

(Note that $F(\mathbf{r}; z) = P(\mathbf{r}; z)/P(\mathbf{0}; z) - \delta_{\mathbf{r},\mathbf{0}}/P(\mathbf{0}; z)$ and that $\sum_{\mathbf{r}} P(\mathbf{r}; z) = 1 + z + z^2 + \cdots = 1/(1-z)$ since $\sum_{\mathbf{r}} P(\mathbf{r}, N) = 1$). The generating function of $\langle S_N \rangle - 1$ thus equals to

$$\frac{z}{(1-z)^2 P(\mathbf{0}; z)}. \tag{11.30}$$

The back transform gives (for $N \gg 1$):

$$\langle S_N \rangle \simeq \begin{cases} \sqrt{\frac{8}{\pi} N} & \text{in } d = 1 \\ \pi N / \ln N & \text{in } d = 2 \\ N / P(\mathbf{0}; 1) & \text{in } d = 3 \end{cases} . \tag{11.31}$$

In $d = 1$ and 2 the number of different visited places grows slower than N, which is a consequence of the fact that each of them is visited repeatedly. On the other hand, in $d = 3$ the number of different visited places grows as N: the sites visited are visited only once or a few times.

These results are also of great importance for the continuous-time lattice random walks, since the subordination transformation, see Section 11.1.2 makes it possible to translate physical time t into N, see the next section.

The Vaks–Balagurov slowing down in trapping can also be reproduced in the approach and has to do with fluctuations in S_N, which are mirrored by the higher moments of this quantity. We refrain here from the detailed discussion of the mathematical approach to the problem.

11.1.2 The continuous-time random walks (CTRW)

The situation of the fixed step time τ corresponds to the so-called simple random walks. This time may however be itself a random variable. For example, the random walk may result from a series of scattering events (in which case the time τ and the displacement s will be strongly correlated). Another situation is exactly the one we discuss in this paragraph: the particle is trapped in some bounded state. From time to time it is released (due to the thermal excitation) and makes a random motion until it gets trapped again. This is exactly the situation Scher and Montroll confronted with when describing the transport in disordered semiconductors (Scher and Montroll, 1975). Having this situation in mind, we shall consider s and τ as independent random variables*.

Thus, let us consider $\psi(\tau, s)$ as a product of the two functions, $\psi(\tau, s) = \psi(\tau)\phi(s)$. The spatial aspect of the problem corresponds to a simple random walk, however, the number of steps now fluctuates: the number of steps performed up to

---

*Although we use the one-dimensional notation here, this decoupling is not exactly what happens in a real one-dimensional system. One may much more consider the situation as a projection of a three-dimensional motion on the x-axis. The genuine one-dimensional situation with its additional correlations introduced by geometrical restrictions is considered e.g. by Bouchaud and Georges (1990).

the time τ is no more the whole part of $t/\bar{\tau}$, but may in principle take any value from 0 to ∞. We note that such a situation corresponds to the case of subordinated Markovian processes: A discrete-time Markovian process X_n, a simple random walk, depends on its number of steps n, which may be called the operational time of the process. The operational time itself is a random process with positive increments which depends on the physical time t. A resulting process may be Markovian or not, depending on the properties of $\psi(\tau)$.

Now, let us suppose that the probability to arrive at point x after n steps is known and is given by a probability density $p_n(x)$. In order to obtain the probability to be at x at time t we have to average $p_n(x)$ over the probability distribution $\chi_n(t)$ to make exactly n steps up to the time t:

$$p(x,t) = \sum_{n=0}^{\infty} p_n(x)\chi_n(t). \tag{11.32}$$

The probability density $p_n(x)$ is known and is given by its characteristic function

$$f(k) = \tilde{\phi}(k)^n. \tag{11.33}$$

Our task is now to find the distribution $\chi_n(t)$ of making exactly n steps up to t. We note that the step times τ are assumed to be independent and identically distributed. The probability to make no steps after beginning is

$$\chi_0(t) = 1 - F_1(t) = 1 - \int_0^t \psi_1(\tau)d\tau, \tag{11.34}$$

where $\psi_1(\tau)$ is the probability density to make the first step during the time τ, and $F_1(t)$ is the corresponding cumulative distribution function. We have to reserve the possibility that the waiting time distribution for the first step may differ from ones of the subsequent steps (Tunaley, 1974); this gives us a clue for understanding some aging phenomena in systems showing the glass-like dynamics (*vide infta*). The probability to make no steps during time t after a step was made is

$$\chi(t) = 1 - F(t) = 1 - \int_0^t \psi(\tau)d\tau. \tag{11.35}$$

Now, the probability that exactly one step was made up to the time t is the one that the first step followed at some time τ and afterwards no steps followed:

$$\chi_1(t) = \int_0^t \psi_1(\tau)\chi(t-\tau)d\tau. \tag{11.36}$$

The probability to make exactly 2 steps is

$$\chi_2(t) = \int_0^t \int_0^t \psi_1(\tau_1)\psi(\tau_2)\chi(t-\tau_1-\tau_2)d\tau_1 d\tau_2, \tag{11.37}$$

etc., so that

$$\chi_n(t) = \int_0^t \cdots \int_0^t \psi_1(\tau_1)\psi(\tau_2)\cdots\psi(\tau_n)\chi(t-\tau_1-\cdots\tau_n)d\tau_1\cdots d\tau_n. \quad (11.38)$$

A multiple convolution structure of these expressions leads to simple structures under the Laplace transform:

$$\hat{\chi}_0(u) = \frac{1-\hat{\psi}_1(u)}{u}$$

$$\cdots$$

$$\hat{\chi}_n(u) = \hat{\psi}_1(u)\hat{\psi}^{n-1}(u)\frac{1-\hat{\psi}(u)}{u}, \quad (11.39)$$

see Blumen et al., 1986. The Fourier–Laplace-transform of Eq. (11.32) thus reads:

$$\hat{f}(k,u) = \sum_{n=0}^{\infty} \phi(k)^n \hat{\chi}_n(u) \quad (11.40)$$

$$= \frac{1-\hat{\psi}_1(u)}{u} + \frac{1-\hat{\psi}(u)}{u}\frac{\hat{\psi}_1(u)}{\hat{\psi}(u)}\sum_{n=1}^{\infty}\left[\phi(k)\hat{\psi}(u)\right]^n. \quad (11.41)$$

In the second term one easily recognizes the geometric series, so that the closed form for the Fourier–Laplace-transform follows:

$$\hat{f}(k,u) = \frac{1-\hat{\psi}_1(u)}{u} + \frac{1-\hat{\psi}(u)}{u}\frac{\phi(k)\hat{\psi}_1(u)}{1-\phi(k)\hat{\psi}(u)}. \quad (11.42)$$

In the case when $\psi_1(\tau) = \psi(\tau)$ (Markovian situation, as well as non-Markovian situations when the time count starts together with the first step, corresponding to the so-called ordinary renewal processes, see Cox (1967)) the result, Eq. (11.42) is further simplified:

$$\hat{f}(k,u) = \frac{1}{u}\frac{1-\hat{\psi}(u)}{1-\phi(k)\hat{\psi}(u)}. \quad (11.43)$$

The corresponding $p(x,t)$ follows by an inverse Fourier and Laplace-transforms.

Let us now discuss the overall behavior of this probability density and concentrate first on the situation when $\phi(s)$ is symmetric and possesses the finite second moment, as it was the case in the Einstein's discussion. Differentiating Eq. (11.43) twice with respect to k we get (assuming that $\phi(k) = 1 - \frac{1}{2}\sigma^2 k^2 + o(k^2)$)

$$\langle x^2(u)\rangle = \frac{\hat{\psi}(u)}{u\left[1-\hat{\psi}(u)\right]}\sigma^2, \quad (11.44)$$

and $\langle x^2(t)\rangle$ is given by the inverse Laplace transform of this expression. We note now that the behavior of $\langle x^2(t)\rangle$ for t large is governed by one of $\langle x^2(u)\rangle$ for small values of u and that, just parallel to Eq. (11.3), we have

$$\hat{\psi}(u) = 1 - \langle \tau \rangle u + \langle \tau^2 \rangle u^2 + \cdots \qquad (11.45)$$

(provided the corresponding moments exist) so that for $u \to \infty$ we have

$$\langle x^2(u) \rangle \simeq \frac{\sigma^2}{\langle \tau \rangle} u^{-2}. \qquad (11.46)$$

The inverse Laplace transform immediately gives:

$$\langle x^2(t) \rangle = \frac{\sigma^2}{\langle \tau \rangle} t. \qquad (11.47)$$

The prefactor can be associated with the diffusion coefficient: $D = \sigma^2/2 \langle \tau \rangle$ and is finite as long as the first moment of the waiting time exists. A simple random walk is just one of the situations when this is the case.

11.1.3 *CTRW and the master equation*

Let us consider an important situation $\psi(\tau) = \langle \tau \rangle^{-1} \exp(-t/\langle \tau \rangle)$ corresponding to the Markovian case. As we proceed to show, on this case the CTRW exactly corresponds to a master-equation scheme with the transition rates $w(x'|x) = \phi(x' - x)/\tau_0$ with $\tau_0 = \langle \tau \rangle$. Let us consider the situation when the particle initially starts at $x = 0$ at time $t = 0$, i.e. for the initial condition $p(x, 0) = \delta(x)$. We moreover denote $p(x, t) = p(x, t|0, 0)$. The Master equation for the case of homogeneous transition rates reads

$$\frac{\partial}{\partial t} p(x, t) = \frac{1}{\tau_0} \int dx' \, \phi(x - x') p(x', t) - \frac{1}{\tau_0} \int dx' \, \phi(x' - x) p(x, t). \qquad (11.48)$$

Note that the second integral in the r.h.s. simplifies due to the homogeneity and to the normalization condition for the step-length distribution $\phi(s)$, $\int dx' \, \phi(x' - x) p(x, t) = p(x, t)$, so that

$$\frac{\partial}{\partial t} p(x, t) = \frac{1}{\tau_0} \int dx' \, \phi(x - x') p(x', t) - \frac{1}{\tau_0} p(x, t). \qquad (11.49)$$

Let us now first make Fourier-transform in the spatial variable x and note that the first integral in the right-hand side has a form of the convolution, so that

$$\frac{\partial}{\partial t} f(k, t) = \frac{1}{\tau_0} \phi(k) f(k, t) - \frac{1}{\tau_0} f(k, t). \qquad (11.50)$$

Now, we take the Laplace transform in the temporal variable: we know that $\frac{\partial}{\partial t} f(k, t) = u \hat{f} - f(k, 0)$. The initial condition is $f(k, 0) = 1$, which is a Fourier-transform of the δ-function. Thus,

$$u \hat{f}(k, u) - 1 = \frac{1}{\tau_0} \left[\phi(k) \hat{f}(k, u) - \hat{f}(k, u) \right]. \qquad (11.51)$$

so that

$$\hat{f}(k,u) = \frac{1}{u - \frac{1}{\tau_0}[\phi(k) - 1]}. \quad (11.52)$$

Now we compare this result with the one of the CTRW approach,

$$\hat{f}(k,u) = \frac{1}{u}\frac{1 - \hat{\psi}(u)}{1 - \phi(k)\hat{\psi}(u)}. \quad (11.53)$$

The expressions, Eq. (11.51) and Eq. (11.52) are equivalent if we take

$$\hat{\psi}(u) = \frac{1}{u\tau_0 + 1}, \quad (11.54)$$

i.e. for the exponential waiting-time probability density $\psi(\tau) = \tau_0^{-1}\exp(-\tau/\tau_0)$. This is the only situation in which the approaches are equivalent[†]. The exponential waiting time distribution corresponds to a Markovian process in time in which the probability to make a jump during the time interval δt equals to $\delta t/\tau_0$. In this case the probability to make a jump to the state x depends only on the actual state of the system, and not on the time, when the previous step was made. In all other cases the probability of the jump depends also on the time of the previous one. However, the CTRW process can be considered as discrete Markovian process in the following sense: At the moment a step is made, the length and the time of the next jump is taken at random, from given probability distributions, independent on the previous history. Being considered at arbitrary time, the process exhibits memory about the instant of the last jump. In this sense the process is non-Markovian. Such processes are sometimes called semi-Markovian. As we proceed to show, many CTRWs are non-stationary processes and exhibit aging.

11.2 Power-law waiting-time distributions

We first stay stuck to the case of uncorrelated spatial and temporal parts (i.e. decoupling $\psi(x,t) = \phi(x)\psi(t)$) and discuss the situation under which $\psi(t)$ lacks the first moment, i.e. when the integral $\int_0^\infty \tau\psi(\tau)d\tau$ diverges. Such an absence of means is not a too-seldom situation in statistical physics, and it is definitely a one, under which usual, close-to-equilibrium thermodynamics fails. However, the situation is experimentally a wide-spread one, so that we first turn to a short discussion of the experimental results which lead Scher and Montroll to the formulation of the CTRW approach.

[†]In general, decoupled continuous-time random walks can be described by generalized Master equations with a memory kernel introduced by Kenkre and Knox (1974), see Klafter and Silbey (1980). We shall not discuss this approach here, but we shall consider continuous limits of a class of such equations later in Section 11.3.

By mid seventies several experiments on photoeffect in disordered semiconductors showed a very peculiar time-dependence of the transient photocurrent. A typical experimental setup consists of a film of material of thickness L kept under voltage between a massive and a thin, semi-transparent electrode. A typical current through the system (at low temperatures) is extremely small, since the thermal activation is not too strong to considerably populate the conduction band. A strong light flash applied through a semi-transparent electrode produces free charge carriers (say, electrons), which are then moving towards the massive electrode, giving a pulse of a current. Essentially we have here to do with a kind of a time-of-flight experiment. One could anticipate (and it is really the case for the temperatures high enough) that a Gaussian packet of electrons moves with a more or less constant velocity $v = \mu e E$ where μ is the mobility and E is the electric field, and broadens according to the diffusive law. Thus, the form of the charge pulse $en(x,t)$ (with n being the electron density) can be calculated via the solution of the corresponding Fokker–Planck equation (under constant field and an absorbing boundary condition at the position of the massive electrode). The instantaneous value of the photocurrent, being the time derivative of the dipole moment of the charge distribution, is $I = e\frac{d}{dt}\int_0^L xn(x,t)dx$. This would give us a current which is practically constant before a considerable part of electrons reaches the electrode, and then decays fast, over the time connected with the width of the Gaussian.

However, experiments at lower temperatures done with a variety of non-organic and organic materials showed a very different picture: The "almost constant" part of the photocurrent and a fast subsequent decay are absent. Instead, one encounters a continuous and very slow decay; a current plotted on the double logarithmic scales shows a crossover between the two linear regimes, denoting the power-laws. Moreover, if the slope of the line at the short-time domain is γ, the slope of one for the long-time domain is $2 - \gamma$; this simple connection was clearly seen in all corresponding experiments. The explanation of the effect was connected with multiple trapping phenomena.

In a disordered semiconductor (compared to an ideal crystal) the density of states in the conduction band has a tail protruding into the energy gap of the ideal one. One can envisage this tail as connected with the local density fluctuations in a disordered solid. Such states are localized, and do not contribute to the electric conductivity. The typical decay of this tail of the density of states into the gap goes as $\rho(E) \propto E_0^{-1} \exp(-E/E_0)$, where E is the energy calculated, say from the localization threshold, and E_0 is the energetic scale characterizing the "fatness" of the tail, and depending on the degree of disorder (the prefactor E_0 is introduced to keep the correct dimension of the density of states). The overall transport process (called "dispersive transport", to stress the difference with the normal diffusive one) can be considered as a sequence of periods during which the particle is trapped and does not move, and ones when it is thermally activated to higher energies, and moves. Neglecting the possibility of multiple trapping in the same state (which does not change anything in the three-dimensional case considered here but is of importance in lower dimensions, see Bouchaud and Georges (1990)) we can obtain the qualitative behavior of the waiting-time in a trap using the following "hand waiving" argument:

A typical life-time of an electron in a trap of depth E follows the Arrhenius law,

$$\tau \simeq \tau_0 \exp(-E/kT). \tag{11.55}$$

To get the probability distribution of τ we note that the probability distribution of the trap's depth is proportional to the corresponding density of states, $p(E) \propto \rho(E) \propto E_0^{-1} \exp(-E/E_0)$. The probability density of the waiting time τ can then be obtained by the change of variables in this expression: $\psi(\tau) = p(E(\tau))\frac{dE}{d\tau}$, where $E(\tau) \simeq kT \ln(\tau/\tau_0)$ according to Eq. (11.55). Using this expression we get

$$\psi(\tau) \propto \exp(-kT \ln(\tau/\tau_0)/E_0) \frac{kT}{E_0 \tau} \propto \tau_0^\gamma \tau^{-1-\gamma} \tag{11.56}$$

with $\gamma = kT/E_0$. Of course, such arguments only give the asymptotic behavior of $\psi(\tau)$ for very large τ, however, as we proceed to show, the exact form of this distribution is of minor importance for what follows. It is clear that the presence or the absence of the higher moments of such distribution depends only on the value of γ: in the low temperature regime (for $T < E_0/k$) the mean value of τ diverges, so that the change of transport regime from diffusion to something else may take place. One of the forms typically used in calculation examples is

$$\psi(\tau) = \frac{\gamma}{\tau_0 \left[1 + \frac{\tau}{\tau_0}\right]^{1+\gamma}}. \tag{11.57}$$

The waiting-time distributions with power-law asymptotics are wide-spread also in other applications, see the review article by Bouchaud and Georges for a detailed discussion. The motion of the particle between the two trapping events is characterized by some displacement probability density $\phi(x)$, and can be adequately modeled through a random walk. In the absence of the external field this walk is unbiased, so that the first moment of $\phi(x)$ vanishes. A realization of a CTRW with fixed step length $x = \pm 1$ is shown in Fig. 11.1. The field introduces a nonzero bias, which can be taken to be proportional to E.

Let us first consider the free diffusion, without external field, in such a system. In this case the mean squared displacement can be obtained from Eq. (11.43) by noting that the second moment of the distribution is given by the second k-derivative of $\hat{f}(k,u)$ (whose first k-derivative vanishes). Noting that for small k one has $\hat{\phi}(k) \simeq 1 - 2\lambda^2 k^2$ one arrives at

$$\hat{f}(k,u) = \frac{1}{u} \frac{1 - \hat{\psi}(u)}{1 - (1 - 2\sigma^2 k^2)\hat{\psi}(u)}. \tag{11.58}$$

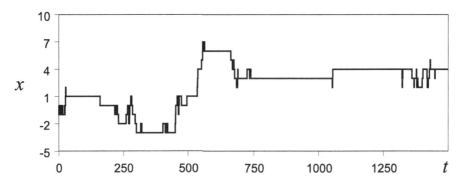

Fig. 11.1 A trajectory of a CTRW with the waiting-time density given by the Lévy–Smirnov distribution $\psi(t) = (2\pi)^{-1/2} t^{-3/2} \exp[-1/(2t)]$ corresponding to $\alpha = 1/2$.

In the case of a power-law, Eq. (11.57) one has

$$\psi(u) \approx 1 - Au^\alpha \qquad (11.59)$$

with $A = \Gamma(1-\alpha)\tau_0^\alpha$ so that

$$\hat{f}(k,u) = \frac{Au^{\alpha-1}}{1 - (1 - \frac{\lambda^2}{2}k^2)(1 - Au^\alpha)} \simeq \frac{1}{Bk^2 u^{1-\alpha} + u} \qquad (11.60)$$

with $B = \lambda^2/2A$. From this form the scaling of the distribution is evident: pdf $p(x,t)$ scales as a function of $\xi = x/t^{\alpha/2}$: $p(x,t) = t^{-\alpha/2}\tilde{p}(\xi)$. We note that the inverse Fourier–Laplace transforms of such forms can be expressed within the class of special functions known as Meijer G-Functions (Prudnikov et al., 1990) or even more general Fox's H-Functions (Fox, 1961), see also a MathWord entry on http://mathworld.wolfram.com/MeijerG-Function.html. In some situations these functions can be reduced to something simpler, however this is typically not the case.

Comparing this situation with the one for the exponential waiting-time distribution, $\psi(t) = \tau^{-1}\exp(-t/\tau)$, whose Laplace-transform is $\hat{\psi}(u) = 1/(1+u\tau) \approx 1 - u\tau$ we get

$$\hat{f}(k,u) \simeq \frac{1}{(\lambda^2/2\tau)k^2 + u} \qquad (11.61)$$

where we easily recognize a Fourier–Laplace-transform of the Gaussian Green's function of the diffusion equation,

$$P(x,t) = \frac{1}{\sqrt{4\pi Dt}} \exp\left(-\frac{x^2}{4Dt}\right) \qquad (11.62)$$

with $D = \lambda^2/2\tau$. This distribution scales as a function of $\xi = x/\sqrt{t}$. In Fig. 11.2 we compare this Gaussian distribution with $D = 1$ with the one following from Eq. (11.60) with $B = 1$ and $\alpha = 1/2$. Note the overall tent-like form of the function

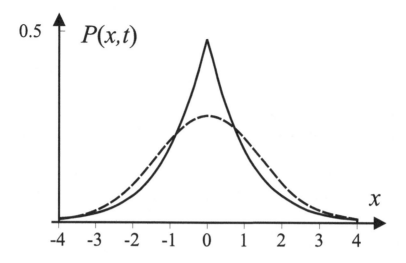

Fig. 11.2 The limiting form of the probability density of a particle performing CTRW (as the inverse Fourier–Laplace transform of Eq. (11.60) (for $B = 1$ and $\alpha = 0.5$). For comparison the Gaussian distribution with the same dispersion is shown as a dashed line.

corresponding to the anomalous diffusion with the cusp at $x = 0$, which mirrors the fact that the initial conditions in such a system never get forgotten.

For biased diffusion (say, in an external field) one has to take $\phi(k) = 1 + ibk + \lambda^2 k^2/2 + \cdots$, where b characterizes the strengths if external bias. In this case the characteristic function reads

$$\hat{f}(k, u) \simeq \frac{1}{(ibk/A + Bk^2)u^{1-\alpha} + u}. \tag{11.63}$$

Using the fact that the characteristic function of a probability distribution is a generating function of its moments, we can calculate for example the mean displacement $M_1(t)$ of particles, through its Laplace-transform:

$$M_1(u) = i\frac{d}{dk}\hat{f}(k, u)|_{k=0} = \frac{b}{A}u^{-1-\alpha}, \tag{11.64}$$

which corresponds asymptotically to $M_1(t) \simeq t^\alpha$. This finding allows us to explain the basic features of photoconductivity experiments discussed at the beginning of the paragraph:

At short time after irradiation the charge carriers do not "feel" the absorbing boundary, so that the mean displacement $\langle X(t) \rangle = M_1(t) \propto t^\alpha$, and $I(t) \propto dX(t)/dt \propto t^{\alpha-1}$: The slope of the initial part of the curve is $\alpha_1 = -\alpha + 1$. This regime (due to the "typical" motion) ends when $\langle X(t) \rangle \sim l$, i.e. at time $\tau(l) \propto l^\alpha$. At even longer times the particles still present within the sample are mostly those which were trapped all the time before. The current due to these particles is proportional to their number, $I \propto \frac{d}{dt}N(t) \propto \psi(t)$ and goes as $t^{-1-\alpha}$ (the tail of the first passage distribution), so that $\alpha_2 = 1 + \alpha$, and thus, $\alpha_1 + \alpha_2 = 2$ in full agreement with the experimental findings.

11.3 Aging behavior of CTRW systems

The subdiffusive CTRW is an intrinsically nonstationary process, and exhibits many effects typical for nonstationarity. It was first considered as a toy model for aging in glasses by Feigelman and Vinokur (1988), and got to be a popular quantitative model for the description of such phenomena afterwards, see Bouchaud and Montus (1996), Laloux and P. Le Doussal (1998). Here we present some simple considerations elucidating the nature of aging due to Sokolov et al (2001). We concentrate on the linear response of the system to a time-dependent external field, and assume the following: We consider an ensemble of random walkers performing the CTRW. During the trapping period the particle does not move, independently on the strength of the external field. However, when the particle jumps, the field biases the direction of the step, so that the mean displacement per step is proportional to the strength of the external field. We again use the one-dimensional notation.

Let us discuss the linear response of an ensemble of the random walkers performing the CTRW to the changing external field and consider some physically short time interval dt. Let dN be the mean number of steps performed during dt. Now, the mean displacement during dt is $\overline{dX} = x dN$ (note that for very short times dt the value of dN can be considered as a proportion of the realizations in which a particle has performed just one step) where x is the mean displacement per step depending on the actual value of the external field E: $x = \kappa E$ with $\kappa \simeq \lambda/k_B T$. We thus get:

$$\overline{dX} = \kappa E(t) dN. \tag{11.65}$$

where $dN = N(t+dt) - N(t) \approx dt \sum_{i=0}^{\infty} n \frac{d}{dt} \chi_n(t)$. Thus, the typical particles' velocity at time t (which is proportional to the particles' current) is given by:

$$\overline{V}(t) = \frac{\overline{dX}}{dt} = \kappa f(t) E(t). \tag{11.66}$$

where

$$f(t) = \sum_{i=0}^{\infty} n \frac{d}{dt} \chi_n(t). \tag{11.67}$$

Now, using Eq. (11.39) and Eq. (11.67) we arrive at the Laplace-transform of $f(t)$, which reads:

$$f(\lambda) = \frac{\psi(\lambda)}{1 - \psi(\lambda)} \tag{11.68}$$

Using Eq. (11.59) and applying the inverse Laplace transform we get the behavior of $f(t)$ for longer times:

$$f(t) = \frac{\sin \pi\alpha}{\pi \tau_0^\alpha} t^{\alpha-1}. \tag{11.69}$$

Thus at longer times one has

$$\overline{V(t)} = \kappa \frac{\sin \pi\alpha}{\pi \tau_0^\alpha} t^{\alpha-1} E(t). \tag{11.70}$$

Note that the current or velocity response of a CTRW-system to the external field is local in time and is explicitly dependent on the time elapsed after the system was prepared. To obtain the particle's position (or the polarization of the medium) we simply have to integrate Eq. (11.70) over time:

$$\overline{X(t)} = \kappa \frac{\sin \pi\alpha}{\pi \tau_0^\alpha} \int_0^t t_1^{\alpha-1} \mathbf{E}(t_1) dt_1. \tag{11.71}$$

Thus, the response of the CTRW-system to a time-dependent field dies out, and its polarization tends to a constant value. Note that the CTRW-system not only ages, but shows a kind of "Freudistic" response: the polarization at time t is mostly due to the early history of the system, immediately after it was prepared in a state corresponding to CTRW. The response of the CTRW system to a pulsed and to the sinusoidal external field is shown in Fig. 11.3.

The behavior for a θ-functional field unveils a very important property of the linear response in such systems. This response depends on the time elapsed since the system was prepared (a time of the 1st step): The system displays *aging*. The response to a constant field E switched at t_w behaves as $\overline{V(t)} \propto (t_w + \Delta t)^{\alpha-1} E \propto t_w^{\alpha-1}(1+\Delta t/t_w)^{\alpha-1} E$, where Δt is the time elapsed after switching the field. Thus, the overall strength of response is proportional to $t_w^{\alpha-1}$ and the characteristic time of this response is exactly t_w. One says, that the situation corresponds to simple or normal aging. It can happen that the characteristic time of relaxation goes as t_w^μ with some $\mu \neq 1$. The situations with $\mu < 1$ are known and are said to correspond to *subaging*.

Let us discuss the reason for aging in a CTRW model in some depth. This reason is exactly the difference between the first step forward waiting time distribution $\psi_1(t)$ and the distribution of all other waiting times.

Let τ_1 be the waiting time for the first renewal of the process after starting observation at time t_w, see Fig. 11.4. The jump immediately preceding t_w (numbered $i-1$), took place at time $t_{i-1} = s$. The forward waiting time distribution $\psi_1(t)$ is

$$\psi_1(\tau \mid t_w) = \int_0^{t_w} p(x)\psi(t_w - x + \tau_1)ds, \tag{11.72}$$

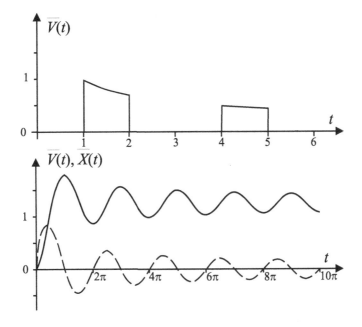

Fig. 11.3 The response of the CTRW system to an external field. The upper panel shows the mean velocity as a response to two pulses of unit height and duration following at $t = 1$ and at $t = 4$. Note that the response to the second pulse is considerably weaker: The system shows aging. The lower panel shows the response to the sinusoidal field. Here the mean velocity and the mean displacement are shown by a dashed and by a solid line, respectively.

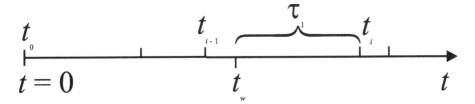

Fig. 11.4 The forward waiting time for the first step of the aging process.

where $p(s)$ is the probability density to make a jump exactly at time s:

$$p(s) = \sum_{n=0}^{\infty} p_n(s). \qquad (11.73)$$

Here $p_n(t)$ is the probability density that it is exactly n-th jump that takes place at time t. This one is given by an n-fold convolution of the waiting time probability density $\psi(t)$ with itself. Under Laplace-transform Eq. (11.73) reads:

$$\hat{p}(u) = 1 + \hat{\psi}(u) + \hat{\psi}^2(u) + \cdots = \frac{1}{1 - \hat{\psi}(u)}. \qquad (11.74)$$

The Laplace-transform of $\psi_1(\tau_1, t_w)$ as a function of t_w is then:

$$\hat{\psi}_1(\tau; u) = \frac{e^{u\tau}\left[\hat{\psi}(u) - \int_0^\tau e^{-ut}\psi(t)dt\right]}{1 - \hat{\psi}(u)}. \tag{11.75}$$

The inverse transform of this expression gives for both t_w and τ_1 large

$$\psi_1(\tau \mid t_w) \simeq C\left(\frac{t_w}{\tau}\right)^\alpha \frac{1}{t_w + \tau}. \tag{11.76}$$

The longer t_w, the longer is typically τ (which is an extreme form of the inspection paradox, see Feller (1991)).

Note that our expression, Eq. (11.76), is only valid for $0 < \alpha < 1$. For $\alpha > 1$, when the mean waiting time $\langle t \rangle$ exists, the distribution of the forward waiting times is given by

$$\psi_1(\tau_1) = \frac{1}{\langle t \rangle}\left[1 - F(\tau_1)\right], \tag{11.77}$$

where F is the cumulative distribution function of the ψ-distribution. Note that for power-law distributions $\langle t \rangle$, if it exists, is the only relevant timescale of the problem, i.e. $\psi(t) \simeq \langle t \rangle^\alpha t^{-1-\alpha}$ (in order to have a correct dimension of the probability density), so that at longer times

$$\psi_1(\tau) \simeq \frac{1}{\langle t \rangle^{1-\alpha}} \tau^{-\alpha}. \tag{11.78}$$

Thus, for $t_w > \langle t \rangle$ $\psi_1(\tau)$ is independent on t_w (and shows no aging!).

Comparing Eq. (11.76) and Eq. (11.78) one readily infers that for $\tau < t_w$

$$\psi_1(\tau \mid t_w) \simeq \frac{1}{t_w^{1-\alpha}} \tau^{-\alpha}, \tag{11.79}$$

i.e. in this case t_w is only typical timescale, so that we take here t_w instead of $\langle t \rangle$! For $\tau \gg t_w$, on the other hand, $\psi_1(\tau \mid t_w)$ follows the behavior similar to one of $\psi(t)$ i.e.

$$\psi_1(\tau \mid t_w) \simeq C t_w^\alpha \tau^{-1-\alpha} \tag{11.80}$$

but again with the typical timescale of t_w. We note that this type of nonstationarity is typical for three-dimensional systems for which CTRW is a valid model describing their asymptotic behavior. In the low-dimensional systems, the situation is more complex and may lead to subaging, see Rinn, Maass and Bouchaud, 2000, 2001.

One more note is here at place. CTRW is a powerful scheme, which describes several kinds of physical processes. One of the important applications of CTRW is connected with intermittent dynamics, as generated, for example by maps. As an example one can consider a map

$$x_{n+1} \equiv F(x_n, z, a) = x_n + a x_n^z, \quad 0 < \tilde{x}_n < \frac{1}{2}, \tag{11.81}$$

where $\tilde{x}_n$ is a fractional part of x_n. For $\frac{1}{2} < \tilde{x}_n < 1$ the map is defined by the mirror symmetry. Parameter $z > 1$ in Eq. (11.81) stands for a degree of non-linearity and a is a control parameter, $a > 1$. If $1 < z < 2$ the diffusion generated by the map is normal, whereas for $z > 2$ the diffusion is anomalous. In this anomalous case, the behavior of the system is excellently described by CTRW, as first shown by Geisel and Thomae (1984), see also the discussion by Zumofen and Klafter (1993). The behavior of such maps shows aging effects typical for CTRW models, which fact was first discussed by Barkai (2003).

11.4 CTRW and fractional Fokker–Planck equations

The description of transport processes within a framework of deterministic Fokker–Planck equations has considerable analytical advantages compared to stochastic approaches. Is it possible to formulate a Fokker–Planck-like approach to processes more complex than normal, Fickian diffusion? The answer to this question is positive, however the corresponding Fokker–Planck equations are somewhat unusual, since they often involve the derivatives of non-integer order with respect to their temporal or spatial variable. Such equations where first postulated on the basis of purely phenomenological considerations (Balakrishnan, 1985, Schneider und Wyss 1987, 1989); their close relation with CTRW schemes was understood much later, see (Metzler and Klafter 2000; Sokolov, Klafter and Blumen 2002). The phenomenological introduction of such equation is based on the following consideration. A usual Fick's diffusion equation is a partial differential equation of a parabolic type, i.e. has a first-order time derivative on its left-hand side and the second-order derivative in coordinate (or a Laplacian, in a multidimensional case) on its right-hand side. It's Green's function solution scales as a function of time: The form of the equation does not change when simultaneously changing the time-scale by a factor of λ and the length-scale by a factor of λ^2. Therefore, such an equation possesses a scaling solution, $P(x,t) = f(t)P(x/\sqrt{t})$, where $f(t)$ is simply a time-dependent normalization factor of the probability density. The Green's function of the equation is exactly such a solution: $P(x,t) = G(x,t;0,0) = (4\pi Dt)^{-1/2} \exp\left[-\frac{1}{4D}(x/\sqrt{t})^2\right]$. From the scaling form of the solution it follows that the characteristic length of the problem scales as a square root of the time, and a characteristic length squared (e.g. the mean squared displacement) grows as the first power of time. Reverting this consideration one can say that the fact that the mean squared displacement scales linearly in time would propose to look for the governing equation being first-order in time and second-order in spatial coordinate, i.e. for a parabolic one. The process, in which the mean squared displacement grows as t^α with $0 < \alpha < 1$ could probably be described by an equation being still of second-order in spatial coordinate but of a fractional α-th order in time. Before we discuss the meaning of such equations, we have to be sure that such an operator as a fractional α-th derivative can be mathematically reasonably defined and possesses a correct physical interpretation, see (Miller and Ross, 1993; Oldham and Spanier, 1974).

The definition of a fractional derivative is a generalization of a definition of "normal" derivative. As already stated, there are several ways of such generalization. We start here from a definition of a *Riemann–Liouville* derivative which starts from the generalization of the repeated integration formula:

$$\frac{d^{-n}}{dx^{-n}}f(x) = \int_a^x \int_a^{y_1} \cdots \int_a^{y_{n-1}} f(y_n) dy_n \cdots dy_1$$
$$= \frac{1}{(n-a)!} \int_a^x (x-y)^{n-1} f(y) dy \quad (x > a). \tag{11.82}$$

This allows to define a fractional integral:

$$(I_{a+}^\alpha f)(x) = \frac{1}{\Gamma(\alpha)} \int_a^x (x-y)^{\alpha-1} f(y) dy \quad (x > a) \tag{11.83}$$

with $0 < \alpha < 1$, which restriction guarantees the convergence of the integral for nonsingular integrable functions $f(y)$. The α-th fractional derivative is then defined through

$$_aD_x^\alpha = \frac{d}{dx} I_{a+}^{1-\alpha} = \frac{d}{dx} \frac{1}{\Gamma(\alpha)} \int_a^x (x-y)^{\alpha-1} f(y) dy. \tag{11.84}$$

The higher derivatives are defined by repeating additional differentiation. We note that the Riemann–Loiuville integrals or derivatives behave under Laplace transform in the same way as the whole-number repeated integrals or derivatives. For the integrals we have

$$\mathcal{L}\{I_{0+}^a f(t)\} = u^{-\alpha} f(t). \tag{11.85}$$

For the fractional derivatives we have correspondingly

$$\mathcal{L}\{_0D_t^\alpha f(t)\} = u^\alpha f(t) - \sum_{j=0}^n u^j c_j \tag{11.86}$$

where $n = [\alpha]$ and c_j are the "quasi-initial values", $c_j = \lim_{t \to 0} {}_0D_t^{\alpha-1-j} f(t)$.

Note also that the Riemann–Liouville definition leads to a trivial generalization of the standard rule for the differentiation of a power: $\frac{d^n}{dx^n} x^m = \frac{m!}{(m-n)!} x^{m-n}$, so that $_0D_x^\alpha x^\beta = \frac{\Gamma(\beta+1)}{\Gamma(\beta-\alpha+1)} x^{\beta-\alpha}$.

Interchanging integration and differentiation in Eq. (11.84) introduces a slightly different operator, called a Caputo derivative, (Caputo, 1969). The Riemann–Liouville and Caputo definitions are the most important ones for the description of subdiffusive CTRW behavior.

Several other definitions are common, for example the *Weyl* definition (1917) starting from

$$(I_+^\alpha f)(x) = \frac{1}{\Gamma(\alpha)} \int_{-\infty}^x (x-y)^{\alpha-1} f(y) dy \quad _{-\infty}D_x^\alpha = \frac{d}{dx} I_+^{1-\alpha} \tag{11.87}$$

so that $_{-\infty}D_x^\alpha = \frac{d}{dx}I_+^{1-\alpha}$ and the *Riesz* definition (1949) leading to a symmetric form:

$$(I^\alpha f)(x) = \frac{1}{2\cos\frac{\alpha\pi}{2}}\left(I_+^\alpha + I_-^\alpha\right); \qquad D^\alpha = \frac{d}{dx}I^{1-\alpha}. \qquad (11.88)$$

These ones are often used in fractional equations describing superdiffusion. The Weyl operator preserves the usual rule for the differentiating of the exponential $_{-\infty}D_x^\alpha e^x = e^x$. The Weyl and Riesz operators are those which reproduce the properties of usual integrals or derivatives under Fourier-transform:

$$\mathcal{F}\{_{-\infty}D_t^\alpha f(t)\} = (i\omega)^\alpha f(\omega) \qquad (11.89)$$

for a Weyl operator and

$$\mathcal{F}\{_{-\infty}D_t^\alpha f(t)\} = -\omega^\alpha f(\omega) \qquad (11.90)$$

for a Riesz one. This one is a very special form generalizing the behavior of the second derivative, and is exactly the one appearing on its place in fractional equations for superdiffusion.

Let us now turn to the fractional generalization of the diffusion or of the Fokker–Planck equation of the form

$$\frac{\partial}{\partial t}P(x,t) = {}_0D_t^{1-\alpha}\left[\nabla\left(-\mu^* f P + K^* \nabla P\right)\right], \qquad (11.91)$$

with the additional Riemann–Liouville fractional derivative operator acting on the "normal" right-hand side of a Fokker–Planck equation, see (Metzler and Klafter, 2000; Sokolov, Klafter and Blumen, 2002). Here μ^* denotes the (generalized) mobility and K^* denotes the (generalized) diffusion coefficient; the change of notation for the diffusion coefficient from D to K is necessary in order not to mix it up with the differential operator. Eq. (11.91) represents a by far the most widely used form of the fractional Fokker–Planck equation. We note that the equation discussed belongs therefore to the class of non-Markovian Fokker–Planck equations, and the additional fractional derivative plays the role of the memory kernel. The corresponding non-Markovian Fokker–Planck equations are often considered as a dangerous theoretical instrument, since they in general do nor guarantee for the non-negativity of their solutions which therefore cannot be interpreted as probability density. However, as we proceed to show, the solution of the fractional Fokker–Planck equation, Eq. (11.91) is probability density, and moreover they correspond to the continuous limit of a CTRW scheme with power-law waiting-time distribution.

Let us first assume the external force to be constant and the initial condition to be δ-functional. Applying the Laplace-transform in temporal variable to the both parts of the equation Eq. (11.91) we get:

$$u\hat{P}(k,u) - P(k,0) = u^{1-\alpha}\left[-ik\mu^* f\hat{P}(k,u) - K^* k^2 \nabla\hat{P}(k,u)\right] \qquad (11.92)$$

with $P(k, 0) = 1$. The solution of Eq. (11.92) then reads:

$$\hat{P}(k, u) = \frac{1}{u^{1-\alpha}(ik\mu^* f + K^* k^2) + u}, \qquad (11.93)$$

exactly the form which immediately follows from the CTRW model, see Eq. (11.63).

Fractional Fokker–Planck equations are however a more versatile tool, giving solutions in many cases when the Fourier transform in spatial coordinate would not simplify the situation due to spatial inhomogeneity.

To see this let us consider a general non-Markovian Fokker–Planck equation with memory operator on the right-hand side:

$$\frac{\partial}{\partial t} P(x, t) = \hat{M} \left[\nabla \left(-\mu^* f P + K^* \nabla P \right) \right], \qquad (11.94)$$

where $\hat{M}$ is an operator of the convolution type acting on the right-hand side of the equation; it is either

$$\hat{M} f(t) = \int_{t_0}^{t} g(t - t') f(t') dt' \qquad (11.95)$$

or

$$\hat{M} f(t) = \frac{d}{dt} \int_{t_0}^{t} g(t - t') f(t') dt'. \qquad (11.96)$$

The corresponding integral is assumed to tend to zero for $t \to t_0$, i.e. the kernel of the transformation may posers only a weak singularity like one of the fractional integral with $0 < p < 1$, so that under the Laplace-transform the action of the operator $\hat{M}$ on the function f can be described by a mononomial $\hat{L}\hat{M} f(t) = \hat{M}(u) \hat{f}(u)$ with $\hat{L}$ denoting the operator of the Laplace transform, i.e. does not introduce additional initial conditions. Note that a fractional derivative is just an operator of the type needed. Let us show now that the formal solution of the non-Markovian Fokker–Planck equation Eq. (11.94) has a following form:

$$P(\mathbf{x}, t) = \int_0^\infty F(\mathbf{x}, \tau) T(\tau, t) d\tau, \qquad (11.97)$$

where $F(\mathbf{x}, \tau)$ is a solution of a Markovian Fokker–Planck equation with the same Fokker–Planck operator,

$$\frac{\partial}{\partial t} F = \nabla \left(-\mu^* f F + K^* \nabla F \right), \qquad (11.98)$$

and for the same initial and boundary conditions, and the function $T(\tau, t)$ is connected with the memory kernel $\hat{M}$ by the following relation: The Laplace-transform of T in its second variable t, $\tilde{T}(\tau, u) = \int_0^\infty T(\tau, t) e^{-ut} dt$ reads:

$$\hat{T}(\tau, u) = \frac{1}{\hat{M}(u)} \exp\left[-\tau \frac{u}{\hat{M}(u)} \right], \qquad (11.99)$$

see (Barkai, 2001; Sokolov 2001, 2003). To show this let us consider the Laplace-transform of $P(\mathbf{x}, t)$ given by Eq. (11.97) with respect to its temporal variable:

$$\hat{P}(\mathbf{x}, u) = \int_0^\infty dt e^{-ut} \int_0^\infty d\tau F(\mathbf{x}, \tau) T(\tau, t)$$

$$= \int_0^\infty d\tau F(\mathbf{x}, \tau) \hat{T}(\tau, u)$$

$$= \int_0^\infty d\tau F(\mathbf{x}, \tau) \frac{1}{\hat{M}(u)} \exp\left[-\tau \frac{u}{\hat{M}(u)}\right]$$

$$= \frac{1}{\hat{M}(u)} \hat{F}\left(\mathbf{x}, \frac{u}{\hat{M}(u)}\right), \qquad (11.100)$$

where $\hat{F}(\mathbf{x}, u)$ is a Laplace-transform of $F(\mathbf{x}, \tau)$ in its second (temporal) variable τ. Let us now note that the Laplace-transform of the non-Markovian FPE, Eq. (11.94) reads:

$$u\hat{P}(\mathbf{x}, u) - P(x, 0) = \hat{M}(u) \mathcal{L} \hat{P}(\mathbf{x}, u), \qquad (11.101)$$

where $\mathcal{L}$ denotes the Fokker–Planck operator acting on the probability density P, and $P(x, 0)$ is the initial condition. Inserting the form, Eq. (11.100), into Eq. (11.101) one gets:

$$\frac{u}{\hat{M}(u)} \hat{F}\left(\mathbf{x}, \frac{u}{\hat{M}(u)}\right) - P(\mathbf{x}, 0) = \mathcal{L} \hat{F}\left(\mathbf{x}, \frac{u}{\hat{M}(u)}\right). \qquad (11.102)$$

Introducing a new variable $s = u/\hat{M}(u)$ we rewrite Eq. (11.102) in a form

$$s\hat{F}(\mathbf{x}, s) - P(\mathbf{x}, 0) = \mathcal{L} \hat{F}(\mathbf{x}, s), \qquad (11.103)$$

in which one readily recognizes the Laplace-transform of an ordinary, Markovian FPE, Eq. (11.98), with the same initial condition $P(\mathbf{x}, 0)$. This completes our proof. Thus, the solution of a non-Markovian Fokker–Planck equation of the type of Eq. (11.94) in the Laplace domain is connected with the solution of the regular Fokker–Planck equation through

$$\hat{P}(\mathbf{x}, u) = \frac{1}{\hat{M}(u)} \hat{F}\left(\mathbf{x}, \frac{u}{\hat{M}(u)}\right). \qquad (11.104)$$

In time domain the solution of Eqs. (11.91) and (11.98) are connected via Eq. (11.97) where $T(\tau, t)$ is given by Eq. (11.99). Now, if the function T can be interpreted as the probability density of the number of steps τ of the random walk (in continuous limit) as a function of the "physical" time t, Eq. (11.97) describes exactly the continuous-time version of the Eq. (11.32), i.e describes CTRW. For the memory kernel corresponding to the fractional derivative of the order between zero and one,

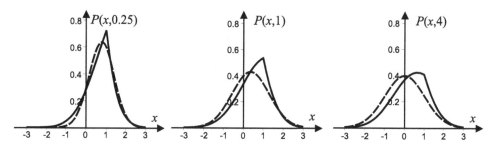

Fig. 11.5 The time evolution of the probability density in a fractional Ornstein–Uhlenbeck process. The initial condition corresponds to $P(x,t) = \delta(x)$. The parameters are $K = \mu = 1$ and $\alpha = 1/2$. For comparison, the results for the simple diffusion are shown as a dashed line.

i.e. for Eq. (11.91), the Laplace-transform of the memory kernel is $M(u) = u^{1-\alpha}$, so that

$$\hat{T}(\tau, u) = \frac{1}{\hat{M}(u)} \exp\left[-\tau \frac{u}{\hat{M}(u)}\right] = u^{\alpha-1} \exp(-\tau u^{-\alpha}), \qquad (11.105)$$

(Barkai, 2001; Sokolov, 2001). This function is a probability density, which can be expressed through the one-sided Lévy probability density,

$$T(\tau, t) = \frac{t}{(\alpha \tau^{1+1/\alpha})} L\left(\frac{t}{\tau^{1/\alpha}}, \alpha, -\alpha\right), \qquad (11.106)$$

where the standard notation (Feller, 1991) is used. This special case corresponds to the limiting distribution of $\chi_n(t)$ for both n and t large, where a continuous variable τ is introduced instead of a discrete n. This special case was mentioned in (Metzler and Klafter, 2000) and is discussed in detail in (Barkai, 2001). We also note here that the procedure showing the connection between the continuous limit of CTRW schemes with power-law waiting-time distributions and fractional Fokker–Planck equation is not the only way to obtain them from the stochastic random walk schemes. Another approach based on the generalization of Kramers–Moyal expansion is discussed by Barkai, Metzler and Klafter (2000).

The physical picture described by the equation closely corresponds to random traps model: The external force biases the direction of jumps but does not influence the waiting-time distribution. This force now can be inhomogeneous in space, moreover, complex boundary condition problems can be discussed.

As an example we show in Fig. 11.5 the time-development of the corresponding probability density of the particle's position for a δ-functional initial condition in a fractional generalization of the Ornstein–Uhlenbeck process, described by Eq. (11.91) with $f(x) = -\kappa x$. Note the persisting cusp at the initial position of particles. This cusp is a consequence of the long-time memory typical for CTRW with power-law waiting time distributions.

Although the integral transformation, Eq. (11.97), essentially solves the equation, the expression is not always the best starting point for the further theoretical analysis. Thus, the eigenfunction decomposition, as used in our second discussion of the Ornstein–Uhlenbeck process may be a better approach. From the form of fractional Fokker–Planck equation it follows that its spatial eigenfunctions are the same as for a normal, Markovian one, so that only the temporal parts differ. The temporal functions are now governed by ordinary fractional differential equations of the type

$$\frac{d\Phi_n(t)}{dt} = -\lambda_{n\,0}D_t^{1-\alpha}\Phi_n(t); \qquad (11.107)$$

λ_n are the eigenvalues of the corresponding Fokker–Planck operator, see Section 8.2.1. The solution of an ordinary fractional differential equation

$$\frac{d\Phi(t)}{dt} = -\tau_0^{-\alpha} D_x^{1-\alpha}\Phi(t) \qquad (11.108)$$

is given by a known special function, a Mittag–Leffler function $E_\alpha\left(-\left(t/\tau\right)^\alpha\right)$ which is defined through the inverse Laplace transform

$$E_\alpha\left(-\left(t/\tau\right)^\alpha\right) = L^{-1}\left\{\frac{1}{u + \tau^{-\alpha}u^{1-\alpha}}\right\}. \qquad (11.109)$$

Its asymptotic behavior is a stretched exponential $E_\alpha\left(-\left(t/\tau\right)^\alpha\right) \simeq \exp\left[-\left(t/\tau\right)^\alpha/\Gamma(1+\alpha)\right]$ for small t and a power-law $E_\alpha\left(-\left(t/\tau\right)^\alpha\right) \simeq -\left(t/\tau\right)^\alpha/\Gamma(1-\alpha)$ for large t. The Mittag–Leffler function is a natural generalization of an exponential function, which corresponds to $E_1(-t/\tau)$. Note that the solutions $\Phi(t)$, except one for $\alpha = 1$, being an exponential, show at long times slow, power-law decay. The Mittag–Leffler functions are ubiquitous in the relaxation patterns governed by subdiffusion. The behavior of the Mittag–Leffler function $E_{1/2}(-(t/\tau)^{1/2})$ is shown in Fig. 11.6.

11.5 Superdiffusion: Lévy flights and Lévy walks

In what follows we shortly discuss what happens if the third Einstein's postulate of the "normal" diffusion is abandoned, namely the one of finite mean free path (i.e. of the finite mean square displacement per step). Here several different models can be discussed.

The simplest one is the uncorrelated simple or continuous-time random walk with the step lengths distributed according, say, to a power-law, $p(x) \simeq x^{-1-\alpha}$ with $0 < \alpha < 2$. According to the generalized central limit theorem due to P. Lévy, after many steps the distribution of the particle's coordinate converges to a one characterized by a probability density corresponding to one of the so-called infinitely divisible stable laws. Apart from trivial translation, these laws are characterized by a family of the probability densities, whose characteristic functions all have very

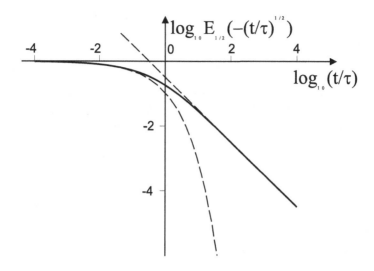

Fig. 11.6 The Mittag–Leffler function as the solution of Eq. (11.108). The dashed lines denote the short- and the long-time asymptotic forms.

simple form:

$$f(k) = \exp\left(-Ak^\alpha e^{i\pi\gamma/2}\right) \qquad (11.110)$$

for $k > 0$, (Feller, 1991). Since the Fourier-transform of this function represents a probability density, which is a real function, the behavior for negative k is given by $f(-k) = f^*(k)$. The parameter γ describes the asymmetry of the distribution; the distributions with $\gamma = 0$ are symmetric. The values of γ change between $-\alpha$ and α for $0 < \alpha \leq 1$ and between $-(2-\alpha)$ and $(2-\alpha)$ for $1 < \alpha \leq 2$. The distributions with γ at either boundaries of the corresponding intervals are called extreme laws. For $0 < \alpha \leq 1$ the corresponding densities vanish identically for all negative (positive) values of x. The Gaussian distribution (the one with $\alpha = 2$) is always symmetric. The parameter A is connected with the width of the distribution, which is proportional to $A^{1/\alpha}$. This can be immediately seen from the overall scaling behavior or from the dimensional arguments. Since all Lévy distributions except for a Gaussian have power-law tails decaying as $x^{-1-\alpha}$, they all have infinite variance, so that the width must be understood either as their interquartile distance or as some fractional moment, $W_q = \langle |x|^q \rangle^{1/q}$ with $q < \alpha$. A simple scaling argument shows, that the parameter A is proportional to the number of steps of the Lévy flight n. Indeed, let us compare a flight of n steps with the one of $2n$ steps. The probability density of positions after $2n$ steps is a convolution of the corresponding identical probability densities after n steps, i.e. its characteristic function is a square of the corresponding function for n steps. This, it is given by a function Eq. (11.110) with twice the corresponding coefficient A. In general, the function after λn steps is given by the characteristic function

$$f(k) = \exp\left(-\lambda A n k^\alpha e^{i\pi\gamma/2}\right), \tag{11.111}$$

($k \leq 0$), i.e. $A \propto n$. The width of the corresponding distribution thus grows as $W \propto n^{1/\alpha}$, i.e. superdiffusively. A trajectory of a Lévy flight (as compared to a simple random walk) is shown in Fig. 11.7 for a one-dimensional process and in Fig. 11.8 in two dimensions.

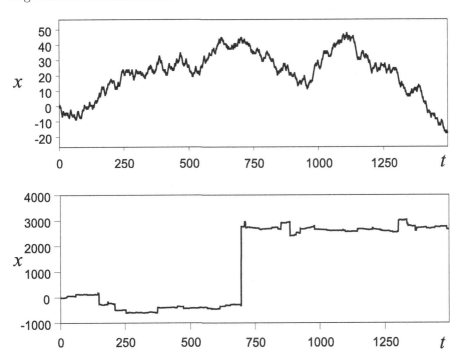

Fig. 11.7 A trajectory of a Cauchy Lévy flight with $p(x) = 1/[\pi(1+x^2)]$ (lower panel) as compared to a one of a simple random walk ($x = \pm 1$) shown in the upper panel; note the difference in vertical scales. Note that while the trajectory of a random walk appears continuous on the scales of the plot, the one of a Lévy flight shows jumps on all scales.

Note that the Lévy flight is a Markovian jump process. When passing to the continuous time variable, the corresponding Chapman–Kolmogorov equation of the process can be put down in a form of a fractional Fokker–Planck (diffusion) equation. For symmetric Lévy distribution the form of this equation is extremely simple,

$$\frac{\partial p(x,t)}{\partial t} = K_\alpha \frac{\partial^\alpha}{\partial |x|^\alpha} p(x,t), \tag{11.112}$$

where the corresponding fractional differential operator has to be considered as a symmetrized Riesz–Weyl derivative, and K_α is the fractional diffusion coefficient having the dimension $[L^\alpha/T]$, $0 < \alpha < 2$. Noting that under the Fourier-

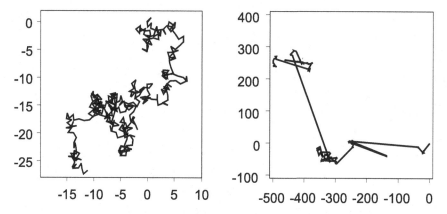

Fig. 11.8 A trajectory of a two-dimensional random walk with fixed step length $x = 1$ and a random direction of each step (left panel) as compared to a trajectory of a two dimensional Cauchy Lévy flight with $p(x) = 1/[\pi(1 + x^2)]$ and a random step direction. Both trajectories correspond to $n = 400$ steps of the process. Note both the tendency of clustering of small steps as well as the existence of very long flights typical for a Lévy process. Again, note the difference in scales.

transform the Riesz–Weyl derivative corresponds to multiplication by $-|k|^\alpha$, we see that Eq. (11.112) corresponds to

$$\frac{\partial f(k,t)}{\partial t} = -K_\alpha |k|^\alpha f(k,t) \qquad (11.113)$$

so that the Green's function solution of Eq. (11.112) corresponding to the initial condition $f(k,t) = 1$, reads

$$f(k,t) = \exp\left(-K_\alpha |k|^\alpha t\right), \qquad (11.114)$$

i.e. is a characteristic function of a symmetric Lévy distribution of type Eq. (11.110) with $\gamma = 0$.

Due to the Markovian nature of a Lévy-flight model, also its formulation within the Langevin scheme is possible: Under the assumption that the deterministic force and the noise term (corresponding to flights) can be separated, one can, for example, write

$$\dot{x} = -\frac{dU}{dx} + Y_\alpha(t) \qquad (11.115)$$

where $U(x)$ corresponds to the potential of the deterministic force and $Y_\alpha(t)$ is a stationary white Lévy noise of index α. It is defined in such a way that the process

$$L(\Delta t) = \int_t^{t+\Delta t} Y_\alpha(t')dt' \qquad (11.116)$$

at nonintersecting Δt intervalls is a stationary α-stabe process with stationary independent increments. Restricting oneself to the symmetric case, one gets that the probability distribution of $L(\Delta t)$ is given by its characteristic function

$f(k, \Delta t) = \exp(-K|k|^\alpha \Delta t)$ where D (of the dimension of $[K] = \text{cm}^\alpha/\text{sec}$) corresponds to the intensity of the Lévy noise. The corresponding fractional Fokker–Planck equation can be derived along the usual lines and reads

$$\frac{\partial p(x,t)}{\partial t} = \frac{\partial}{\partial x}\frac{dU}{dx} p(x,t) + K_\alpha \frac{\partial^\alpha}{\partial |x|^\alpha} p(x,t). \qquad (11.117)$$

This equation can be solved analytically for some special cases and numerically for any potential. Two stationary solutions are shown in Fig. 11.9. Note that Eq. (11.117), with a drift term proportional to the deterministic force, assumes linear friction, so that, combined with non-Gaussian noise it gives rise to solutions which violate fluctuation-dissipation theorem. This means that it is pertinent to essentially nonequilibrium situation. We note therefore, that this is not the only possible type of the fractional Fokker–Planck equation for Lévy flights in potential.

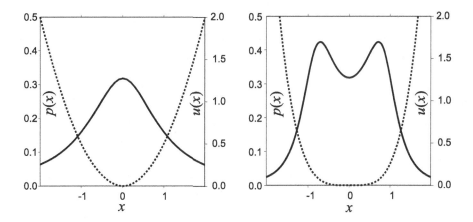

Fig. 11.9 Stationary solutions of Eq. (11.117) for $K = 1$ and $\alpha = 1$ (symmetric Cauchy noise of unit intensity) as obtained analytically by Chechkin et al. (2004) for a quadratic and for a quartic potential. The left panel shows the situation in a harmonic potential $U(x) = x^2/2$, shown as a dotted line. The corresponding solution is simply a Cauchy distribution: $p(x) = 1/[\pi(1 + x^2)]$ (full line). The right panel shows the situation in a quartic potential $U(x) = x^4/4$ (dotted line). The corresponding solution reads $p(x) = 1/[\pi(1 - x^2 + x^4)]$ (full line). This pdf is bimodal. Note that both solutions do not correspond to Boltzmann distributions in the corresponding potentials, which stresses the nonequilibrium nature of the process described by Eq. (11.117).

The genuine Lévy flight model might seem unphysical since it allows for indefinitely large jumps, i.e. the model does not possesses a limiting velocity. However, as long as x is not a Euclidean coordinate, this is not a real problem. As an example let us consider the following system. Imagine a (quasi)particle (say an exciton) jumping between the monomers of a long polymer chain. The monomers of the chain are numbered consecutively starting from the chain's beginning, the number n of the monomer is considered as a coordinate in the one-dimensional chemical space. The jumps of the particle might take place along the chain or between the monomers which are far away in the chemical sequence of the chain, but are by chance close to each other in the Euclidean space under a chain's particular conformation (Sokolov, 1996). If the conformational changes in a chain are fast enough compared with the typical time between the particle's jumps, the corresponding process in the chemical space is the genuine Lévy flight. The behavior of such a process in the force-free case is governed by Eq. (11.112). Incorporation of the external forces into a problem was considered by Brockmann and Sokolov (2002) (it does not follow Eq. (11.117) since the model describes a system posessing Boltzmann equilibrium distribution). The corresponding equations was used by Brockmann and Geisel (2003) for the description of flights in different potential fields. We also note that the theory of the equations like Eq. (11.112) is not fully developed. Thus it is not yet clear how the boundary condition problems for them have to be uniquely formulated (Chechkin et al., 2003).

Another important model is a Lévy walk, introduced by Shlesinger, West and Klafter (1987). A simplest realization of the Lévy walk is a process in which the particle moves with a constant velocity $\vec{v}$ during a period of time t_i. After time t_i the direction of the velocity is chosen anew (a Drude-like model). The duration periods t_i in given by the waiting-time probability density $\psi(t)$. If $\psi(t) \simeq \tau_0^\alpha t^{-1-\alpha}$ (τ_0 is the characteristic time of a step) and the second moment of the waiting-time t is absent ($\alpha < 2$) the mean free path diverges, so that departure from the simple diffusion can be anticipated. The initial model was formulated as a toy model of diffusion in turbulent flows; however in this case the approach seems to be oversimplified. The behavior closely described by this model arises however in several other physical situations. One of the experimental realizations (a particles' transport in a flow between two rotating cylinders) and some simulation results can be found on the webpage http://chaos.ph.utexas.edu/research.html. Another realization corresponds to a Hamiltonian system in which a particle moves in a two-dimensional egg-crate potential

$$V(x,y) = B(\cos x + \cos y) + C \cos x \cos y, \qquad (11.118)$$

at energies above the one at which infinite motion gets possible, Klafter and Zumofen (1993), see Fig. 11.10.

The difference between the Lévy walk model and the random walk models considered so far is that in the present case the temporal and spatial aspects are coupled: a single motion event is described by a probability density $\psi(r,t)$ to make a step of length r and of duration t. In our simple one dimensional case where between the scattering events the particle moves with the constant velocity either to the left or to the right we have, for example $\psi(r,t) = \frac{1}{2}\psi(t)\left[\delta(r-vt) + \delta(r-vt)\right]$. A trajectory of a Lévy walk is shown in the lower panel of Fig. 11.11.

Let $\eta(r,t)$ be the probability density of *just arriving* at r in time t, i.e. the probability density of the particle's position just after completing the step. This

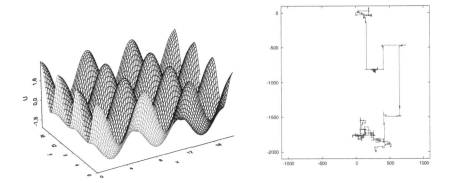

Fig. 11.10 The left panel shows the potential, Eq. (11.118) with $B = 1.5$ and $C = 0.5$. The existence of the "valleys" between the potential peaks allows (for high enough energies) for the particle's motion in form of long straights interrupted by distinct scattering events, i.e. the one of the Lévy walk type. A typical trajectory as obtained by the integration of the equations of motion for a particle of mass $m = 1$ with initial conditions $x(0) = y(0) = \pi/2, v_x(0) = 1.2$ and $v_y(0) = 0.001$ up to the time $t = 10^4$ is shown in the right panel. This Lévy-walk-type behavior is typical for energies which correspond to infinite motion but are lower than the maximal value of the potential.

probability density fulfills the following recursion equation

$$\eta(r,t) = \int dr' \int_0^t d\tau\, \eta(r',\tau)\psi(r-r', t-\tau) + \delta(t)\delta_{r,0}, \tag{11.119}$$

where the last term introduces the initial condition. Here we use the fact that our process is semi-Markovian: the probability density of the random walker's position after completing the next step depends only on its position at the beginning of the step. In the Fourier–Laplace representation $\psi(r,t) \to \psi(k,u)$, we have:

$$\eta(k,u) = \eta(k,u)\psi(k,u) + 1 \tag{11.120}$$

i.e.

$$\eta(k,u) = \frac{1}{1-\psi(k,u)} \quad \left(= \psi^0 + \psi^1 + \cdots + \psi^n + \cdots\right). \tag{11.121}$$

We have also to include the possibility that a walker didn't complete a full step up to time t. Let us consider the position of a random walker which completed n steps up to time $t - \tau$ and is moving freely with a constant velocity during the rest time τ. The probability density of positions at time t is given by

$$p(r,t) = \int dr' \int_0^t d\tau\, \eta(r-r', t-\tau)\Psi(r',\tau) \tag{11.122}$$

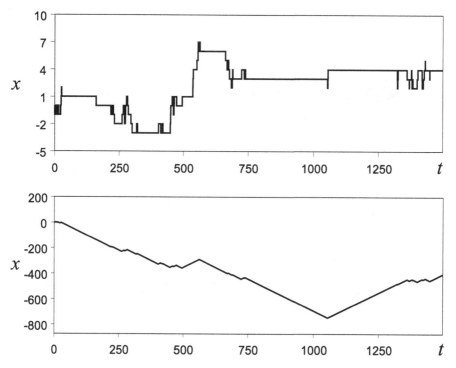

Fig. 11.11 A Lévy walk with $v = \pm 1$ (lower panel) as compared with the CTRW (upper panel, essentially the same as Fig. 11.1). The processes are in fact closely related: in CTRW at the beginning of each step i of duration t_i a jump of length $x = \pm 1$ is performed. In a Lévy walk, at the beginning of each step a particle assumes a velocity $v = \pm 1$ and moves with this velocity untill the step is completed. The two panels of the figure correspond not only to the same realization of step times but also the the same directions of jumps resp. velocities.

where $\Psi(r', \tau)$ the probability to move exactly at the distance r' at time τ. For a Lévy walk it reads

$$\Psi(r, t) = \frac{1}{2} \delta\left(|r| - vt\right) \int_t^\infty \psi(t') dt' \qquad (11.123)$$

since it corresponds to the motion with a constant velocity under the condition that no scattering events took place before time t. The Fourier–Laplace-transform of $p(r, t)$ is thus

$$f(k, u) = \frac{\Psi(k, u)}{1 - \psi(k, u)}. \qquad (11.124)$$

From here on we take for simplicity $v = 1$. The back transforms are made with use of Tauberian theorems.

Let us first concentrate of the behavior of the second moment of the displacement. The calculations show that the behavior depends on the range of the parameter α of the distribution of the times of free motion and are summarized as follows

for $\alpha < 1$ one has $\langle R^2(t) \rangle \propto t^2$ (a ballistic regime)
for $1 < \alpha < 2$ $\langle R^2(t) \rangle \propto t^{3-\alpha}$ (a subballistic superdiffusion)
for $\alpha > 2$ $\langle R^2(t) \rangle \propto t^1$ (a normal diffusion).

The asymptotic form of a Lévy-walk-propagator ($v = 1$) for $\alpha > 1$ is

$$p(r,t) \simeq \begin{cases} t^{-\beta} \exp\left[-c\left(\dfrac{r}{t^\beta}\right)^2\right] & \text{for } |r| < t^\beta \\ t/|r|^{1+\alpha} & \text{for } t^\beta < |r| < t \\ t^{1-\alpha}\delta(|r|-t) & \text{a } \delta\text{-function at } |r| = t \\ 0 & \text{for } |r| > t \end{cases} \quad (11.125)$$

with $\beta = 1/\alpha$ for $1 < \alpha < 2$ and $\beta = 1/2$ for $\alpha > 2$ (in this case the middle part is a normal Gaussian; it (very) slowly overweighs a still heavy tail). Here the characteristic time τ_0 is taken as a unit time and $v\tau_0$ as a unit length. One can thus distinguish four characteristic domains in such a distribution:

(1) a Gaussian middle part due to multiple scattering
(2) a power-law (Lévy) tail
(3) δ-"horns" due to particles which never got scattered
(4) cutoff due to the finite velocity

Note that the overall distribution *does not scale as a whole.*

For $0 < \alpha < 1$ only one exemplary analytical form (namely the arcsine-law for $\alpha = 1/2$) is known, which is exactly

$$p(x,t) = \frac{1}{\pi\sqrt{(vt+x)(vt-x)}}. \quad (11.126)$$

Numerical simulations show that all distributions look more or less the same (with flat middle part and singularities at $\pm vt$). The corresponding distributions scale as a whole, see (Zumofen and Klafter, 1993).

Lévy walks can also be described within the framework of fractional kinetic equations. The corresponding equation was derived by Sokolov and Metzler (2003), see also discussion in (Uchaikin, 2003; Becker–Kern, Meerschaert and Scheffler, 2004). The equation (reducing to a telegrapher's equation for the case of normal diffusion) does not have a form of a Fokker–Planck equation. Other physical situations lead to many other different problems, being combinations or variants of ones considered here. The classification of all of them might be a topic of a separate review so that we refrain from giving full details in our introductory discussion.

Chapter 12
Active Brownian Motion

12.1 The general model

12.1.1 *Self-propelling of Brownian particles*

When the British botanist *Robert Brown* discovered in 1827 the erratic motion of small particles immersed in a liquid, he considered them first as living entities. He addressed even a letter to *Charles Darwin*, asking him about his opinion concerning these creatures. A legend is saying that Darwin, wise by long experience, answered in a rather indefinite way. Indeed, only after the turn of the century, Einstein, Smoluchowski, Langevin and others have shown that the behaviour of Brownian particles are due to physical effects only. As we have demonstrated in the previous chapters, the behavior of usual Brownian particles is completely due to the (passive) stochastic collisions, the particles suffer from the surrounding medium. There is no active transfer of energy to the particles. The energetic equilibrium between particles and surrounding medium is expressed by the fluctuation–dissipation theorem.

In this Chapter, we want to generalize the idea of Brownian particles including an energy input from the surrounding. This way we will be able to derive a simplified model of active biological motion. We introduce active Brownian particles which are Brownian bodies with the ability to take up energy from the environment. Simple models of active Brownian particles were studied already in Chapter 5 of this book and in in many earlier works (Schienbein & Gruler,1993; Steuernagel et al., 1994; Klimontovich, 1995; Derenyi & Viscek,1995; Bier & Astumian, 1996; Mikhailov & Calenbuhr, 2002; Schweitzer, 2003). Here we will study in a more systematic way models of many Brownian particles with negative friction and will investigate in more details the depot model for particles which are able to store the inflow of energy in an internal depot and to convert internal energy to perform different activities (Schweitzer, Ebeling & Tilch, 1998; Ebeling, Schweitzer & Tilch, 1999). Other versions of active Brownian particle models (Schimansky-Geier et al., 1995, 1997; Schweitzer et al., 1997; Schweitzer, 2003) consider more specific activities, such as environmental changes and signal–response behavior. In these models, the active Brownian particles (or active walkers, within a discrete approximation) are able to generate a self-consistent field, which in turn influences their further movement and physical or chemical behavior. This non-linear feedback between the particles and the field generated by themselves results in an interactive structure formation process on the macroscopic level. Hence, these models have been used to simulate a broad variety of pattern formations in complex systems, ranging from physical to biological and social systems (Schweitzer & Schimansky-Geier, 1994; Schimansky-Geier et al., 1995, 1997; Schweitzer et al., 1997; Helbing, 1997, 2001, Mikhailov & Cahlenbuhr, 2002; Schweitzer, 2003).

The plan of the Chapter is in brief as follows: At first we will develop several models of systems of active Brownian particles in 2-d systems including energy input and noise. The energy input is modeled

(i) by Rayleigh–Helmholtz-type velocity–dependent friction,
(ii) by models of self-propelling based on internal energy depots,
(iii) by more general partially negative space–dependent friction laws.

The Rayleigh–Helmholtz friction was originally developed to model the complex energy input in musical instruments. In the framework of depot models, the Brownian particles have explicitly the ability to take up energy from the environment, to store it in an internal depot and to convert internal energy into kinetic energy. Considering also internal dissipation, we consider this as a simplified model of active biological motion. For the take-up of energy several examples we will discuss here only a spatially homogeneous supply of energy. The case of supply of energy at spatially localized sources (food centers) generates a more complicated dynamics (Ebeling et al., 1999).

The motion of the particles is described here by a Langevin equation which includes an acceleration term resulting from the pumping. The corresponding Fokker–Planck equations are derived. Simulations of the Brownian particles are compared with analytical solutions of the Fokker–Planck equation. The velocity distributions show a crater-like shape which strongly deviate from Maxwell distributions. In the presence of external parabolic forces, the system develops a limit cycle in the 4-d phase space, the corresponding distribution has the form of a hoop or a tire in the 4-d space.

Summarizing, our basic assumption is to add to the dynamics of simple physical Brownian particles the new mechanism of pumping with free energy, which may be realized in several steps as by energy take-up, storage and conversion of energy, and energy consuming motion. This way, the particles become more complex, which result in new dynamical features that may resemble active biological motion. Hence, the basic idea can be formulated as follows: how much of physics is needed to achieve a degree of complexity which gives us the impression of motion phenomena found in biological systems? However, we will study here only the physical aspects of the problem. In particular we are interested in the question how known types of Hamiltonian motion or Brownian motion could be extended by mechanisms of energy take-up, storage and conversion. These new elements should contribute to the development of a microscopic theory of active biological motion. In the present model we restrict ourselves to take into account specific aspects of energy balances that are related to the mechanisms of energy pumping and energy dissipation.

We will show, that in comparison with simple Brownian particles, the active particles become much more complex, which result in new dynamical features as e.g.:

(i) New diffusive properties with with large mean square displacement,
(ii) Unusual velocity distributions with craterlike shape,
(iii) Formation of limit cycles corresponding to the motion on circles in space.

Some of these features may resemble active biological motion. Hence, the basic idea can be formulated as follows: how much of physics is needed to achieve a degree of complexity which gives us the impression of motion phenomena found in biological systems?

In order to avoid misunderstandings we underline again, that we do not intend here to model any particular biological or social object but instead to analyze particular physical nonequilibrium systems which show new types of dynamics which might be interest for a later more concrete approach.

12.1.2 *Equations of motions*

The motion of Brownian particles with general velocity- and space-dependent friction in a space-dependent potential $U(r)$ can be described by the Langevin equation:

$$\frac{dr}{dt} = v\,; \; m\frac{dv}{dt} = F - \nabla U(r) + \mathcal{F}(t)\,. \tag{12.1}$$

Here **F** is a dissipative force which is in the simplest case given by a friction law

$$F = -m\gamma(r,v)v \tag{12.2}$$

where $\gamma(r,v)$ is the friction function of the particle with mass m at position r, moving with velocity v. The friction $\gamma(r,v)$ may depend on space and time. $\mathcal{F}(t)$ is a stochastic force with strength S and a δ-correlated time dependence

$$\langle \mathcal{F}(t) \rangle = 0\,; \; \langle \mathcal{F}(t)\mathcal{F}(t') \rangle = 2S\,\delta(t-t')\,. \tag{12.3}$$

The noise strength S for the momentum is connected with the previously used noise strength for the velocities D_v by the simple relation $S = m^2 D_v$. In the case of thermal equilibrium systems, with $\gamma(r,v) = \gamma_0 = \text{const.}$, we may assume that the loss of energy resulting from friction, and the gain of energy resulting from the stochastic force, are compensated in the average. In this case the fluctuation-dissipation theorem (Einstein relation) is saying:

$$S = D_v\, m^2 = m k_B T \gamma_0\,. \tag{12.4}$$

T is the temperature and k_B is the Boltzmann constant, and D_v is a scaled expression for the strength of the stochastic force in the velocity space. In the following we often will choose units in which $m \equiv 1$ what leads to $S = D_v$. We are interested mainly in statistical descriptions, i.e. in the probability $P(r, v, t)$ to find the particle at location r with velocity v at time t. As shown earlier the distribution function, $P(r, v, t)$, which corresponds to the Langevin equation (12.1), can be described by a Fokker–Planck equation of the form:

$$\frac{\partial P(r,v,t)}{\partial t} + v\,\frac{\partial P(r,v,t)}{\partial r} + \nabla U(r)\,\frac{\partial P(r,v,t)}{\partial v} \tag{12.5}$$

$$= \frac{\partial}{\partial v}\left[\gamma(r,v)\,v\,P(r,v,t) + D_v\,\frac{\partial P(r,v,t)}{\partial v}\right]. \tag{12.6}$$

As discussed already previously (see Chapter 6, section 6.1), in the special case $\gamma(r,v) = \gamma_0$ the stationary solution of eqn.(12.6), $P_0(r,v)$, is known to be the Boltzmann distribution:

$$P_0(r,v) = \mathcal{N}\,\exp\left\{-\frac{1}{k_B T}\left[\frac{m}{2} v^2 + U(r)\right]\right\}. \tag{12.7}$$

The major question discussed throughout this Chapter is, how this known picture changes if we add a new activity to the model by considering that Brownian particles can be also pumped with energy from the environment. While for usual Brownian motion the dissipation of energy caused by friction is compensated by the stochastic force, we now discuss the case of an additional influx of energy, which may be used to accelerate the particle's motion. In our model, this will be considered by a more complex friction function which now can be a space- and/or velocity-dependent function, $\gamma(\mathbf{r}, \mathbf{v})$. To gain more insight, we restrict the discussion to cases where the friction depends only on $\mathbf{v}$.

Let us consider now several models of the self-propelling mechanism, in part repeating and generalizing the results from Chapter 5. First we consider velocity-dependent friction as a mechanism accelerating the Brownian motion. Velocity-dependent friction plays an important role e.g. in certain models of the theory of sound developed by Rayleigh and Helmholtz. In the simplest case we may assume the following friction force of the individual Brownian particle:

$$\gamma(r,v) = -\gamma_1 + \gamma_2 v^2 = \gamma_1\left(\frac{v^2}{v_0^2} - 1\right) = \gamma_2(v^2 - v_0^2)\,. \tag{12.8}$$

This Rayleigh–Helmholtz-model is a standard model studied in many papers on Brownian dynamics (Klimontovich, 1995; Erdmann et al., 2000). We note that $v_0^2 = \gamma_1/\gamma_2$ defines a special value of the velocities where the friction is zero. Another standard model for active friction with a zero point v_0 was detected empirically in experiments with moving cells and analyzed by Schienbein and Gruler (1973)

$$\gamma(v) = \gamma_0 \left(1 - \frac{v_0}{|v|}\right). \tag{12.9}$$

It was shown by the mentioned authors that this model allows to describe the active motion of several cell types as e.g. granulocytes (Schienbein & Gruler, 1993; Erdmann et al., 2000). A disadvantage of this model is the singularity of the friction function at $v = 0$. On the other hand we may consider as an advantage that the friction function converges at large v to the constant of passive friction.

Now we will consider the so-called depot model (see also Chapter 5) for the friction function which is well behaved in the full velocity range. This friction function is based on the idea of an energy depot of the particles (Schweitzer et al., 1998; Ebeling et al., 1999; Schweitzer, 2003). We assume that the Brownian particle itself should be capable of taking up external energy storing some of this additional energy into an internal energy depot, $e(t)$. This energy depot may be may be altered by three different processes:

(1) take-up of energy from the environment; where $q(\mathbf{r})$ is the space-dependent pump rate of energy
(2) internal dissipation, which is assumed to be proportional to the internal energy. Here the rate of energy loss, c, is assumed to be constant.
(3) conversion of internal energy into motion, where $d(v)$ is the rate of conversion of internal to kinetic degrees of freedom. This means that the depot energy may be used to accelerate motion on the plane.

This extension of the model is motivated by investigations of active biological motion, which relies on the supply of energy, which is dissipated by metabolic processes, but can be also converted into kinetic energy. The resulting balance equation for the internal energy depot, e, of a pumped Brownian particle is then given by:

$$\frac{d}{dt}e(t) = q(\mathbf{r}) - c\, e(t) - d(v)\, e(t). \tag{12.10}$$

A simple ansatz for $q(\mathbf{r})$ and $d(\mathbf{v})$ reads:

$$q(\mathbf{r}) \equiv q_0 \qquad d(\mathbf{v}) = d v^2 \tag{12.11}$$

where $d > 0$ is the conversion rate of internal into kinetic energy. Under the condition of stationary depots we get

$$e_0 = \frac{q_0}{c + d\mathbf{v}^2}. \tag{12.12}$$

The energy conversion may result in an additional acceleration of the Brownian particle in the direction of movement. This way we get for the dissipative force including the usual passive friction and the acceleration on the cost of the depot

$$F = -m\gamma v + mde(t)v. \tag{12.13}$$

Correspondingly we find a Langevin equation, which contains an additional driving force, $de(t)v$:

$$m\dot{v} + m\gamma_0 v + \nabla U(r) = mde(t)v + \mathcal{F}(t). \tag{12.14}$$

Hence, the Langevin Eq. (12.14) is now coupled with the equation for the energy depot, Eq. (12.10). The energy loss of the depot is fully converted into kinetic energy of motion of the Brownian particle. In most cases we will assume in the following that the energy bag is stationary $\dot{e}(t) = 0$. This allows the adiabatic elimination of the energy and leads to an effective dissipative force (see Fig. 12.1):

$$F(v) = -m\left[\gamma_0 - \frac{dq}{c + dv^2}\right]v. \tag{12.15}$$

The corresponding friction function is

$$\gamma(v) = \gamma_0 - \frac{dq}{c + dv^2}. \tag{12.16}$$

The behavior of the force and the friction changes qualitatively in dependence on the bifurcation parameter (Erdmann et al., 2000; Erdmann & Ebeling, 2003)

$$\zeta = \frac{dq}{c\gamma_0} - 1. \tag{12.17}$$

For positive ζ-values we observe that the force disappears for 3 values of the velocity. Let us now consider several special cases in more detail: In the case that the velocities are rather small we get for the friction law

$$\gamma(v) = \left(\gamma_0 - \frac{dq}{c}\right) - \frac{q\,d}{c^2}v^2 + \mathcal{O}\left(v^4\right) \tag{12.18}$$

which corresponds with

$$\gamma_1 = \frac{dq}{\gamma_0} - \gamma_0; \qquad \gamma_2 = \frac{qd}{c^2} \tag{12.19}$$

to the Rayleigh–Helmholtz model discussed above.

For positive ζ, due to the pumping with free energy, slow particles are accelerated and fast particles are damped. At definite conditions our active friction functions have a a zero corresponding to stationary velocities v_0, where the friction function and the friction force disappear. The deterministic trajectory of our system moving on a plane is in both cases attracted by a cylinder in the 4-d space given by

$$v_1^2 + v_2^2 = v_0^2 \tag{12.20}$$

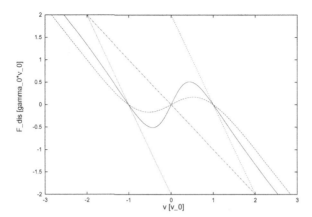

Fig. 12.1 Friction force driving active particles corresponding to the depot model (SET-model): (i) Passive friction force $q = 0, \zeta = -1$ (dash-dotted straight line crossing the center). (ii) Depot model for positive values of the strength of driving: $\zeta = 0.5$ (dashed line); $\zeta = 2$ full line; $\zeta = \infty$ (dash-dotted line with a step at zero).

where v_0 is the value of the stationary velocity which is for the Rayleigh-model or the depot model respectively

$$v_0^2 = \frac{\gamma_1}{\gamma_2}; \qquad v_0^2 = \frac{q_0}{\gamma_0} - \frac{c}{d}. \tag{12.21}$$

Before we conclude this section let us discuss briefly several other closely related depot models. We may define a second variant of the depot model by the assumptions:

$$\boldsymbol{F}(\boldsymbol{v}) = m \left[\frac{\boldsymbol{v}}{v} de - \gamma_0 \boldsymbol{v} \right] \tag{12.22}$$

where again e is the energy content of a depot and d a conversion parameter. In difference to the standard depot model (SET-model) the first term which expresses an acceleration in the direction of $\boldsymbol{v}$ is not dependent on the modulus $|\boldsymbol{v}|$. The acceleration depends only on the energy content e. The corresponding balance of the depot energy reads

$$\frac{de}{dt} = q - ce - d|\boldsymbol{v}|e. \tag{12.23}$$

Within this second variant of the depot model we get assuming $q > 0$ and requiring that the internal energy depot relaxes fast compared to the motion of the particle in adiabatic approximation

$$\boldsymbol{F} = m\boldsymbol{v} \left(\frac{dq}{cv + dv^2} - \gamma_0 \right). \tag{12.24}$$

Now for any $q > 0$, a root $v_0 > 0$ exists, and the Schienbein–Gruler law follows with the correct derivative in the limit $|\boldsymbol{v}| \gg v_0$. On the other hand the Rayleigh law cannot be obtained from the present depot model in a simple way. There is one point which seems to be unrealistic in both depot models discussed so far: The dissipative force increases linearly with the energy content. For real systems one would expect a saturation with increasing e. This leads us to a third variant of the depot model:

$$F(v) = m\left[\frac{v}{v}\frac{de}{1+ge} - \gamma_0 v\right]. \quad (12.25)$$

The parameter g leads to the wanted saturation for $ge \gg 1$. The corresponding balance of the depot energy reads

$$\frac{de}{dt} = q - ce - d|v|\frac{de}{1+ge}. \quad (12.26)$$

Within this third variant of the depot model we get assuming $q > 0$ and requiring that the internal energy depot relaxes fast compared to the motion of the particle in adiabatic approximation

$$F = -mv\left(\frac{G(v)}{2gv^2} - \gamma_0\right) \quad (12.27)$$

where the function $G(v)$ is defined by

$$G(v) = dv + c + gq - \sqrt{(dv + c - gq)^2 + 4qgc}. \quad (12.28)$$

The behavior of this third variant of depot models introduced here is not essentially different from the standard depot model (SET-model) or from the Schienbein–Gruler law. Which of the models is more realistic has to be decided from case to cae, based on experimental data.

12.2 Force-free motion of active particles and mean square displacement

12.2.1 *Distribution functions*

Let us study first the stationary solutions of the equation for the Rayleigh-model of active friction. For the case of free motion (no external forces) we get the stationary solution

$$P_0(v) = \mathcal{N}\exp\left[\frac{\gamma_1}{2D_v}v^2 - \frac{\gamma_2}{4D_v}v^4\right]. \quad (12.29)$$

The shape of this distribution (Eq. (12.29)) can be seen in Fig. 12.2. With $\gamma_2 = 1$ the normalization constant is (Erdmann et al., 2000):

$$\mathcal{N}^{-1} = \pi\sqrt{\pi D_v}\exp\left(\frac{\gamma_1^2}{4D_v}\right)\left[1 + \mathrm{erf}\left(\frac{\gamma_1}{2\sqrt{D_v}}\right)\right]. \quad (12.30)$$

For the Schienbein–Gruler model the solution is of particular simplicity (Schienbein & Gruler, 1993)

$$P_0(v) = \mathcal{N} \exp\left[\frac{\gamma_0}{2D_v}(|v| - v_0)^2\right]. \qquad (12.31)$$

For the depot-model (SET-model) the stationary solution reads

$$P_0(v) = \mathcal{N}\left(1 + \frac{d}{c}v^2\right)^{\frac{q}{2D_v}} \exp\left[-\frac{\gamma_0}{2D_v}v^2\right]. \qquad (12.32)$$

The Fig. 12.2 shows a cross section of the probability distribution for Rayleigh-Helmholtz and Schienbein–Gruler friction function For strong noise corresponding

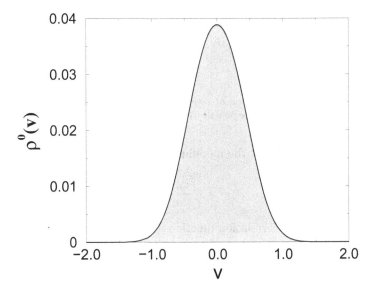

Fig. 12.2 Velocity distribution function of active Brownian particles for the depot model with undercritical values of the parameters (passive regime $d = 1$).

to high temperatures $D \sim T \to \infty$ we get by using Eq. (12.4) the Maxwell distribution

$$P_0(v) = \left(\frac{m}{2\pi k_B T}\right) \exp\left[-\frac{mv^2}{2k_B T}\right]. \qquad (12.33)$$

This limit case is well known, it corresponds to the standard Brownian motion. Many characteristic quantities are explicitly known and were given already in earlier Chapters, as e.g. the dispersion of the velocities

$$\langle v^2 \rangle = 2\frac{k_B T}{m}, \qquad (12.34)$$

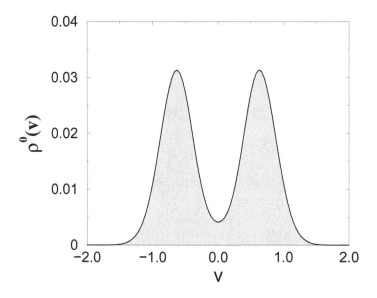

Fig. 12.3 Velocity distribution function of active Brownian particles for the depot model with overcritical parameter values (active regime $d = 10$).

and the most probable value of the modulus of the velocity

$$\tilde{v} = \sqrt{\frac{k_B T}{m}}. \tag{12.35}$$

Further we know the autocorrelation functions

$$\langle \boldsymbol{v}_i(t)\boldsymbol{v}_j(0)\rangle = \delta_{ij}\frac{k_B T}{m}. \tag{12.36}$$

In the opposite case of weak noise we get a hat-like distribution, see Fig. 12.3, and in the limit $D_v \sim T \to 0$ and strong pumping we find a δ-distribution of the velocities

$$P_0(\boldsymbol{v}) = \mathcal{N}\delta\left(\boldsymbol{v}^2 - \boldsymbol{v}_0^2\right). \tag{12.37}$$

In this case of strong pumping the distribution function is maximal on the cylinder discussed above. The cross-section with the $v_1 - v_2$-plane has the shape of a hat.

12.2.2 Mean-square displacement

Near equilibrium the following well-known expression holds for the mean square displacement in two-dimensional Brownian motion

$$\langle (\boldsymbol{r}(t) - \boldsymbol{r}(0))^2 \rangle = \frac{2kT}{m\gamma_0}\left[t + \gamma_0^{-1}(\exp(-\gamma_0 t) - 1)\right]. \tag{12.38}$$

which gives in the limit $t \to \infty$ the Einstein formula

$$\langle (\boldsymbol{r}(t) - \boldsymbol{r}(0))^2 \rangle = 2D_r t. \tag{12.39}$$

Here

$$D_r = \frac{kT}{m\gamma_0} \tag{12.40}$$

is the coefficient of spatial diffusion. Following Schienbein and Gruler (1993) or Mikhailov and Meinköhn (1997) we get in this case the following formula for the mean-square displacement

$$\langle (\boldsymbol{r}(t) - \boldsymbol{r}(0))^2 \rangle = \frac{2v_0^4}{D_v} t + \frac{v_0^6}{D_v^2} \left[\exp\left(-\frac{2D_v t}{v_0^2} \right) - 1 \right]. \tag{12.41}$$

Following recent work (Ebeling, 2004) we will derive a formula for the mean square displacement which contains the Einstein formula and the Mikhailov/Meinköhn formula. We consider the mean-square displacement on the plane $d = 2$ for a dynamics according to the depot model (SET). We will apply a procedure which generalizes the methods developed by Stratonovich (1961, 1963, 1967), Klimontovich (1982, 1986) and Mikhailov & Meinköhn (1997). A particle starting at $t = 0$ in $\boldsymbol{r}(0)$ will at time t at the coordinate vector

$$\boldsymbol{r}(t) = \int_0^t dt_1 \boldsymbol{v}(t_1). \tag{12.42}$$

The general expression for the mean square displacement reads

$$\langle (\boldsymbol{r}(t) - \boldsymbol{r}(0))^2 \rangle = \int_0^t dt_1 \int_0^t dt_2 \langle \boldsymbol{v}(t_1) \boldsymbol{v}(t_2) \rangle. \tag{12.43}$$

The correlation function of the velocities may be calculated exactly if the velocities are Maxwell distributed or δ-distributed (see above). We will apply here the here more general assumption that the velocities have a rather narrow distribution around the most probable value $\tilde{v}$. For the depot model the the most probable velocity $\tilde{v}$ is the positive root of the bi-quadratic equation

$$\frac{D_v}{\tilde{v}^2} + \frac{qd}{c + d\tilde{v}^2} = \gamma_0. \tag{12.44}$$

Following Klimontovich (1982, 1984) we introduce now radial and angle variables for the velocities

$$v_1(t) = \rho(t) \cos \phi(t); \qquad v_2(t) = \rho(t) \sin \phi(t). \tag{12.45}$$

We consider $\rho(t)$ as a slow variable which is decoupled from the fast dynamics of the angle $\phi(t)$. For the correlation function we find

$$K(t) = \langle \boldsymbol{v}(t_1) \boldsymbol{v}(t_2) \rangle = \langle \rho(t_1) \rho(t_2) \cos(\phi(t_1 - t_2)) \rangle. \tag{12.46}$$

Assuming that the dynamics is decoupled we derive

$$K(t) \simeq \langle \rho(t_1)\rho(t_2)\rangle \langle \cos(\phi(t_1 - t_2))\rangle \simeq \tilde{v}^2 \langle \cos(\phi(t_1 - t_2))\rangle. \qquad (12.47)$$

Here we assumed that the absolute values of the velocities are always near to the value of maximal probability. Let us now study the correlation of the angles. Following again Klimontovich (1982, 1986) we have to study the Fokker–Planck equation

$$\frac{\partial f(\phi,t)}{\partial t} = D_\phi \frac{\partial^2 f}{\partial \phi^2} \qquad (12.48)$$

with the diffusion coefficient $D_\phi = D_v/(2\rho^2)$. Replacing here ρ^2 by its mean value we get

$$D_\phi = \frac{D_v}{2\langle v^2\rangle} = \frac{1}{t_0}. \qquad (12.49)$$

The characteristic time t_0 determines the relaxation of the angle-angle correlations. Using the distribution

$$f(\phi,\tau|0,0) = \frac{1}{\sqrt{4\pi t/t_0}} \exp\left[-\frac{t_0\phi^2}{4\tau}\right] \qquad (12.50)$$

we get finally after carrying out the time integration (Ebeling, 2004)

$$\langle (\boldsymbol{r}(t) - \boldsymbol{r}(0))^2\rangle = \tilde{v}^2 \left[tt_0 - t_0^2 + t_0^2 \exp(-t/t_0)\right]. \qquad (12.51)$$

In the limit of passive Brownian motion with Maxwell-distributed velocities $q = 0$ with $\tilde{v}^2 = kT/m$ and $\langle v^2\rangle = 2kT/m$ this leads us back to the classical Einstein formula Eq. (12.38). In the opposite case of strong driving $q \to \infty$ and δ-distributed velocities we come back to the Mikhailov–Meinköhn formula Eq. (12.41). This way we have shown that our approximate expression Eq. (12.51) is correct in the limits $q \to 0$ and $q \to \infty$. For intermediate values of the driving parameter q our result Eq. (12.51) is a useful approximation. This has been confirmed by several simulations (Erdmann et al., 2004). Going to the limit of infinite time $t \gg t_0$ we find the mean-square displacement

$$\langle (\boldsymbol{r}(t) - \boldsymbol{r}(0))^2\rangle = \tilde{v}^2 t_0 t. \qquad (12.52)$$

In the case of strong noise or weak driving we get in agreement with the Einstein-formula

$$D_{\text{eff}} = \frac{2kT}{m\gamma_0} \qquad (12.53)$$

and accordingly for weak noise or strong driving the effective spatial diffusion coefficient of Mikhailov and Meinköhn

$$D_{\text{eff}} = \frac{v_0^4}{D_v}. \qquad (12.54)$$

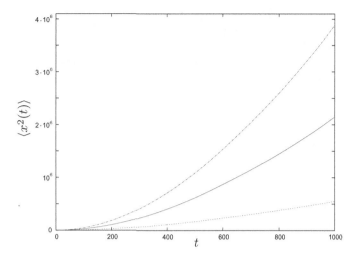

Fig. 12.4 Mean square displacement as a function of time for the parameters $\gamma_0 = 0.5, c = d = 1$ and the values of the driving parameter $q = 1.0$ (dotted line), $q = 2.0$ (fill line) and 3.0 (dashed-dotted line).

For the depot model the expression for v_0^2 provides for large values of the bifurcation parameter ζ the formula

$$D_{\text{eff}} = \frac{1}{D_v}\left(\frac{q}{\gamma_0} - \frac{c}{d}\right)^2. \tag{12.55}$$

Several numerical experiments (Erdmann et al., 2004) correspond in a reasonable way to this asymptotic theory for large q-values, see Fig. 12.4.

12.3 Deterministic motion in external potentials

12.3.1 Parabolic confinement

In the following, we continue to discuss the particle's motion in a two-dimensional space, $r = \{x_1, x_2\}$. The case of constant external forces was already treated by Schienbein and Gruler (1993). Here we specify the potential $U(r)$ as a symmetric parabolic potential:

$$U(x_1, x_2) = \frac{1}{2}a\left(x_1^2 + x_2^2\right). \tag{12.56}$$

First, we study the deterministic motion, which is described by four coupled first-order differential equations:

$$\begin{aligned}\dot{x}_1 &= v_1, & m\dot{v}_1 &= -\gamma(v_1, v_2)\,v_1 - ax_1, \\ \dot{x}_2 &= v_2, & m\dot{v}_2 &= -\gamma(v_1, v_2)\,v_2 - ax_2.\end{aligned} \tag{12.57}$$

For the one-dimensional Rayleigh-model it is well known that this system processes a limit cycle corresponding to sustained oscillations with the energy $E_0 = \gamma_1/\gamma_2$. For the 2-d case we can show by simulation and theoretical considerations that a limit cycle in the 4-d space is developed (Ebeling et al., 1999). The projection of this periodic motion to the $\{v_1, v_2\}$ plane is the circle

$$v_1^2 + v_2^2 = v_0^2 = \text{const.} \tag{12.58}$$

The projection to the $\{x_1, x_2\}$ plane also corresponds to a circle

$$x_1^2 + x_2^2 = r_0^2 = \text{const.} \tag{12.59}$$

Due to the condition of equilibrium between centripetal and centrifugal forces on the limit cycle we have

$$\frac{mv_0^2}{r_0} = ar_0 = m\omega_0^2. \tag{12.60}$$

Therefore the radius of the limit cycle is given by

$$r_0 = \frac{v_0}{\omega_0}. \tag{12.61}$$

The energy for motions on the limit cycle is

$$E_0 = \frac{m}{2}(v_1^2 + v_2^2) + \frac{a}{2}(x_1^2 + x_2^2), \tag{12.62}$$

$$= \frac{m}{2}v_0^2 + \frac{a}{2}r_0^2. \tag{12.63}$$

From Eq. (12.60) follows

$$\frac{m}{2}v_0^2 = \frac{m}{2}r_0^2. \tag{12.64}$$

This means we have equal distribution of potential and kinetic energy on the limit cycle (Ebeling et al., 1999). As for the harmonic oscillator in 1-d, both parts of energy contribute the same amount to the full energy. The energy of motions on the limit cycle, which is asymptotically reached, is double the kinetic energy

$$H \longrightarrow E_0 = mv_0^2. \tag{12.65}$$

The energy is a slow (adiabatic) variable which allows a phase averaging with respect to the phases of the rotation (Ebeling et al., 1999). In explicit form we may represent a solution representing a cycle in the 4-d space by the 4 equations

$$\begin{aligned} x_1 &= r_0 \cos(\omega t + \Phi) & v_1 &= -r_0 \omega \sin(\omega t + \Phi), \\ x_2 &= r_0 \sin(\omega t + \Phi) & v_2 &= r_0 \omega \cos(\omega t + \Phi). \end{aligned} \tag{12.66}$$

By insertion into the dynamic equations we can prove easily, that this is an exact solution (for zero noise) if $\omega = \omega_0$. The frequency is given by the time the particle needs for one period moving on the circle with radius r_0 with constant speed v_0. This leads to the relation

$$\omega = \frac{r_0}{v_0} = \left(\frac{m}{a}\right)^{1/2} = \omega_0. \tag{12.67}$$

Correspondingly, the particle rotates even at strong pumping with the frequency given by the linear oscillator frequency ω_0. The trajectory defined by the above 4 equations looks like a hoop in the 4-d space. Most projections to the 2-d subspaces are circles or ellipses however there are to subspaces namely $x_1 - v_2$ and $x_2 - v_1$ where the projection is like a rod.

A second limit cycle is obtained by reversal of the sense of rotation

$$t \to -t; \quad v_1 \to -v_1; \quad v_2 \to -v_2. \tag{12.68}$$

This leads to the solution

$$\begin{aligned} x_1 &= r_0 \cos(\omega t - \Phi) & v_1 &= -r_0\omega \sin(\omega t - \Phi), \\ x_2 &= -r_0 \sin(\omega t - \Phi) & v_2 &= -r_0\omega \cos(\omega t - \Phi). \end{aligned} \tag{12.69}$$

This second cycle forms also a hula hoop which is different from the first one, however both l.c. have the same projections to the $x_1 - x_2$- and to the $v_1 - v_2$-plane. The projection to the $x_1 - x_2$-plane has the opposite direction of rotation in comparision with the first limit cycle. The projections of the two hula hoops on the $x_1 - x_2$-plane or on the $v_1 - v_2$-plane are 2-d rings (Fig. 12.5) The hula hoops distribution intersect perpendicular the $x_1 - v_2$-plane and the $x_2 - v_1$-plane (see Fig. 12.5). The projections to these planes are rod-like and the intersection manifold with these planes consists of two ellipses located in the diagonals of the planes (see Fig. 12.5). In order to construct solutions for stochastic motions we will need beside $H = mv_0^2$ other appropriate invariants of motion. Looking at the first solution we see, that the following relation is valid

$$v_1 + \omega_0 x_2 = 0; \quad v_2 - \omega_0 x_1 = 0. \tag{12.70}$$

In order to characterize the first l.c. we introduce in accordance with Chapter 5 the invariant

$$J_+ = H - \omega_0 = \frac{m}{2}(v_1 + \omega_0 x_2)^2 + \frac{m}{2}(v_2 - \omega_0 x_1)^2. \tag{12.71}$$

We see immediately that $J_+ = 0$ holds on the first limit cycle which correspond to positive angular momentum. In order to characterize the second l.c. we use the invariant

$$J_- = H + \omega_0 L = \frac{m}{2}(v_1 - \omega_0 x_2)^2 + \frac{m}{2}(v_2 + \omega_0 x_1)^2. \tag{12.72}$$

Correspondingly on the second l.c., which corresponds to negative angular momentum, holds $J_- = 0$.

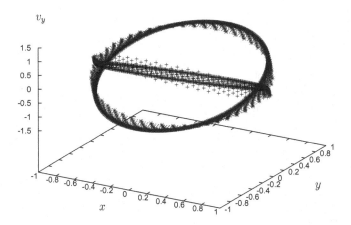

Fig. 12.5 Stroboscopic plot of the 2 limit cycles for driven Brownian motion. We show projections of trajectories for the parameters $v_0 = v_1 = 1$ to the subspace $x_1 - x_2 - v_1$.

As pointed out by Deng and Zhu (2004), there are two other simple but important invariants of motion

$$H_1- = \frac{m}{2}v_1^2 + \frac{m}{2}\omega_0 x_1^2, \qquad H_y- = \frac{m}{2}v_2^2 + \frac{m}{2}\omega_0 x_2^2. \qquad (12.73)$$

As shown by the above mentioned authors, these invariants may also be used for deriving explicit solutions for problems of the stochastic dynamics in parabolic fields.

12.3.2 Perturbed and transient limit cycles

The most important result obtained so far, is the existence of limit cycles, corresponding to stable rotational excitations. In the present section we will discuss briefly several extensions of the theory developed in the previous section:

- effects of anharmonicity of the potential,
- effects of rotational asymmetry of the potential,
- effects due to transients to the stationary energy state.

At first we will discuss how anharmonicity of the potentials influences the dynamics. For the general case of radially symmetric but anharmonic potentials $U(r)$, the equal distribution between potential and kinetic energy $mv_0^2 = ar_0^2$ does not hold. Therefore the relation for the frequencies $\omega_0 = v_0/r_0 = \omega$ is no more valid. It has to be replaced by more general relations. For example the physical condition that on the limit cycle the attracting radial forces are in equilibrium with the centrifugal forces should still be valid. This condition reads in the general case

$$\frac{mv_0^2}{r_0} = |U'(r_0)|. \tag{12.74}$$

If v_0 is given, the equilibrium radius r_0 may be found from the implicit relation

$$v_0^2 = \frac{r_0}{m}|U'(r_0)|. \tag{12.75}$$

Then the frequency of the limit cycle oscillations is given by

$$\omega_0^2 = \frac{v_0^2}{r_0^2} = \frac{|U'(r_0)|}{mr_0}. \tag{12.76}$$

In the case of linear oscillators this leads us back to to $\omega_0 = \sqrt{(a/m)}$. For the case of quartic oscillators

$$U(r) = \frac{k}{4}r^4 \tag{12.77}$$

we get the limit cycle frequency

$$\omega_0 = \frac{k^{1/4}}{v_0^{1/2}}. \tag{12.78}$$

If the equation (12.75) has several solutions, the dynamics might be much more complicated. A second interesting application is the case of attracting Coulomb forces

$$U(r) = -\frac{Ze^2}{r} \tag{12.79}$$

where Ze is the charge of a positive nucleus in the center of the coordinate system and $-e$ a negative charge in distance r. Then the balance condition (12.75) leads us to the a stable radius and the corresponding frequency (Schimansky-Geier et al., 2005)

$$r_0 = \frac{Ze^2}{mv_0^2}; \qquad \omega_0 = \frac{mv_0^3}{Ze^2}. \tag{12.80}$$

The energy on the limit cycle and the angular momentum read

$$H_0 = -\frac{m}{2}v_0^2; \qquad L_0 = \pm\frac{Ze^2}{v_0}. \tag{12.81}$$

Applications to charged grains in dusty plasmas were given in a recent paper (Dunkel et al., 2005). Another interesting application of the theoretical results given above, is the following: Let us imagine a system of Brownian particles which are pairwise bound by a Lennard–Jones-like potential $U(r_1-r_2)$ to dumb-bell-like configurations. Then the motion consists of two independent parts, the free motion of the center of mass, and the relative motion under the influence of the potential. The motion of the center of mass is described by the equations for free motion and the relative motion is described by the equations given in this section. As a consequence, the center of mass of the dumb-bell will make a driven Brownian motion but in addition the dumb-bells are driven to rotate around there center of mass. What we observe then is a system of pumped Brownian molecules which show driven translations with respect to their center of mass. On the other side the internal degrees of freedom may also be excited and we may observe then driven rotations and possibly also driven oscillations. In other words, the mechanisms described here may be used to excite the internal degrees of freedom of Brownian molecules (Erdmann et al., 2000). Further the theory may be applied to the dynamics of clusters of molecules. Similar as in the case of the dumb-bells, the clusters will be driven to make spontaneous rotations. Finally a stationary state will be reached which is a mixture of rotating clusters or droplets similar as described by Mikhailov and Calenbuhr (2002) in the framework of a different model. We will come back to the dynamics of clusters in a later section.

Let us study now another interesting effect, connected with broken rotational symmetry of the potential. So far we studied only external potentials depending on the radius r. In reality this symmetry might be broken and we should study the more general case of asymmetric external potentials (Erdmann et al., 2000). Let us assume that the former symmetric parabolic potential is a bit stretched, elliptically. We introduce a small asymmetry. In nature, systems with exact radial symmetry cannot be found. In the general case of elliptic symmetry we find after turning to the main axis:

$$U(x_1, x_2) = \frac{1}{2} \left(a_1 x_1^2 + a_2 x_2^2 \right). \tag{12.82}$$

We introduce the parameter of asymmetry

$$\Delta = a_1 - a_2 = \omega_1^2 - \omega_2^2, \tag{12.83}$$

what can be understood as small detuning of the frequencies of two oscillators. The deterministic dynamics is then given by the equations

$$\dot{x}_1 = v_1 \qquad \dot{v}_1 = \left[\alpha - \beta \left(v_1^2 + v_2^2 \right) \right] v_1 - \omega_1^2 x_1, \tag{12.84}$$

$$\dot{x}_2 = v_2 \qquad \dot{v}_1 = \left[\alpha - \beta \left(v_1^2 + v_2^2 \right) \right] v_2 - \omega_2^2 x_2. \tag{12.85}$$

We investigate now the stability of the existing limit cycles in the parameter space $\{\Delta, \alpha\}$. Apart from the limit cycles described above there exist an infinite number of limit cycles if one starts with radial initial conditions. This can be understood if one looks at the system Eq. (12.84) in complex representation ($z = x_1 + ix_2$ and $\omega_1 = \omega_2 = \omega_0$):

$$\ddot{z} - \beta \left(\frac{\alpha}{\beta} - \dot{z}^2 \right) + \omega_0^2 z = 0. \tag{12.86}$$

Assuming

$$z(t) = z\, e^{i\Omega_0 t} = |z|\, e^{i\Phi} e^{i\Omega_0 t} \qquad (12.87)$$

one can see that for angles Φ between zero and 2π, stable oscillations are possible. As far as one increases the detuning parameter Δ, the symmetric cycle will be destroyed. For asymmetric initial conditions one can observe two limit cycles as in the case of the parabolic potential. Compared to the parabolic case, for a detuned potential the two limit cycles are just stable within a region of finite size. In simulations one can observe a region within the existing limit cycles stay stable even with a relatively high amount of the detuning Δ. Outside that region (so-called Arnold tongues) the limit cycle can go trough certain bifurcation scenarios like period doubling of the cycle (Neimark–Sacker bifurcations). More details on the bifurcation scenario may be found in (Erdmann et al., 2002).

Finally we want to discuss the role of transient processes. We studied so far the dynamics in the 4-dimensional space $x_1 - x_2 - v_1 - v_2$ and neglected the dynamics of the depot variable. This however, is already a simplification and sometimes the dynamics of the additional 5th variable $e(t)$ might be complicated (Tilch et al., 1999). As a rule however, this dynamics is smooth, exponentially approaching the limit values. In order to show this we will study here the transition of the energy of the depot to its stationary state. Including the full depot dynamics, we have in addition to the previous 4 dynamic variables an additional variable $e(t)$ which is the content of the depot at time t. This variable has of course its own dynamics (Ebeling et al., 1999; Ebeling, 2003). In order to study this dynamics let us assume at first that the depot is full at the initial time $t = 0$ and that there is no feeding $q = 0$. A numerical solution of the simplified depot equation

$$\frac{d}{dt}e(t) = q - c\, e(t) - dv^2\, e(t) \qquad (12.88)$$

together with the dynamic equations for the coordinates and momenta (for $D = 0$) gives transient limit cycles corresponding to left/right rotations. One of them is shown in Fig. 12.6. We see that in the first period a transient limit cycle is formed which then decays since soon the depot energy is exhausted. Therefore stationary or quasistationary processes need a permanent energy support. In later applications of the theory we will study for simplicity only an adiabatic approximation for the depot dynamics.

12.4 Stochastic motion in symmetric external potentials

Since the main effect of noise is the spreading of the deterministic attractors we may expect that the two hoop-like limit cycles are converted into a distribution looking like two embracing hoops with finite size, which for strong noise converts into two embracing tires in the 4-d space. In order to get the explicit form of the

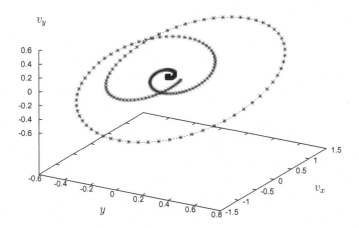

Fig. 12.6 Projection of the 5-d trajectory of the SET depot model corresponding to a transient limit cycle to the 3-d space. We show a numerical solution of Eqs. (12.1), (12.13), (12.56) and (12.88) for the parameters: $e(0) = 20; q = 0; c = .05; d = .1; \gamma_0 = 1)$.

distribution we may introduce the amplitude–phase representation

$$\begin{aligned} x_1 &= \rho(t)\cos(\omega_0 t + \phi(t)) & v_1 &= -\rho(t)\omega_0 \sin(\pm\omega_0 t + \phi(t)), \\ x_2 &= \rho(t)\sin(\omega_0 t + \phi(t)) & v_2 &= \rho(t)\omega_0 \sin(\pm\omega_0 t + \phi(t)), \end{aligned} \tag{12.89}$$

where the radius $\rho(t)$ is now a slow and the phase $\phi(t)$ is a fast stochastic variable (Erdmann et al., 2000). Again the row signs $\pm\omega t$ correspond to the two directions of the angular momentum (right or left rotations). On the basis of this "ansatz" we get for the Hamiltonian

$$H = \omega_0^2 \cdot \rho(t)^2. \tag{12.90}$$

The angular momenta $L = xv_y - yv_x$ corresponding to the two limit cycles are

$$L = +L_0; \qquad L = -L_0; \qquad L_0 = mv_0^2/\omega_0. \tag{12.91}$$

Both limit cycles are located on the sphere $H = mv_0^2$ where H is the Hamiltonian. The dynamics may be treated by using the standard procedure of averaging with respect to the fast phases (Erdmann et al., 2000). A simpler approach is based on the procedure developed in Section 6.1. Considering harmonic oscillators and using equipartition of potential and kinetic energy (see Eq. (12.65) we find for motions on the limit cycle

$$v^2 = \frac{H}{m}. \tag{12.92}$$

Assuming that $v^2 \simeq H/m$ holds also near to the limit cycle, the dynamic system is converted to a canonical dissipative system with

$$\gamma(v^2) \simeq \gamma(H/m). \tag{12.93}$$

This way we come to the following solution for the Rayleigh-model

$$P_0(x_1, x_2, v_1, v_2) = \mathcal{N} \exp\left[\frac{\gamma_1 H - \gamma_2 H^2/2}{D_v}\right] \tag{12.94}$$

with the most probable value of the energy

$$\tilde{H} = H_0 = \frac{\gamma_1}{\gamma_2} = mv_0^2. \tag{12.95}$$

Here H_0 is the energy on the limit cycle.

A different way to derive this distribution is to start from the relation

$$\frac{dH}{dt} = -m\gamma(v^2)v^2 + \sqrt{2D_v}v \cdot \xi(t). \tag{12.96}$$

This leads for the Rayleigh model and near to the limit cycle to the linearized Langevin equation

$$\frac{d\delta H}{dt} = -\gamma_2 v_0^2 \delta H + \sqrt{2D_H}\xi(t) \tag{12.97}$$

with

$$\delta H = H - H_0; \qquad D_H = D_v m v_0^2. \tag{12.98}$$

The stationary distribution of this linearized problem reads

$$P_0(H) = C \exp\left[-\frac{\gamma_2}{2m^2 D_v}\delta H^2\right]. \tag{12.99}$$

We see that the Langevin method leads to the same stationary distribution of the Hamiltonian, the probability is in fact distributed on the surface of a 4-dimensional sphere.

By using Eq. (12.90) we get for the Rayleigh-model of pumping in our approximation the following distribution of the radii:

$$P_0(\rho) \simeq \exp\left[\frac{\gamma_1 \omega_0^2}{D_v}\rho^2\left(1 - \frac{\rho^2}{2r_0^2}\right)\right], \qquad r_0^2 = \omega^2 v_0^2 = \frac{m\gamma_1}{a\gamma_2}. \tag{12.100}$$

We see in Fig. 12.7 that the probability crater is located above the deterministic limit cycles. This way the maximal probability corresponds indeed to the deterministic limit cycle. So far we represented only a projection on the $x_1 - x_2$-plane. The full probability distribution in the 4-d space is not constant on the 4-d sphere $H = mv_0^2$ as suggested by Eq. (12.94) but should be concentrated around the limit cycles which are closed curves on the 4-d sphere $H = mv_0^2$. This means, only a subset of this sphere is filled with probability. The correct stationary probability has

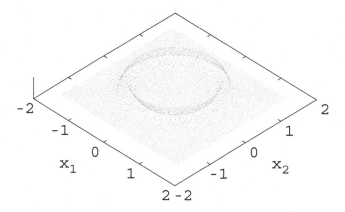

Fig. 12.7 Probability density for the Rayleigh-model represented over the $x_1 - x_2$-plane.

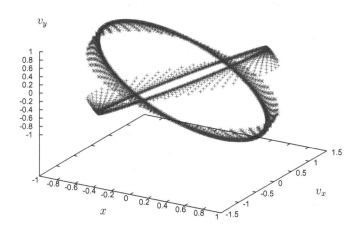

Fig. 12.8 Stroboscopic plot of the 2 limit cycles for driven Brownian motion in parabolic confinement, including weak noise. We show projections of the l.c. to the subspace $y - v_x - v_y$.

the form of two noisy distributions in the 4-d space, which look like hula hoops. This characteristic form of the distributions was confirmed also by simulations (see Fig. 12.8) The projections of the distribution to the $\{x_1, x_2\}$ plane and to the $\{v_1, v_2\}$ plane are noisy tori in the 4-d space (see Fig. 12.8). The hula hoop distribution intersects perpendicular the $\{x_1, v_2\}$ plane and the $\{x_2, v_1\}$ plane. The projections to these planes are rod-like and the intersection manifold with these planes consists of two elipses located in the diagonals of the planes. In order to find the distribution of the angular momenta we start from the Langevin equation

$$\frac{dL}{dt} = -\gamma(v^2)L + \sqrt{2D_L}\xi(t) \quad (12.101)$$

with

$$D_L = m^2 r_0^2 D_v . \quad (12.102)$$

This leads for the Rayleigh model to the following equation for the deviations $\delta L = L \pm m r_0 v_0$:

$$\frac{d\delta L}{dt} = -\gamma_2 v_0^2 \delta L + \sqrt{2D_L}\xi(t) . \quad (12.103)$$

The stationary solution reads

$$P_0(L) = C \exp\left[-\frac{\gamma_2 \cdot \omega_0^2}{2m^2 D_v}(\delta L)^2\right] . \quad (12.104)$$

The given method does not provide a complete solution in the 4-d space, but gives us a good idea about the projections on different planes.

In order to represent the distribution in the 4-d space we come to the following "ansatz":

$$P_0(x_1, x_2, v_1, v_2) = C \exp\left[-\frac{\gamma_2}{2S}(H - H_0)^2\right] \quad (12.105)$$

$$\cdot \left(\exp\left[-\frac{\gamma_2}{2S}J_+^2\right] + \exp\left[-\frac{\gamma_2}{2S}J_-^2\right]\right) \quad (12.106)$$

with $S = m^2 D_v$. We may convince ourselves that this formula agrees with all projections studied above. Furthermore, it is in agreement with the general "ansatz" derived in Chapter 5 from information theory. Since the new expression for the stationary distribution depends only on invariants of motion and does not depend on any parameter of the concrete potential which is investigated, we may apply Eq. (12.106) to any radially symmetric potential (in particular even to Coulomb potentials) (Schimansky-Geier et al., 2005).

In a recent paper Deng and Zhu (2004) derived a different approximative solution. We briefly sketch the result of Deng and Zhu for Rayleigh-type driving, which is based on the Hamiltonian framework for stochastic systems developed by Zhu (2003). This well-developed method for the treatment of certain class of stochastic mechanical systems, is related to the formalism of canonical-dissipative systems (Ebeling, 2000, Schweitzer et al., 2001). Deng and Zhu (2004) start from the amplitude-phase representation

$$\begin{aligned}
x_1 &= \rho_1(t)\cos(\omega_0 t + \phi_1(t)) & v_1 &= -\rho_1(t)\omega_0 \sin(\pm\omega_0 t + \phi(t)), \\
x_2 &= \rho_2(t)\sin(\omega_0 t + \phi_2(t)) & v_2 &= \rho_2(t)\omega_0 \sin(\pm\omega_0 t + \phi_2(t)).
\end{aligned} \quad (12.107)$$

By using the variables

$$\rho_1^2 = H_1 = \frac{m}{2}v_1^2 + \frac{1}{2}\omega_0^2 x_1^2,$$
$$\rho_2^2 = H_2 = \frac{m}{2}v_2^2 + \frac{1}{2}\omega_0^2 x_2^2,$$
(12.108)

and

$$\Phi = \phi_2 - \phi_1.$$
(12.109)

Deng and Zhu derived the following expression for the stationary distribution

$$P_0(x_1, x_2, v_1, v_2) = C \exp\left[-\frac{\lambda(\rho_1, \rho_2, \Phi)}{D_v}\right]$$
(12.110)

with the stochastic potential

$$\lambda(\rho_1, \rho_2, \Phi) = -\gamma_1(\rho_1^2 + \rho_2^2) + \gamma_2\left[\frac{3}{4}(\rho_1^4 + \rho_2^4) + \left(1 + \frac{1}{2}\cos 2\Phi\right)\rho_1^2 \rho_2^2\right].$$
(12.111)

A necessary condition for the correctness of the solution is, that the stochastic potential $\lambda(\rho_1, \rho_2, \Phi)$ disappears on the two attractors of our system (the two hula hoops, representing the limit cycles). We may easily convince ourselves that due to Eq. (12.70), we have on the attractors the relation

$$\Phi = \pm\frac{\pi}{2}; \qquad \cos\Phi = -1$$
(12.112)

and further

$$\rho_1^2 = \rho_2^2 = r_0^2; \qquad r_0^2 = v_0^2/\omega_0^2 = \gamma_1/\gamma_2.$$
(12.113)

This way we have shown that the distribution function is indeed maximal on the attractors of motion. Evidently there exist close relations between both approaches. In order to clarify these relations we write the Deng–Zhu-formula as a product:

$$P_0(x_1, x_2, v_1, v_2) = C \exp\left[-\frac{\gamma_2}{D_v}\left(\frac{3}{4}(H_1^2 + H_2^2) + \frac{1}{2}H_1 H_2 - 2H_0^2\right)\right] \cdot P_2(H_1, H_2, \Phi).$$
(12.114)

The first term is in close relation to Eq. (12.106), but is not identical with it. The second factor is

$$P_2(H_1, H_2, \Phi) = C \exp\left[-\frac{\gamma_2}{2D_v}H_1 H_2 (1 + \cos(2\Phi))\right].$$
(12.115)

The second part has maxima (zero of exponent) for $\Phi = \pm\pi$. In the limit of small D_v we may represent it as

$$P_2(H_1, H_2, \Phi) = \exp\left[-\frac{\gamma_2}{D_v} H_1 H_2 (\Phi - \pi)^2\right] + \exp\left[-\frac{\gamma_2}{D_v} H_1 H_2 (\Phi + \pi)^2\right]. \tag{12.116}$$

Again we see the similarities between both results. The quantitative relation is still not yet clear. However we find identical agreement in the limit $D_v = 0$ where both results yield two delta-functions localized exactly on the limit cycles.

12.5 Collective dynamics of clusters and swarms

12.5.1 *Dynamics of active clusters and swarms*

This paragraph covers in brief a wide range of applications reaching from the dynamics of molecular clusters in nonequilibrium to moving swarms of animals. We introduce the general notation of "swarms" for confined systems of particles (or more general objects) in nonequilibrium. The study of nonequilibrium clusters of molecules begins more than 70 years ago with the pioneering papers of Farkas, Becker and Döring. However most of these studies are restricted to near equilibrium phenomena, as moving over a threshold of the free energy (Schmelzer et al., 1999). Much recent interest was devoted to the theory of Coulomb clusters (Bedanov & Peeters, 1994; Bonitz et al., 2002, Dunkel et al., 2005). Compared to the theory of clusters, the study of objects like swarms of animals is a rather young field of physical studies (see e.g. Helbing, 2001; Vicsek, 2001; Mikhailov & Cahlenbuhr, 2002; Schweitzer, 2003; Erdmann et al., 2005).

Since the dynamics of swarms of driven particles has captured the interest of theorists, many interesting effects have been revealed and in part already explained. We mention the comprehensive survey of Okubo and Levin (2002) on swarm dynamics in biophysical and ecological respect. Further we mention the survey of Helbing (2001) covering traffic and related self-driven many-particle systems and the comprehensive books of Vicsek (2001), Mikhailov and Calenbuhr (2002) and of Schweitzer (2003). In the book of Okubo and Levin (2001) we find a classification of the modes of collective motions of swarms of animals. It is discussed that animal groups have three typical modes of motion:

- translational motions,
- rotational excitations and
- amoeba-like motions.

For example, Ordemann, Moss and Balazci (2003) studied the modes of motion of daphnia. Depending on the existence of a external light source a whole swarm of these animals switches from a uncorrelated type of motion to a very correlated type. The whole swarm starts to rotate then.

At present it seems to be impossible to describe all the complex collective motions of swarms observed in nature. Instead we study in the following the collective modes and the distribution functions of a simple model. We investigate finite systems of particles confined by attracting forces which are self-propelled by active friction and have some hydrodynamic interaction. This is considered as a rough model for the collective motion of nonequilibrium clusters and of swarms of cells and organisms as well (Schweitzer et al., 2001; Ebeling & Schweitzer, 2003). For alternative models based on velocity-velocity interactions see (Vicsek et al., 1995; Czirok & Vicsek, 2000; Vicsek, 2001). From the point of view of statistical mechanics the main problem to study is the dynamics of active Brownian particles including interactions. The self-propelling of the particles is modelled by active friction. The interaction between the particles is modelled by harmonic (linear) forces or by Morse potentials. The consideration is restricted here to 2-d models.

So we will consider many particle systems with interactions, active friction and noise by means of analytical tools and simulations. Driving the system by negative friction we may bring the system to far from equilibrium states. Earlier studies of 1-d models of "swarms" of particles have shown that driven interacting systems may have many attractors (Ebeling et al., 2000; Dunkel et al., 2002; Chetverikov & Dunkel, 2003). For example one finds for Toda chains $N+1$ attractors of deterministic motion including attractors describing dissipative solitons (Chetverikov et al., 2004, 2005). The existence of dissipative soliton-like excitations in 1-d systems found several applications including transport and conductance phenomena (Velarde et al., 2005; Chetverikov et al., 2005).

For applications to more realistic models of swarms we need an extension to 2-d or 3-d systems. We will show, that several phenomena observed in 1-d persist in higher dimensions. In particular, the existence of rotational and soliton-like excitations corresponding to several attractors is not not restricted to 1-d Toda or Morse lattices, but persists at least qualitatively also in more realistic 2-d and 3-d models with Morse- or Lennard–Jones interactions. The superposition of solitons in 1-d systems corresponds to multiple collisions in higher dimensions. Further, in higher dimensions a weak localisation of potential energy was observed which is connected with the collimation of several soliton-like excitations in the region of transitions between different states/configurations (Ebeling et al., 1995, 1996, 1998; Chetverikov et al., 2005).

The most characteristic phenomenon in higher dimensions is the excitation of rotational degrees of freedom connected with angular momenta which may have two directions in 2-d motions and many directions in 3-d motions. As shown above, the generation of angular momenta is connected with specific attractors of the dynamics. Noise may lead to transitions between the deterministic attractors (Erdmann et al., 2005). In the case of two-dynamical motion of interacting particles, positive or negative angular momenta may be generated. This may lead to left/right rotations of pairs, clusters and swarms. We will show that the collective motion of large clusters of driven Brownian particles reminds very much the typical modes of parallel motions in swarms of living entities.

12.5.2 Confined systems of active particles

The study of the dynamics of N interacting active particles is a rather difficult task. We will consider here only simple models and approximations. Let us first introduce forces and interactions described by the potential

$$U(\mathbf{r}_1,\ldots,\mathbf{r}_N) = \sum_i U(\mathbf{r}_i) + \sum_{ij} \Phi(r_{ij}). \tag{12.117}$$

Here the first term describes the external forces and the second one the pair interactions depending on the distance r_{ij} only. We will assume in this section that the external forces are strong enough to take care for the confinement of the swarm. The interactions may be attractive ore repulsive. The dynamics of Brownian particles is determined by the Langevin equation:

$$\dot{\mathbf{r}}_i = \mathbf{v}_i\,;\quad m\dot{\mathbf{v}}_i = -\gamma(v_i)\mathbf{v}_i - \sigma^2(\mathbf{v}_i - \mathbf{V}(t)) - \nabla U(\mathbf{r}_1,\ldots,\mathbf{r}_N) + \mathcal{F}(t) \tag{12.118}$$

where $\mathcal{F}(t)$ is a stochastic force with strength D and a δ-correlated time dependence as defined above. The term proportional to $(v_i - V(t))$ is specific for the swarm dynamics. Here $\mathbf{V}(t)$ is the average velocity of the swarm which in 2-d is the vector

$$V_1 = \frac{1}{N}\sum_i v_{i1};\quad V_2 = \frac{1}{N}\sum_i v_{i2}. \tag{12.119}$$

The small parameter σ measures a possible global coupling of the particles in the swarm to the average velocity of the swarm. This term should express a tendency to parallelize the motion of the particles. We start from the results for the case of one particle $N = 1$ demonstrated in earlier sections. The case $N = 1$ and constant external forces was already treated by Schienbein et al. (1993, 1994); parabolic external forces for $N = 1$ were studied in the previous sections. Here we are going to study the dynamics of 2-d systems of $N \geq 2$ interacting particles forming a swarm as demonstrated in Fig. 12.9. First we study a specific (solvable) approximation, the rigid body model. We assume that the swarm is like a rigid body, i.e. the relative positions of the particles do not change in time. Then the problem is restricted to the dynamics of the center of mass

$$X_1 = \frac{1}{N}\sum_i x_{i1};\quad X_2 = \frac{1}{N}\sum_i x_{i2}. \tag{12.120}$$

The relative motion under the influence of the interactions is described by the coordinates

$$\tilde{x}_{i1} = (x_{i1} - X_1);\quad \tilde{x}_{i2} = (x_{i2} - X_2). \tag{12.121}$$

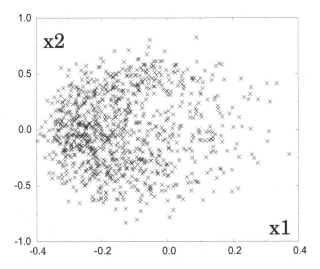

Fig. 12.9 Snapshot of 1000 Brownian particles rotating in a parabolic self-consistent field.

The motion of the center of mass M is described by the equations

$$\dot{X}_1 = V_1 \qquad M\dot{V}_1 = -M\gamma(V_1, V_2) V_1 + \mathcal{F}_1(t),$$
$$\dot{X}_2 = V_2 \qquad M\dot{V}_2 = -M\gamma(V_1, V_2) V_2 + \mathcal{F}_2(t). \qquad (12.122)$$

In the rigid body approximation the stationary solutions of the corresponding Fokker–Planck corresponds to our earlier findings and reads for the depot model

$$P_0(\mathbf{V}) = C\left(1 + dV^2\right)^{(q/2D_v)} \exp\left[-\frac{\gamma_0}{2D_v} V^2\right] \qquad (12.123)$$

corresponding to the driven motion of a free particle with the mass M located at the center of mass. For simplification we specify the potential $U(\mathbf{r})$ as a symmetric parabolic potential:

$$U(x_1, x_2) = \frac{1}{2} a \left(x_1^2 + x_2^2\right). \qquad (12.124)$$

Then the deterministic motion is described by four coupled equations corresponding to the absolute motion of 1 particle in an external field:

$$\dot{X}_1 = V_1 \qquad m\dot{V}_1 = -M\gamma(V_1, V_2) V_1 - aX_1,$$
$$\dot{X}_2 = V_2 \qquad m\dot{V}_2 = -M\gamma(V_1, V_2) V_2 - aX_2. \qquad (12.125)$$

We have shown in the previous sections that a limit cycle in the 4-d space is developed, which corresponds to left/right rotations with the frequency ω_0. As demonstrated in Section 12.4, the projections of this periodic motion are cilcles

$$V_1^2 + V_2^2 = v_0^2 = \text{const}. \qquad (12.126)$$

$$X_1^2 + X_2^2 = r_0^2 = \frac{v_0^2}{\omega_0^2}. \qquad (12.127)$$

The trajectories converge to a limit cycle and any initial value of the energy converges to

$$H \longrightarrow E_0 = mv_0^2. \qquad (12.128)$$

This corresponds to an equal distribution between kinetic and potential energy as known for the harmonic oscillator in 1-d. The energy is a slow (adiabatic) variable which allows a phase average with respect to the phases of the rotation. In explicit form we may represent the motion on the limit cycle in the 4-d space by the 4 equations

$$X_1 = r_0 \cos(\omega_0 t + \Phi); \quad V_1 = -r_0 \omega_0 \sin(\omega_0 t + \Phi), \qquad (12.129)$$

$$X_2 = r_0 \sin(\omega_0 t + \Phi); \quad V_2 = r_0 \omega_0 \cos(\omega_0 t + \Phi). \qquad (12.130)$$

The frequency is given by the time the particle need for one period moving on the circle with radius r_0 with constant speed v_0 leading to

$$\omega_0 = \frac{r_0}{v_0} = \left(\frac{m}{a}\right)^{1/2}. \qquad (12.131)$$

The particle oscillates (in our approximation) with the frequency given by the linear oscillator frequency. The trajectory on the limit cycle defined by the above 4 equations is like a hula hoop in the 4-d space. The projections to the subspaces X_1–V_2 and X_2–V_1 are like a rod. A second limit cycle is obtained by reversal of the velocity. This second limit cycle forms also a hula hoop which is different from the first one, however both l.c. have the same projections to the $\{x_1, x_2\}$ and to the $\{v_1, v_2\}$ plane. The motion in the $\{x_1, x_2\}$ plane has the opposite sense of rotation in comparision with the first limit cycle. Therefore both limit cycles correspond to opposite angular momenta. $L_3 = +\mu r_0 v_0$ and $L_3 = -\mu r_0 v_0$. Applying similar arguments to the stochastic problem we find that the two hoop-rings are converted into a distribution looking like two embracing hoops with finite size, which for strong noise converts into two embracing tyres in the 4-d space. In order to get the explicit form of the distribution we may introduce the amplitude–phase representation (Erdmann et al., 2000)

$$x_1 = \rho \sin(\omega_0 t + \phi); \quad v_1 = -\rho \omega_0 \cos(\omega_0 t + \phi), \qquad (12.132)$$

$$x_2 = \rho \cos(\omega_0 t + \phi); \quad v_2 = -\rho \omega_0 \sin(\omega_0 t + \phi), \qquad (12.133)$$

where the radius ρ is now a slow and the phase ϕ is a fast stochastic variable. By using the standard procedure of averaging with respect to the fast phases we get the distribution of the radii (Erdmann et al., 2000):

$$P_0(\rho) = C \left(1 + d\rho\omega_0^2\right)^{(q/2D_v)} \exp\left[-\frac{\gamma_0}{2D_v}\rho\omega_0^2\right]. \quad (12.134)$$

The probability crater is located above the two deterministic limit cycles on the sphere $r_0^2 = v_0^2/\omega_0^2$. Strictly speaking, not the whole spherical set is filled with probability, but just two circle-shaped subsets on it, corresponding to a narrow region around the limit sets. The full stationary probability has the form of two hula hoop distributions in the 4-d space. This was confirmed by simulations (Erdmann et al., 2000). The projections of the distribution to the $\{x_1, x_2\}$ plane and to the $\{v_1, v_2\}$ plane are smoothed 2-d rings. The distributions intersect perpendicular the $\{x_1, v_2\}$ plane and the $\{x_2, v_1\}$ plane. Due to the noise the swarm of Brownian particles may switch between the two limit cycles, this means inversion of the total angular momentum (direction of rotation of the center of mass around the center of the confinement).

We will study now the rotational excitations of swarms confined to a fixed center. The approximations to be used now are opposite to the approximations used before. We study a finite swarm with the center of mass fixed at the origin of the confinement $\mathbf{R}(t) = 0; \mathbf{V}(t) = 0$. A possible global coupling of the velocities will be neglected ($\sigma = 0$). The pair interaction is assumed either to be weak, or of mean field type. The total confinement potential is approximated oround the origin by a parabolic potential. For this case the N-particle distribution factorizes and we come back to the harmonic one-particle problem solved already in the Section 12.4. For the one-particle Langevin equation of motion in a parabolic potential we have

$$\frac{d\mathbf{v_i}}{dt} + \omega_0^2 \mathbf{r}_i + \gamma(\mathbf{v}_i^2) = (2D_v)^{1/2}\xi_i(t). \quad (12.135)$$

The one-particle distribution function is at the same time the relative density of a swarm of non-interacting particles moving in the external field. Let us first consider the deterministic motion. Under stationary conditions the particles have to observe the requirement of balance between centrifugal and attracting forces $v^2/r = |U'(r)|$. For the harmonic potential this leads to the stationary radius $r_0 = v_0/\omega_0$. Actually the particles are moving in the neighborhood of two limit cycle orbits which have the projections given above and are located on two surfaces in the 4-d space corresponding to the angular momenta $L = \pm v_0^2/\omega_0$. This way the probability is concentrated on two closed curves in the four-dimensional phase space which are similar to tires (see Section 12.4 and Erdmann et al., 2000). Observing this condition we get in accordance with Section 12.4 the approximate solution for the stationary distribution function of the depot model

$$f_0(x_1, x_2, v_1, v_2) = C\left[1 + \frac{dH}{c}\right]^{\frac{q}{D_v}} \exp\left[-\frac{H}{kT}\right]$$
$$\times \left\{\exp\left[-\alpha_1 J_+^2\right] + \exp\left[-\alpha_1 J_-^2\right]\right\} \quad (12.136)$$

with the invariants

$$H = \frac{m}{2}(x_1^2 + x_2^2 + \omega_0^2 x_1^2 + \omega_0^2 x_2^2); \quad J_+ = H - \omega_0 L; \quad J_- = H + \omega_0 L. \quad (12.137)$$

Here

$$L = m(x_1 v_2 - x_2 v_1) \quad (12.138)$$

is the angular momentum. On the limit cycles it has the values $L = \pm m r_0 v_0$. By studying all possible projections of this distribution we we may convince ourselves that in the case $D_v \to 0$ this approximative solution is concentrated around the two limit cycles. On the limit cycles $L = \pm H/\omega_0$ holds. In the stochastic case the angular momentum is distributed around $L \simeq \pm H/\omega_0$. In our approximation this distribution is Gaussian.

Now we will consider more general laws of interactions making use of a mean field approximation. In particular we will consider Morse forces (Erdmann et al., 2000). We consider the case that all particles form one cluster which is hold together by relatively long range forces of Van der Waals-like character with a relatively long range range. For example we may use the interaction model proposed by Morse (1929)

$$\Phi(r) = A \left[\exp(-ar) - 1 \right]^2 - A. \quad (12.139)$$

Due to the attracting tail the particles may form clusters. The individual particles move then in the collectice (self-consistent) field of the other molecules which might be represented by a mean field approximation (Erdmann et al., 2002)

$$V(\tilde{\mathbf{r}}) = \int d\mathbf{r}' \phi(\tilde{\mathbf{r}} - \mathbf{r}') \rho(\mathbf{r}') \quad (12.140)$$

where $\tilde{\mathbf{r}} = (\tilde{x}_1, \tilde{x}_2)$ is the radius vector counted from the center of mass and $\rho(\mathbf{r}')$ is a mean density in the cluster. Approximating the sum of forces in a mean field approximation we may represent the local field near to the center of the swarm by an anharmonic oscillator potential centered around the center of mass. In the normal representation the potential reads

$$U(x_1, x_2) = \frac{1}{2} \left[a_1 (x_1 - X_1) + a_2 (x_2 - X_2) \right]. \quad (12.141)$$

In arbitrary representations we have a second-rank tensor $\mathbf{a}$ with the eigenvalues a_1, a_2, which determine the normal modes of oscillations around the center. The coefficients of the tensor depend on the force law and on the shape of the swarm. In this linear approximation the Langevin equation for an individual Brownian particle in an N-particle system bears some similarity to the symmetrical parabolic case studied above

$$\dot{\mathbf{r}}_i = \mathbf{v}_i, \quad m\dot{\mathbf{v}}_i + \mathbf{a}\mathbf{r}_i = -m\gamma(\mathbf{v}_i^2)\mathbf{v}_i + \sqrt{2D}\xi_i(t). \quad (12.142)$$

The difference to the case studied earlier is the anharmonicity. For small asymmetry of the cluster we may expect to get the same solutions as above for the symmetric parabolic case. However with increasing asymmetry of the shape we find a bifurcation of Arnold tongue type. As a result rotating clusters are getting unstable similar to asymmetric driven oscillators (Erdmann et al., 2002, 2005). In order to check these predictions at least qualitatively we have carried out several simulations for swarms with Morse interactions (Fig. 12.10). We see that the Morse clusters show

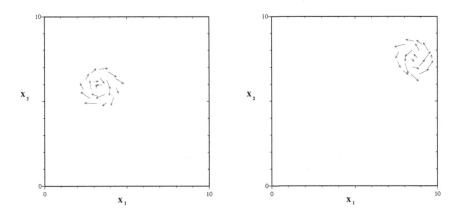

Fig. 12.10 The two possible stationary states of a rotating cluster of 20 particles. The arrows correspond to the velocity of the single particle. In the presence of noise the cluster changes from time to time the direction of rotation.

the expected behavior: We observe rotations changing from time to time the sense of rotations due to stochastic effects. Further we see a very slow drift, which we try to explain in the next section.

12.5.3 *Dynamics of self-confined driven particles*

In the general case the study of the self-confinement of interacting driven particles is a rather difficult task. We begin with the study of 2 driven Brownian particles which are self-confined by attracting forces. In this relatively simple case we observe already several basic features of the self-confined dynamics of swarms. Let us consider two Brownian particles which are pairwise bound by a radial pair potential $U(r_1 - r_2)$. The pair of particles will form dumb-bell like configurations. Then the motion consists of two independent parts: The free motion of the center of mass and the relative motion. The center of mass $\boldsymbol{R} = (\boldsymbol{r}_1 + \boldsymbol{r}_2)/2$ moves with the mass velocity $\boldsymbol{V} = (\boldsymbol{v}_1 + \boldsymbol{v}_2)/2$. The corresponding coordinates are $X_1 = (x_{11} + x_{21})/2$ and $X_2 = (x_{12} + x_{22})/2$. The relative motion under the influence of the forces is described by the relative radius vectors $\boldsymbol{r} = (\boldsymbol{r}_1 - \boldsymbol{r}_2)/2$ and the relative velocity

$v = (v_1 - v_2)/2$. The relative coordinates are $x_1 = (x_{11} - x_{12})$ and $x_2 = (x_{12} - x_{21})$. The deterministic motion of the center of mass is described by the equations

$$m\frac{d}{dt}\mathbf{V} = \frac{1}{2}\left[\mathbf{F}(\mathbf{V}+\mathbf{v}) + \mathbf{F}(\mathbf{V}-\mathbf{v})\right]. \tag{12.143}$$

The relative motion is described by

$$m\frac{d}{dt}\mathbf{v} + U'(r)\frac{\mathbf{r}}{r} = \frac{1}{2}\left[\mathbf{F}(\mathbf{V}+\mathbf{v}) - \mathbf{F}(\mathbf{V}-\mathbf{v})\right] - \sigma^2 \mathbf{v}. \tag{12.144}$$

This system possesses two types of attractors. The first one correspond corresponds to the rotational excitations studied above. We find in linear approximation for the relative velocities

$$m\frac{d}{dt}\mathbf{v} + U'(r)\frac{\mathbf{r}}{r} = -\mathbf{\Gamma} \cdot \mathbf{v} + \mathcal{O}(v^2). \tag{12.145}$$

Here $\mathbf{\Gamma}$ is a tensor which we call friction tensor

$$\mathbf{\Gamma} = \left[(\sigma^2/m + \gamma(V^2))\delta + 2\gamma'(V^2)(\mathbf{V}\mathbf{V})\right]. \tag{12.146}$$

From the dynamical equation for $\mathbf{v}$ we get for the relative energy

$$\frac{d}{dt}\left(\frac{m}{2}\mathbf{v}^2 + U(r)\right) = -\mathbf{v} \cdot \mathbf{\Gamma} \cdot \mathbf{v} + \mathcal{O}(v^2). \tag{12.147}$$

The second attractor corresponds to a translational motion were the two particles move nearly parallel and we have $v^2 \ll V^2 \simeq v_0^2$. With this assumption we find in quadratic approximation in v:

$$\frac{d}{dt}\mathbf{V} = -\left[\gamma(V^2) + \gamma'(V^2)v^2\right]\mathbf{V} - 2\gamma'(V^2)(\mathbf{V}\cdot\mathbf{v})\mathbf{v} + \ldots. \tag{12.148}$$

Assuming that the term of second order $\mathcal{O}(v^2)$ is bounded and remains small we may conclude from this dynamical equation that all velocity states $V^2 \simeq v_0^2$ will converge to the attractor state $V^2 = v_0^2$. Since $\gamma'(v_0^2) > 0$ the coupling to the relative motion may lead to an enhancement of the translation due to an energy flow from the relative to the translational mode. For translational velocities near to the root i.e. $V^2 \simeq v_0^2$ the tensor $\mathbf{\Gamma}$ is positive, having only positive eigenvalues. In this case the r.h.s. of the energy equation is negative i.e. the energy tends to zero, both particles collapse to the minimum of the potential. In other words the attractor of motion is

$$\mathbf{V} = v_0 \mathbf{n}; \qquad \mathbf{R}(t) = v_0 \mathbf{n} t + \mathbf{R}(0). \tag{12.149}$$

In the attractor state $V^2 = v_0^2$ itself the (positive) tensor reads

$$\mathbf{\Gamma} = \sigma^2/m + 2\gamma'(v_0^2)(\mathbf{V}\mathbf{V}). \tag{12.150}$$

This means, excitations with **v** perpendicular to **V** (i.e. (**Vv**) = 0) are only weakly damped and excitations with **v** parallel to **V** show a stronger damping. With decreasing $V^2 < v_0^2$ the tensor Γ shows a bifurcation. This happens, when the first eigenvalue crosses zero, at this point the translation mode is getting unstable. As a consequence the terms $\mathcal{O}(v^2)$ may increase unboundedly and the translational mode breaks down. A similar bifurcation has been found for the 1-d case in (Mikhailov & Zanette, 1999).

Let us consider now the influence of noise. Including noise we expect some distribution around the attractors. In the stable translational regime we find in some approximation the stationary distribution

$$f^{(0)}(\mathbf{V},\mathbf{v},\mathbf{r}) = C\left(1+\frac{d}{c}\mathbf{V}^2\right)^{\frac{q}{2D}} \exp\left[-\frac{1}{kT}\left(m\mathbf{V}^2 + \frac{m}{2}\mathbf{v}^2 + U(r)\right)\right]. \quad (12.151)$$

This corresponds to a driven motion of a free particle located in the center of mass supplemented by a small oscillatory relative motion against the center of mass. The solutions for the rotational model are similar to what we have found for the case of external fields. The probability is distributed around two limit cycles corresponding to left or right rotations.

Summarizing our findings we may state: For two interacting active particles there exists a rotational mode. In this mode the center of the dumb-bell is at rest and the system is driven to rotate around the center of mass. Only the internal degrees of freedom are excited and we observe driven rotations. In the translational mode of the dumb-bell the center of mass of the dumb-bell makes a driven Brownian motion similar to a free motion of the center of mass.

In the case of N-particle systems of interacting active particles the dynamics of the system is given by the equations of motion

$$\frac{d}{dt}\mathbf{r}_i = \mathbf{v}_i \, ; \quad m\frac{d}{dt}\mathbf{v}_i + \sum_j U'(\mathbf{r}_{ij})\frac{\mathbf{r}_{ij}}{|\mathbf{r}_{ij}|} = \mathbf{F}_i(\mathbf{v}_i) - \sigma^2(\mathbf{v}_i - V) + \sqrt{2D}\xi_i(t). \quad (12.152)$$

The result of several simulations is shown in Fig. 12.11 and Fig. 12.13.

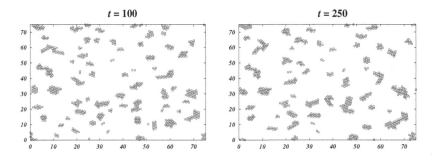

Fig. 12.11 Rotating and drifting clusters of 625 particles with Morse interactions.

An extension of the theory of pairs leads to a theory of the motion of clusters of active molecules. This will be explained in the rest of this Chapter. In order to simplify the simulations we assumed that the interaction of the molecules in the cluster is given by a van der Waals-like interaction with a relatively long range range. For example we may use the interaction model of Morse defined above. Due to the attracting tail the molecules form clusters. The individual molecules move then in the collectice (self-consistent) field of the other molecules which might be represented by a mean field approximation (Erdmann et al., 2002) Approximating the mean field by quadratic terms only we get

$$V(\tilde{x}_1, \tilde{x}_2) = V_0 + \frac{1}{2}\left(a_1 \tilde{x}_1^2 + a_2 \tilde{x}_2^2\right) + \cdots . \qquad (12.153)$$

The problem reduces to that of particles' motion in a harmonic potential discussed above. We may conclued that the individual molecules in the cluster move at least in certain approximation in a parabolic potential. From this follows that they will make rotations in the field. We have performed simulations with 1000 particles moving in a self-consistent potential of parabolic shape. The driving function has a zero at $v_0^2 = 1$. In Figs. 12.9 one sees a snapshot of the moving cluster formed by the molecules. Since the indidual particles move in an effective parabolic potential angular momenta are generated, the swarm starts to rotate. Similar as in the case of the dumb-bells the clusters are driven to make spontaneous rotations. Finally a stationary state will be reached which corresponds to a rotating cluster with nearly constant angular momentum (see Fig. 12.12). Under the influence of noise the cluster may switch to the opposite angular momentum, i.e. to the opposite sense of rotation, the system occurs to be bistable. In Fig. 12.12 is shown that in noisy systems of active Brownian particles the two values $L_3 = -mr_0v_0, +mr_0v_0$ have the maximal probability. Strong coupling of the particles leads to synchronization of the angular momenta, for weak coupling the cluster may be decomposed into groups with different angular momentum (Erdmann et al., 2001). The basic results of our observations may be summarized as follows:

cluster drift We see in the simulations clusters drifting clusters rotating very slowly and clusters without rotations which move rather fast (see Fig. 12.11). The latter state corresponds to the translational mode studied in the previous section for $N = 2$. Here most of the energy is concentrated in the kinetic energy of translational movement.
generation of rotations As we see from the simulations, small Morse clusters up to $N \simeq 20$ generate left/right rotations around their center of mass. The angular momentum distribution is bistable. This corresponds to the rotational mode studied above for $N = 1, 2$.
breakdown of rotations The rotation of clusters may come to a stop due to several reasons. The first is the anharmonicity of clusters. As shown in Erdmann et al. (2002), *strong anharmonicity destroys* the rotational mode.

Another reason are noise induced transitions, this needs som special study.

shape distribution Under special conditions the shape of the clusters is amoeba-like and is getting more and more complicated (Fig. 12.13). A theoretical interpretations of the shape dynamics is still missing.

cluster composition With increasing noise we observe a distribution of clusters of different size. Again a theory of clustering in the two-dimensional case is still missing. For the case of one-dimensional rings with Morse interactions several theoretical and numerical results are available (Dunkel et al., 2001, 2005).

The rotating swarms simulated in our numerical experiments remind very much the dynamics of swarms studied in papers of Viscek and collaborators (Vicsek et al., 1995, Vicsek, 2001) and in other recent work (Erdmann et al., 2002; Schweitzer et al., 2001; Ebeling & Schweitzer, 2001).

We will study now a particular simple case, the model of harmonic swarms with global coupling. Within this model all interactions are reduced to a global coupling ot the particles. We consider two-dimensional systems of N point masses m with the numbers $1, 2, \ldots, i, \ldots, N$ and assume that the masses m are connected by linear pair interactions $\omega_0^2 (\mathbf{r}_i - \mathbf{r}_j)$. The dynamics of the system is given by the following equations of motion

$$\frac{d}{dt}\mathbf{r}_i = \mathbf{v}_i \; ; \; m\frac{d}{dt}\mathbf{v}_i + \omega_0^2 (\mathbf{r}_i - \mathbf{R}(t)) = \mathbf{F}_i(\mathbf{v}_i) - \sigma^2(\mathbf{v}_i - \mathbf{V}) + \sqrt{2D}\xi_i(t) \, . \quad (12.154)$$

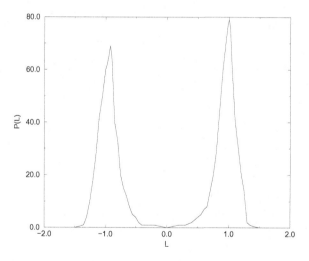

Fig. 12.12 Numerical results for the probability distribution of the angular momentum for active Brownian particles with $v_0^2 = r_0^2 = 1$.

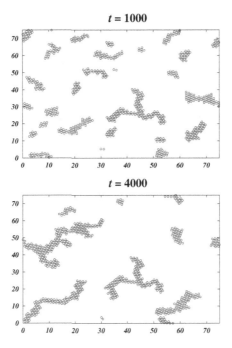

Fig. 12.13 Amoeba-like structures of clusters of 625 particles with Morse interactions.

The center of mass is defined by

$$\mathbf{R}(t) = \frac{1}{N} \sum_{i=1}^{N} \mathbf{r}_j \,. \tag{12.155}$$

The dissipative forces are expressed in the form

$$\mathbf{F}_i(\mathbf{v}_i) = -m\gamma(\mathbf{v}_i^2)\mathbf{v}_i \,. \tag{12.156}$$

The function γ denotes as above the velocity-dependent friction, which possibly has a negative part. The second term on the r.h.s. models a weak global coupling to the average velocity of the swarm

$$\mathbf{V} = \frac{1}{M} \sum_i m_i \mathbf{v}_i \,. \tag{12.157}$$

This way the dynamics of our Brownian particles is determined by Langevin equations with dissipative contributions. The Langevin equations contain as usually a stochastic force with strength D and a δ-correlated time dependence.

$$\langle \xi_i(t) \rangle = 0 \,; \quad \langle \xi_i(t)\xi_j(t') \rangle = \delta(t-t')\delta_{ij} \,. \tag{12.158}$$

In the case of thermal equilibrium systems we have $\gamma(\mathbf{v}) = \gamma_0 =$ const. In the general case where the friction is velocity dependent we will assume that the friction is monotonically increasing with the velocity and converges to γ_0 at large velocities. In the following we will use the Rayleigh ansatz

$$\gamma(\mathbf{v}^2) = -\left(\gamma_1 - \gamma_2 v^2\right). \tag{12.159}$$

Here γ_1, γ_2 are certain constants characterizing the energy flows to the particle. Dependent on the parameters the friction force function may have one zero at $\mathbf{v} = 0$ or two more zeros. For the Rayleigh model we get for positive values of γ_1 which plays the role of a bifurcation parameter the zero

$$v_0^2 = \frac{\gamma_1}{\gamma_2}. \tag{12.160}$$

In the case that such a finite characteristic velocity v_0 exists, we speak about active particles. We see that for $|\mathbf{v}| < v_0$, i.e. in the range of small velocities the dissipative force is positive, i.e. the particle is provided with additional free energy. Hence, slow particles are accelerated, while the motion of fast particles is damped. The exact form of the friction function is not relevant for the dynamics. However it is relevant that a zero of the friction $\gamma(v^2) = 0$ exists, which is an attractor of motion in the velocity space.

We start again with an investigation of the translational mode of this system. For the mean velocity we find by summation and expanding around $\mathbf{V}$ in a symbolic representation

$$\frac{d}{dt}\mathbf{V} = \mathbf{F}(\mathbf{V}) + \frac{1}{2}(\delta \mathbf{v}) * \mathbf{F}''(\mathbf{V}) * (\delta \mathbf{v}) + \ldots. \tag{12.161}$$

For the relative motion $\delta \mathbf{v}_i = \mathbf{v}_i - \mathbf{V}$ we get in first order approximation

$$\frac{d}{dt}\delta \mathbf{v}_i + \omega_0^2 \delta \mathbf{r}_i = -\mathbf{\Gamma} * \delta \mathbf{v}_i + \sqrt{2D}\xi_i(t). \tag{12.162}$$

In the translational mode of this system all the particles form a noisy flock which moves with nearly constant velocity modulus

$$\mathbf{V}(t) = \dot{\mathbf{R}}(t) = v_0 \mathbf{n} \qquad \mathbf{r}_i(t) - \mathbf{R}(t) = 0; \qquad i = 1, \ldots, N. \tag{12.163}$$

The direction $\mathbf{n}$ may change from time to time due to stochastic influences. The distribution function of the flock is Boltzmann-like. However this distribution is stable only in the region where the friction tensor $\mathbf{\Gamma}$ has only positive eigen values. This is for sure if $V^2 = v_0^2$. Our solution breaks down if the dispersion δv^2 is so large that the linearization around V is no more possible. With increasing noise we find a bifurcation. This corresponds to the findings of Mikhailov and Zanette for equivalent 1-d systems (Mikhailov & Zanette, 1999, Erdmann et al., 2005). We note that in the 2-d system the dispersion of the relative velocity $\delta \mathbf{v}$ is not isotropic. The dispersion in the direction of the flight $\mathbf{V}$ is smaller than perpendicular to it. We introduce here an isotropic approximation which allows for explicite solutions of the bifurcation problem. Beyond the region of stability of the translational mode the swarm converges to a rotating swarm at rest (Erdmann et al., 2005). This second stationary state of the swarm corresponds to left/right rotating ring configurations with a center at rest (Schweitzer, Ebeling, Tilch, 2001, Ebeling and Schweitzer, 2001a). In order to describe the numerical results semi-quantitatively, we introduce further approximations. At first we simplify the equation for the mean momentum **V** similar as in the work by Mikhailov and Zanette (1999), assuming

$$\frac{d}{dt}\mathbf{V} = (\gamma_1 - \gamma_2 \mathbf{V}^2)(\mathbf{V}) - \frac{\beta}{N}\sum_i (\delta v_i)^2 - 2\frac{\beta}{N}\sum_i (\mathbf{V}\delta\mathbf{v}_i)\,\delta\mathbf{v}_i + \ldots + \sqrt{2D_v}\xi(t)\,. \tag{12.164}$$

In order to find explicit solutions we decouple the center of mass motion from the relative motion. By averaging with respect to δv_i and neglecting the tensor character of the coupling to the relative motion we get (Ebeling & Erdmann, 2005)

$$\frac{d}{dt}\mathbf{V} = (\alpha - \beta \mathbf{V}^2)\mathbf{V} + \ldots + \sqrt{2D}\xi(t)\,. \tag{12.165}$$

Here the effective driving strength α (which strictly speaking is a tensor) is approximated by

$$\alpha = \gamma_1 - s\frac{\beta}{N}\sum_i (\delta v_i)^2; \qquad \beta = \gamma_2\,. \tag{12.166}$$

The factor s is between 1 (corresponding to strictly perpendicular fluctuations) and 3 (corresponding to only parallel fluctuations). As some reasonable avarage we will assume $s \simeq 2$. The corresponding velocity distribution is

$$f^{(0)}(\mathbf{V}) = C \exp\left[\frac{1}{2D_v}\left(\alpha_1 \mathbf{V}^2 - \beta \mathbf{V}^4\right)\right]\,. \tag{12.167}$$

This way we find the most probable velocity

$$\mathbf{V_1}^2 = \frac{1}{b}\left(\alpha - s\frac{\beta}{N}\sum_i (\delta v_i)^2\right)\,. \tag{12.168}$$

The most probable velocity of the swarm is shifted to values smaller than for the free motion. The shift with respect to the free mode V_0 is proportional to the noise strength D_v. For the fluctuations around the center om mass of the swarm we find

$$\frac{d}{dt}\delta\mathbf{v}_i + \omega^2 \delta\mathbf{r}_i = -\Gamma\delta\mathbf{v}_i + \sqrt{2D}\xi_i(t)\,. \tag{12.169}$$

Here $\Gamma = 2\gamma_2 V_1^2 - \gamma_1$ follows from a diagonal approximation of the tensor $\boldsymbol{\Gamma}$. In this way the relative distribution can be approximated as

$$f_{rel}^{(0)}(\mathbf{v}_i, \mathbf{r}_i) = C \exp\left[-\frac{\Gamma}{2D_v}\left((\delta\mathbf{v}_i)^2 + \omega^2(\delta\mathbf{r}_i)^2\right)\right]\,. \tag{12.170}$$

Now we get a quadratic equation for the dispersion

$$\langle \delta v^2 \rangle = \frac{D_v}{\alpha - 2s\beta \langle \delta v^2 \rangle} \qquad (12.171)$$

with the solution

$$\langle \delta v^2 \rangle = \frac{\alpha}{4s\beta} \left[1 - \sqrt{1 - \frac{8s\beta D_v}{\alpha^2}} \right]. \qquad (12.172)$$

The corresponding effective friction reads

$$\Gamma = \frac{\alpha}{2} \left[1 + \sqrt{1 - \frac{8s\beta D_v}{\alpha^2}} \right]. \qquad (12.173)$$

At the critical noise strength $D_v = \alpha^2/8s\beta$ the dispersion has its maximal and the effective friction its minimal value, for larger noise strength the dispersion and the effective frioction get complex. In simulations we found for $\alpha = \beta = 1$ a critical noise strength $D_{cr} \approx 0.067$. This is close to our theoretical estimate with $s = 2$ which gives $D_{cr} = 1/16$. This way we gave a simple explanation for the transition from translational to rotational modes. A more advanced theory is presented elsewhere (Erdmann et al., 2005).

12.5.4 The influence of hydrodynamic Oseen-type interactions

One could imagine that the particles do not just interact in a direct way but mediately trough a liquid-like medium. Then, the local hydrodynamic field $\mathbf{v}_F$ in which an individual is moving is influenced by the other moving particles. This field might be very complicated, but will have in general certain simple symmetry properties. For simplicity we will assume that it is composed by additive Oseen-contributions similar as used in the theory of electrolytes and macromolecules (Falkenhagen, 1971):

$$\mathbf{v}_F = \sum_j \frac{R}{r_{ij}} \left[\delta + \frac{\mathbf{r}_{ij} \otimes \mathbf{r}_{ij}}{r_{ij}^2} \right] \mathbf{v}_j, \qquad (12.174)$$

where R is a kind of hydrodynamic radius of the particles. The parameter measures the strength of hydrodynamic interactions. Strictly speaking the Oseen-law is valid only asymptotically for $r_{ij} \gg R$ but we will use it here for all distances.

The equations of motion of a active Brownian particle moving in a liquid can written as follows

$$\dot{\mathbf{v}}_i = \mathbf{F}_i(\mathbf{v}_i)\mathbf{v}_i + \kappa_F \mathbf{v}_F + \sqrt{2D}\xi_i(t), \qquad (12.175)$$

where κ_F is a constant which describes the strength of the hydrodynamic coupling. Therefore, $\kappa_F \mathbf{v}_F$ is the hydrodynamic force which acts on every particle induced by the surrounding ones. We do not claim that this is the only way to introduce hydrodynamic interactions but just state that the Oseen-law is the simplest hydrodynamic interaction field with the right symmetry properties. This is to be seen if one looks at the alternative way of writing Eq. (12.174)

$$\mathbf{v}_F(r_{ij}) = \sum_j \frac{R}{r_{ij}} \mathbf{v}_j + \sum_j \frac{R(\mathbf{r}_{ij} \cdot \mathbf{v}_j)}{r_{ij}^3} \mathbf{r}_{ij} \quad \text{if } r_{ij} > R. \qquad (12.176)$$

The new form of writing shows that the Oseen-flow consists of a parallel and a radial component. With respect to collective (in particular synchronization) effects the parallel components are of special relevance. We have shown above that active Brownian particles without hydrodynamic interactions develop rotations with full right/left symmetry. This way we find without hydrodynamic interactions symmetrical distributions of the angular momenta. Including hydrodynamic interactions the situations changes dramatically as we have shown by simulations. The Oseen-type interaction has the tendency to parallelize the motion of two interacting particles. This breaks the the symmetry of left-right in the limit $N \to \infty$. In Fig. 12.14, it is shown that depending on the initial conditions the system develops either left or right rotations corresponding to unsymmetrical distributions of the angular momentum. The initial distribution corresponds to zero angular momentum, then a bistable distribution develops with a small preference to negative L. In this intermediate stage the particles move on a limit cycle with the radius $r_0 = v_0/\omega_0$ with a nearly equal distribution. Finally du to the parallelizing effect of the hydrodynamic interactions the symmetry is broken and the whole swarm makes a right rotation corresponding to negative L. At the same time the particles are concentrating in a noisy flock moving on the limit cycle. The situation is quite similar to what we observe in ferromagnetic systems or in more general Ising-type systems without external fields. Here due to the interactions a spontaneous magnetization (orientation) of the system is developed pointing to a direction with a small preference. In our case the interacting acts in the velocity space and tries to parallelize the directions of the "velocity spins". A "spin-model" of active Brownian particles has been developed by Vicsek and coworkers (Vicsek et al., 1995; Czirok & Vicsek, 2000; Vicsek, 2001).

This way we have shown that an ensemble of pumped Brownian particles with hydrodynamic interactions can produce a *collective swarm dynamics*. Hence, by means of an appropriate *asymmetric potential* and an additional mechanism to drive the system into *non-equilibrium*, we are able to convert the genuinely *non-directed* Brownian motion into *directed motion*. In this respect, our result agrees with other models of swarm dynamics which have been proposed to reveal the microscopic mechanisms resulting in collective movement. In our earlier studies based only on conservative interactions we typically found full symmetry of left/right rotations.

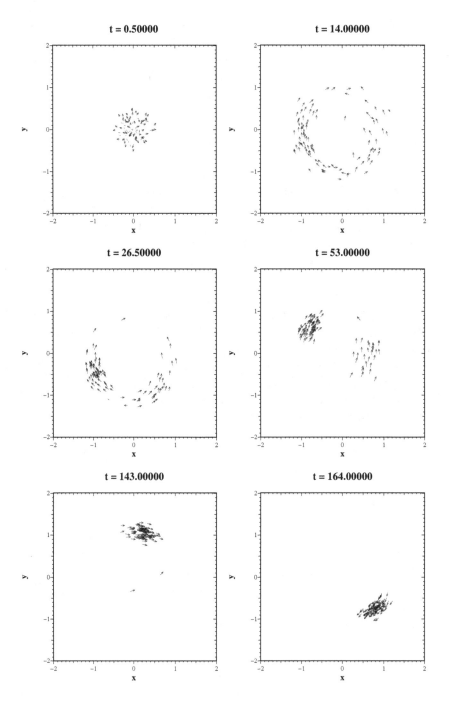

Fig. 12.14 Time evolution of a swarm with Oseen interactions. The globule-like initial distribution without net angular momentum first forms two rings with opposite sense of rotation. Later the symmetry is broken and only one of the rings survives.

This means that in one swarm in a stationary state half of the particles perform left rotations and the other half performs right rotations. As a consequence we found in the stationary state symmetrical distributions of the angular momentum.

Let us draw some conclusions:

We studied in this section the active Brownian dynamics of a finite number $N \geq 2$ of confined particles with velocity-dependent friction and global coupling. Confinement was created

(i) by pair-wise linear attracting forces leading to global coupling, or

(ii) by attracting Morse (or more general Van der Waals-type)interactions.

We have given first an analysis of several simple approximation as the rigid-body, and the fixed-center approximations. Based on these analytical investigations we have made several predictions for the behavior of Morse clusters. We could identify several qualitative modes of movement. Further we have made a numerical study of special N particle Morse systems. In particular we investigated the rotational and translational modes of the swarm and the clustering phenomena. Further we investigated the effect of global couplings. Finally we introduced interactions in the velocity-space. This should model the fact that many creatures swimming in water are able to react on the flow patterns created by other nearby swimming animals. We have modeled this interaction — which in nature might be very complicated — by a simple Oseen-type ansatz. The Oseen-law which originally is based on hydrodynamic studies of the interaction of spheres supports parallel swimming. This is a realistic feature in spite of the fact that the detailed form of the Oseen-law is an oversimplification. The introduction of velocity-dependent interactions breaks (at least in the limit of big swarms $N \gg 1$) the symmetry of the left/right rotations. As shown in our simulations the preference of parallel motions leads to either left or right vortices. Which of the two directions of the angular momentum is selected, depends on initial conditions or small perturbations. Spontaneous transitions between the two directions are still in principle possible, due to stochastic influences, however in practice nearly impossible since the transition time drastically increases with the number of particles in the swarm. We consider the breaking of the symmetry of left/right rotations as an essential new feature of our model since in reality full left/right symmetry of rotations in one swarm is never observed. A real swarm shows either right or left rotations, the particles synchronize the sense of rotations.

We did not intend here to model any particular problem of biological or social collective movement. However may be, that the study of dynamic modes of collective movement of swarms may be of some relevance for the understanding of biological and social collective motions (Ebeling & Schweitzer, 2003; Hänggi et al., 2003). To support this view we may refer again to the book of Okubo and Levin (2001) where the modes of collective motions of swarms of animals are classified in way which reminds very much the theoretical finding for the model investigated here. In particular we mention also the motion of animals in water, for example the collective motion of daphnia (Ordemann et al., 2003).

Concluding the present book on statistical thermodynamics of systems far from equilibrium we tried to give some hints on modern applications to physical and non-physical systems. We express finally the hope that the further investigation of simple model system of the type investigated here will give support to the understanding of complex biological and social motions.

References

B.J. Alder, T.E. Wainwright (1967): "Velocity Autocorrelations for Hard Spheres", *Phys. Rev. Lett.* **18**, 988.

B.J. Alder, T.E. Wainwright (1969): "Enhancement of diffusion by vortex-like motion of classical hard particles", *J. Phys. Soc. Japan (Suppl.)* **26**, 267.

B. Andersen, P. Salamon, R.S. Berry (1984): "Thermodynamics in finite time", *Phys. Today* **37**(9): 62–70.

A.A. Andronov (1929): "Les cycles limites de Poincare et la theorie des oscillations autoentretenues", Comptes Rendus *Acad. Sci. Paris* **189**, 559.

A.A. Andronov, S.E. Chaikin (1937): "Theory of oscillations", (in Russian), Moscow.

A.A. Andronov, E.A. Witt, S.E. Chaikin (1959): "Theory of oscillations" (in Russian), FisMatGis, Moscow.

A.A. Andronov, E.A. Witt, S.E. Chaikin (1965, 1967): "Theorie der Schwingungen", Akademie-Verlag, Berlin.

A.A. Andronov, E.A. Witt, S.E. Chaikin (1966): "Theory of oscillations", Pergamom Press, Oxford.

V.S. Anishchenko (1989): "Dynamical chaos in physical systems", Teubner, Leipzig.

V.S. Anishchenko (1990): "Complex oscillations in simple systems", (in Russian), Nauka, Moskva.

V.S. Anishchenko (1995): "Dynamical chaos — models and experiments", World Scientific, Singapore.

V.S. Anishenko, A.B. Neiman, F. Moss, L. Schimansky-Geier (1999): "Nonlinear dynamics of chaotic and stochastic systems", *Physics — Uspekhi* **42**, 7–77.

V.S. Anishenko, V.V. Astakhov, A.B. Neiman, T.E. Vadivasova, L. Schimansky-Geier (2002): "Nonlinear dynamics of chaotic and stochastic systems", Springer, Heidelberg.

M. Araujo, H. Larralde, S. Havlin, H.E. Stanley (1993): "Scaling anomalies in reaction front dynamics in confined systems", *Phys. Rev. Lett.* **71**(21), 3592–3595.

V.I. Arnold, A. Avez (1968): "Ergodic problems of classical dynamics", Benjamin, New York.

L. Arnold (1998): "Random dynamical systems", Springer, Berlin.

R.D. Astumian, P. Hänggi (2002): "Brownian Motors", *Physics Today* **55** (Nov.), 33.

L. Bachelier (1900): "Theorie de la speculation", *Ann. Sci. Col. Norm. Sup.* **17**, 21.

P. Bak (1996): "How Nature Works", Springer, New York.

P. Bak, L. Chen, M. Creutz (1989): "Self-Organized Criticality in the Game of Life", *Nature* **342**, 780.

P. Bak, C. Tang, K. Wiesenfeld (1987): "Self-organized criticality - an explanation of 1/f noise", *Phys. Rev. Lett.* **59**, 381, see also "Self-organized criticality", *Phys. Rev.* **A38**, 364 (1988).

V. Balakrishnan (1985): "Anomalous diffusion in one dimension", *Physica A* **132**, 569–580.

B.Y. Balagurov, V.G. Vaks (1973): "Random-walk of a particle in a lattice with traps", *J. Eksp. Teor. Phys.* **65**(5), 1939–1946.

R. Balescu (1975): "Equilibrium and nonequilibriun statistical mechanics", Wiley, New York.

R. Balescu (1997): "Statistical dynamics. Matter out of equilibrium", Imperial College Press, London.

J.D. Bao, Y.Z. Zhuo (2003): "Ballistic diffusion induced by a thermal broadband noise", *Phys. Rev. Lett.* **91**, Art. No. 138104, Sep. 26, 2003.

E. Barkai, R. Metzler, J. Klafter (2000): "From continuous time random walks to the fractional Fokker–Planck equation", *Phys. Rev. E* **61** 132-138.

E. Barkai (2001): "Fractional Fokker–Planck equation, solution, and application", *Phys. Rev. E* **63** 046118.

E. Barkai (2003): "Aging in subdiffusion generated by a deterministic dynamical system", *Phys. Rev. Lett.* **90**, 104101.

H. Barkhausen (1907): "Das Problem der Schwingungserzeugung", Hirzel, Leipzig.

F. Beck, M.T. Hütt, U. Lüttge, eds. (2003): "Nonlinear dynamics and the spatiotemporal principles of biology", *Nova Acta Leopoldina NF* **88**, No. 322.

P. Becker-Kern, M.M. Meerschaert, H.P. Scheffler (2004): "Limit theorems for coupled continuous time random walks", *Annals of Probability* **32**(1B), 730–756.

V.M. Bedanov, F.M. Peeters (1994): "Ordering and phase-transitions of charged-particles in a classical finite 2-dimensional system", *Phys. Rev. B* **49** (1994) 2667.

H. Berg (1993): "Random walks in biology", Princeton Univ. Press, Princeton.

P. Berge, Y. Pomeau, Ch. Vidal (1984): "Ordre dans le chaos", Hermann, Paris.

J. Bernasconi, T. Schneider, Eds. (1981): "Physics in one dimension", Springer-Verlag, Berlin–Heidelberg–New York.

P. Bliokh, V. Sinitsin, V. Yaroshenko (1995): "Dusty and Self-Gravitational Plasmas in Space", Kluwer, Dordrecht.

A. Blumen, J. Klafter, G. Zumofen (1986), in "Optical Spectroscopy of Glasses", edited by I. Zschokke (Reidel, Dordrecht) pp. 199–265.

H. Bolterauer, M. Opper (1981): "Solitons in the statistical mechanics of the Toda lattice", *Z. Phys. B* **42**, 155–161.

M. Bonitz, V. Golubnychiy, A.V. Filinov, Yu.E. Lozovik (2002): "Single-electron control of Wigner crystallization", *Microelectronic Engineering* **62**, 141.

J.P. Bouchaud, A. Georges (1990): "Anomalous diffusion in disordered media — statistical mechanisms, models and physical applications", *Phys. Rep.* **195** (4-5), 127–293.

P.W. Bridgman (1928): "Note on the principle of detailed balancing", *Phys. Rev.* **31**, 101–102.

L. Brillouin (1950): "Can the rectifier become a thermodynamical demon?", *Phys. Rev.* **78**, 627–628.

L. Brillouin (1956): "Science and information theory", Academic Press, New York.

M. Bramson, J.L. Lebowitz (1988): "Asymptotic behavior of densities in diffusion-dominated annihilation reactions", *Phys. Rev. Lett.* **61** (21) 2397–2400.

R. Brown (1829): "Additional remarks on active molecules", http://dbhs.wvusd.k12.ca.us/Chem-History/Brown-1829.html.

E. Bruckner, W. Ebeling, A. Scharnhorst (1989): "Stochastic dynamics of instabilities in evolutionary systems", *System Dynamics Review* **5**, 176.

E. Brunet, B. Derrida (1997): "Shift in the velocity of a front due to a cutoff", *Phys. Rev. E* **56**(3), 2597–2604.

E. Brunet, B. Derrida (2001): "Effect of microscopic noise on front propagation", *J. Stat. Phys.* **103**(1–2), 269–282.

S.G. Brush (1965): "Kinetic theory", Vols. 1–2, Pergamon Press, Oxford.

L.A. Bunimovich, Ya.G. Sinai (1980): "Markov partitions for dispersed billiards", *Comm. Math. Phys.* **78**(2), 247.

L.A. Bunimovich, Ya.G. Sinai (1988): "Spacetime chaos in coupled map lattices", *Nonlinearity* **1**, 491.

N.W. Butenin, Ju.L. Neimark, N.A. Fufaev (1976): "Introduction to nonlinear oscillations" (in Russian), Nauka, Moscow.

M. Buttiker (1987): "Transport as a consequence of state-dependent diffusion", *Z. Phys. B — Condensed Matter* **68**(2–3), 161–167.

V. Cahlenbuhr, J.-L. Deneubourg (1991): "Chemical communication and collective behaviour in social and gregarious insects", in: "Models of Self-organization in Complex Systems", Akademie-Verlag, Berlin.

H.C. Callen (1960): "Thermodynamics", Wiley, New York.

M. Caputo (1969): "Elasticità e dissipazione", Zanichelli Printer, Bologna.

S. Chandrasekhar (1943): "Stochastic problems in physics and in astronomy", *Rev. Mod. Phys.* **15**, 823.

D. Chandler (1987): "Introduction to modern statistical mechanics", Oxford University Press, New York.

S. Chandrasekhar (1943): "Stochastic problems in physics and astronomy", *Rev. Mod. Phys.* **15**, 3–100.

A.V. Chechkin, R. Metzler, V.Y. Gonchar, J. Klafter, L.V. Tanatarov (2003): "First passage and arrival time densities for Levy flights and the failure of the method of images", *J. Phys. A: Math. Gen.* **36**(41), L537–L544.

A. Chetverikov, W. Ebeling, M.G. Velarde (2005): "Thermodynamics and phase transitions in dissipative Morse chains", *Eur. Phys. J. B*, in press.

A. Chetverikov, W. Ebeling, M.G. Velarde (2005): "Nonlinear ionic excitations, dynamic bound states and non-linear currents in one-dimensional plasmas", *Conts. Plasma Phys.* **45**(3–4), in press.

A. Chetverikov, W. Ebeling, M.G. Velarde (2005): "Dissipative solitons and complex currents in active lattices", *Int. J. Bifurc. Chaos*, in press.

A.Yu. Chikishev, W. Ebeling, A.V. Netrebko, N.V. Netrebko, Yu.M. Romanovsky, L. Schimansky-Geier (1996): "Stochastic cluster dynamics of macromolecules", in: "Nonlinear Dynamics and Structures in Biology and Medicine: Optical and Laser Technologies", V.V. Tuchin, editor, *Proc. SPIE* **3053**, 54–70.

A.Yu. Chikishev, W. Ebeling, A.V. Netrebko, N.V. Netrebko, Yu.M. Romanovsky, L. Schimansky-Geier (1998): "Stochastic cluster dynamics of macromolecules", *Int. Journal of Bifurcation & Chaos* **8**, 921–926.

O. Chichigina, W. Ebeling, A.V. Netrebko, A. Makarov, Yu.M. Romanovsky, L. Schimansky-Geier (2003): "Stochastic cluster dynamics of macromolecules", *Proc. SPIE* **5110**, 28–40.

G.I. Christov, M.G. Velarde (1995): "Dissipative solitons", *Physica D* **86**, 323–347.

R. Clausius (1865): "Die mechanische Wärmetheorie", Vieweg, Braunschweig.

P. Collet, J.-P. Eckmann (1980) "Iterated Maps on the interval as dynamical Systems", Birkhäuser, Basel.

F.C. Collins, G.E. Kimball (1949): "Diffusion-controlled reaction rates", *J. Colloid Sci.* **4**, 425–437.

M. Conrad (1983): "Adaptability. The significance of variability from molecules to ecosystems", Plenum Press, New York.

D.R. Cox (1967): "Renewal theory", Chapman and Hall, London.

F.L. Curzon, B. Ahlborn (1975): "Efficiency of a Carnot engine at maximum power output", *Am. J. Phys.* **43**, 22–24.

A. Czirok, E. Ben-Jacob, I. Cohen, T. Vicsek (1996): "Formation of complex bacterial colonies via self-generated vortices", *Phys. Rev. E* **54**, 1791–1799.

T. Dauxois, M. Peyrard & A.R. Bishop (1993): "Dynamics and thermodynamics of a nonlinear model for DNA denaturation", *Phys. Rev. E* **47**, 648–695.

P. Debye (1942): "Reaction rates in ionic solutions", *Trans. Electrochem. Soc.* **82**, 265.

S.R. DeGroot, P. Mazur (1984): "Non-equilibrium thermodynamics", Dover, New York.

K.G. Denbigh, J.S. Denbigh (1985): "Entropy in relation to incomplete knowledge", Cambridge University Press, Cambridge.

M.L. Deng, W.Q. Zhu (2004): "Stationary motion of active Brownian particles", *Phys. Rev. E* **69**, 046105-1-9.

T. Dittrich, P. Hänggi, G.-L. Ingold, B. Kramer, G. Schön, W. Zwerger (1989): "Quantum Transport and Dissipation", Wiley, Weinheim-Berlin.

M.D. Donsker, S.R.S. Varadhan (1979): "Number of distinct sites visited by a random walk", *Commun. Pure Appl. Math.* **32**(6), 721–747.

J.R. Dorfman (1999): "An introduction to chaos in nonequilibrium statistical mechanics", Cambridge Univ. Press, Cambridge.

J. Dunkel, W. Ebeling, U. Erdmann (2001): "Thermodynamics and transport in an active Morse ring chain", *Eur. Phys. J.* **24**, 511–524.

J. Dunkel, W. Ebeling, U. Erdmann, V.A. Makarov (2002): "Coherent motions and clusters in a dissipative Morse ring chain", *Int. J. Bifurcations & Chaos* **12**, 2359–2377.

J. Dunkel, W. Ebeling, S. Trigger (2004): "Active and passive Brownian motion of charged particles in two-dimensional plasma models" *Phys. Rev. E* **70**(4), 046406.

A. Dvoretzky, P. Erdös (1951): in J. Neumann, ed., "Proceedings of the second Berkeley Symposium", Univ. Cal. Press, Berkeley, p. 33.

W. Ebeling (1965): "Statistisch-mechanische Ableitung von verallgemeinerten Diffusionsgleichungen", *Ann. Physik* **16**, 147–159.

W. Ebeling, R. Feistel (1977): "Stochastic theory of molecular replication processes", *Ann. Physik* **34**, 91.

W. Ebeling (1981): "Structural stability of stochastic systems", in "Springer Series in Synergetics" (ed. H. Haken) **11**, 188–198.

W. Ebeling, R. Feistel (1982): "Physik der Selbstorganisation und Evolution", Akademie-Verlag, Berlin (2nd ed. Berlin, 1986).

W. Ebeling, Yu.L. Klimontovich (1984): "Selforganization and turbulence in liquids", Teubner, Leipzig.

W. Ebeling, A. Engel, B. Esser, R. Feistel (1984): "Diffusion and reactions in random media", *J. Stat. Phys.* **37**, 369–378.

W. Ebeling, Yu.M. Romanovsky (1985): "Energy Transfer and Chaotic Oscillations in Enzyme Catalysis", *Z. Phys., Chem. Leipzig* **266**, 836–843.

W. Ebeling, I. Sonntag (1986): "Stochastic description of evolutionary processes in underoccupied systems", *BioSystems* **19**, 91.

W. Ebeling, M. Jenssen (1988): "Trapping and fusion of solitons in a nonuniform Toda lattice", *Physica D* **32**, 183–193.

W. Ebeling (1989): "On the Entropy of Dissipative and Turbulent Structures", *Physica Scripta* **T25**, 238.

W. Ebeling, M.V. Volkenstein (1990): "Entropy and the Evolution of Information", *Physica A* **163**, 398.

W. Ebeling, H. Engel, H. Herzel (1990): "Selbstorganisation in der Zeit", Akademie-Verlag, Berlin.

W. Ebeling, A. Engel, R. Feistel (1990): "Physik der Evolutionsprozesse", Akademie-Verlag, Berlin.

W. Ebeling, L. Schimansky-Geier (1990): "Transition phenomena in high-dimensional systems and models of evolution", in "Noise in nonlinear dynamical systems", ed. by F. Moss and P.V.E. McClintock, Cambridge University Press, Cambridge.

W. Ebeling, M. Jenssen (1991): "Soliton-assisted activation processes", *Ber. Bunsenges. Phys. Chem.* **95**, 356–362.

W. Ebeling, M. Jenssen (1992): "Trapping and fusion of solitons in in a nonuniform Toda lattice", *Physica A* **188**, 350–355.

W. Ebeling, G. Nicolis (1991): "Entropy of symbolic sequences: the role of correlations", *Europhys. Lett.* **14**, 191.

W. Ebeling, D. Hoffmann (1991): "The Berlin school of thermodynamics founded by Helmholtz and Clausius", *Eur. J. Phys.* **12**, 1–9.

W. Ebeling (1992): "On the relation between entropy concepts", in "From Phase Transitions to Chaos", G. Györgyi, I. Kondor, L. Asvari, T. Tel, eds., World Scientific, Singapore.

W. Ebeling (1992): "On the relation between various entropy concepts and the valoric interpretation", *Physica A* **182**, 108.

W. Ebeling, G. Nicolis (1992): "Word frequency and entropy of symbolic sequences: a dynamical perspective", *Solitons & Fractals* **2**, 635–640.

W. Ebeling, W. Muschik, eds. (1993): "Statistical physics and thermodynamics of non-linear nonequilibrium systems", World Scientific, Singapore.

W. Ebeling, V. Podlipchuk, A.A. Valuev (1995): "Molecular dynamics simulation of the activation of soft molecules solved in condensed media", *Physica A* **217**, 22–37.

W. Ebeling, V. Podlipchuk, M. G. Sapeshinskyy (1996): "Microscopic models and simulations of local activation processes", *Int. J. Bifurcation & Chaos* **8**, 755–765.

W. Ebeling (1997): "Prediction and entropy of sequences with LRO", *Physica D* **109**, 42–50.

W. Ebeling, K. Rateitschak (1998): "Symbolic dynamics, entropy and complexity of the Feigenbaum map at the accumulation point", *Discrete Dyn. in Nat. & Soc.* **2**, 187–194.

W. Ebeling, M. Sapeshinsky, A.A. Valuev (1998): "Simulations of activation processes", *Physica A* **2**, 635–640.

W. Ebeling, J. Ortner (1998): "Quasiclassical theory and simulations of strongly coupled plasmas", *Physica Scripta* **T75**, 93–98.

W. Ebeling, F. Schweitzer, B. Tilch (1999): "Active Brownian motion with energy depots modelling animal mobility", *BioSystems* **49**, 17–29.

W. Ebeling, M.Jenssen (1999): 'Brownian particles with Toda interactions — a model of nonlinear molecular excitations', *SPIE* **3726**, 112–124.

W. Ebeling, A. Chetverikov, M. Jenssen (2000): "Statistical thermodynamics and nonlinear excitations of Toda systems", *Ukrainian. J. Phys.* **45**, 479–487.

W. Ebeling, U. Erdmann, J. Dunkel, M. Jenssen (2000): "Nonlinear dynamics and fluctuations of dissipative Toda chains", *J. Stat. Phys.* **A101**, 443–457.

W. Ebeling (2000): "Canonical nonequilibrium statistics and applications to Fermi-Bose systems", *Cond. Matter Phys.* **3**, no. 2, 285–293.

W. Ebeling, P. Landa, V. Ushakov (2000): "Self-oscillations in ring Toda chains with negative friction", *Phys. Rev. E* **63**, 046601-1-8.

W. Ebeling, F. Schweitzer (2001): "Swarms of particle agents with harmonic interactions", *Theory BioSci.* **120** (2001) 1–18.

W. Ebeling, R. Steuer, M.R. Titchener (2001): "Partition-based entropies of deterministic and stochastic maps", *Stochastics and Dynamics* **1**, 45–61.

W. Ebeling (2002): "Nonequilibrium statistical mechanics of swarms of driven particles", *Physica A* **314**, 93–97.

W. Ebeling, L. Schimansky-Geier, Yu.M. Romanovsky, eds. (2002): "Stochastic Dynamics of Reacting Biomolecules", World Scientific, Singapore.

W. Ebeling (2003): "Synchronization of stochastic motions in swarms of active Brownian particles with global coupling", *Fluctuation and Noise Letters* **3**, L137–L144.

W. Ebeling, U. Erdmann (2003): "Nonequilibrium statistical mechanics of swarms of driven particles", *Complexity* **8**, No. 4, 25–35.

W. Ebeling, G. Röpke (2004): "Statistical mechanics of confined systems with rotational excitations", *Physica D* **187**, 268–280.

W. Ebeling (2004): "Nonlinear Brownian motion — mean square displacement", *Condensed Matter Physics* **7**, 539–550.

W. Ebeling, U. Erdmann (2005): "Active Brownian motion — stochastic dynamics of swarms", Proc. Conf. Erice July, 2004, A. Rapisarda (ed.) World Scientific.

J. P. Eckmann (1981): "Roads to turbulence in dissipative dynamical systems", *Rev. Mod. Phys.* **53**, 643–654.

J.P. Eckmann, D. Ruelle (1985): "Ergodic theory of chaos and strange attractors", *Rev. Mod. Phys.* **57**, 617–656.

M. Eigen (1971): "The selforganization of matter and the evolution of biological macromolecules", *Naturwissenschaften* **58**, 465.

M. Eigen, P. Schuster (1977): "The hypercycle", *Naturwissenschaften* **64**, 541; (part II **65**, 341).

A. Einstein (1905): "Über die von der molekularkinetischen Theorie der Wärme geforderte Bewegung von in ruhenden Flüssigkeiten suspendierten Teilchen", *Ann. der Physik*, 4. Folge, Band **17**, 549–560.

A. Einstein (1906): "Zur Theorie der Brownschen Bewegung", *Ann. der Physik*, 4. Folge, Band **19**, 371–381.

H. Engel-Herbert, W. Ebeling (1988): "The behavior of the entropy during transitions far from thermodynamic equilibrium", *Physica* **149A**, 182, 195.

A. Engel, W. Ebeling (1987): "Diffusion in a random potential", *Phys. Rev. Lett.* **39**, 1979.

A. Engel, W. Ebeling (1987): "Interaction of moving interfaces with obstacles", *Phys. Lett. A* **122**, 20–25

U. Erdmann, W. Ebeling, L. Schimansky-Geier, F. Schweitzer (2000): "Brownian particles far from equilibrium", *Eur. Phys. J.* **15**, 105–113.

U. Erdmann, W. Ebeling (2003): "Collective motion of Brownian particles with hydrodynamic interactions", *Fluctuation and Noise Letters* **3**, L145–L150.

U. Erdmann, W. Ebeling, L. Schimansky-Geier (2004): "Generalzation of Einsteins theory to active Brownian motion", Preprint Sfb 555, Humboldt University.

U. Erdmann, W. Ebeling (2005): "On the attractors and bifurcations of two-dimensional Rayleigh oscillators including noise", *Int. J. Bifurc. & Chaos*, submitted.

U. Erdmann, W. Ebeling, A.S. Mikhailov (2005) "Noise-induced transition from translational to rotational motion of swarms", *Phys. Rev. E* **71**, 051904.

G. Falk, W. Ruppel (1976): "Energie und Entropie", Springer-Verlag, Berlin.

H. Falkenhagen (1971): "Theorie der Elektrolyte", Hirzel-Verlag, Leipzig.

H. Falkenhagen, W. Ebeling (1965): "Statistical derivation of diffusion equations according to the Zwanzig method", *Phys. Lett.* **15**, 131.

M.V. Feigelman, V.M. Vinokur (1988): "On the stochastic transport in disordered systems", *J. de Physique* **49**, 1731–1736.

M. J. Feigenbaum (1979): "The universal metric properties of nonlinear transformations", *J. Stat. Phys.* **21**, 669–706.

M.J. Feigenbaum (1978): "Quantitative universality for a class of nonlinear transformations", *J. Stat. Phys.* **19**, 25.

R. Feistel, W. Ebeling (1989): "Evolution of Complex Systems", Kluwer, Dordrecht.

W. Feller (1991): "An introduction to probability theory and its applications", Wiley, N.Y. (3rd edition).

E. Fermi (1965): "Collected papers", Univ. Chicago Press, Chicago.

A. Fick (1855): "Über Diffusion", *Ann. Phys.* **94**, 59.

R.A. Fisher (1937): "The advance of advantageous genes", *Ann. Eugenics* **7**, 355–369.

A. Fokker (1914): "Die mittlere Energie rotierender elektrischer Dipole im Strahlungsfeld", *Ann. d. Phys.* **43**, 810.

G. Ford, M. Kac, P. Mazur (1965): "Statistical mechanics of ensembles of coupled oscillators", *J. Math. Phys.* **6**, 504.

B.J. Ford (1992): "Bownian movement in *clarkia* pollen: A reprise of the first observations", *The Microscope* **40**(4), 235–241.

C. Fox (1961): "The G and H-functions as symmetrical Fourier kernels", *Trans. Amer. Math. Soc.* **98**, 395–429.

H. Frauenfelder, P.G. Wolynes (1985): "Rate theories and the puzzles of protein kinetics", *Science* **229**, 337–345.

J. Freund, W. Ebeling, K. Rateitschak (1996): "Self similar sequences and universal scaling of dynamical entropies", *Phys. Rev. E* **54** (1996); *Int. J. Bifurc. & Chaos* **6**, 611–620.

J. Freund, T. Pöschel, T., eds. (1999): "Stochastic Dynamics", Springer, Berlin.

E. Frey (2002): "Physics in cell biology: On the physics of biopolymers and molecular motors", *ChemPhysChem* **3**(3), 270–275.

A. Fulinski (1998): "On Marian Smoluchowski's life and contribution to physics", *Acta Physica Polonica* **29**, 1523.

C.W. Gardiner (1983): "Handbook of stochastic methods", Springer, Berlin.

C. Gardiner (1991): "Quantum Noise", Springer, Berlin.

L. Garrido, ed. (1989): "Systems far from equilibrium", Springer, New York.

P. Gaspard (1998): "Chaos, scattering and statistical mechanics", Cambride Univ. Press, Cambridge.

T. Geisel, S. Thomae (1984): "Anomalous diffusion in intermittent chaotic systems", *Phys. Rev. Lett.* **52**, 1936.

L.E. Gendenstein, I.N. Krive (1985): "Supersymmetry in quantum-mechanics", *Uspekhi Fiz. Nauk* **146**, 553–590.

J.W. Gibbs (1902): "Collected works of J.W. Gibbs", Yale Univ. Press, New Haven; reprinted Longmans, New York, 1931.

P. Glansdorff, I. Prigogine (1971): "Thermodynamics of Structure, Stability and Fluctuations", Wiley-Interscience Publ., New York.

I.V. Gopich, A. Szabo (2002): "Kinetics of reversible diffusion influenced reactions: The self-consistent relaxation time approximation", *J. Chem. Phys.* **117**, 507–517.

G. Gouy (1889): "Sur la mouvement Brownien", *Comptes Rendus* **109**, 102.

R. Graham (1973): "Stochastic models", in "Springer Tracts in Modern Physics", Vol. 66, Springer, Berlin.

R. Graham (1978): "Models of stochastic behaviour", in "Scattering Techniques", ed. by S.H. Chen, B. Chu, and R. Nossal, Plenum Press, New York.

R. Graham (1990): "Macroscopic potentials, bifurcations and noise in dissipative systems", in "Noise in Nonlinear Dynamical Systems", ed. by F. Moss, P.V.E. McClintock, Cambridge University Press, Cambridge.

W.T. Grandy, Jr. (1988): " Foundations of Statistical Mechanics", Reidel Publ. Co., Dordrecht.

W.T. Grandy, L.H. Schick, eds. (1991): "Maximum entropy and Bayesian methods", Kluwer, Dordrecht.

P. Grassberger (1986): "Entropy and complexity", Int. J. Theor. Phys. **25**, 907–915.

P. Grassberger, I. Procaccia (1982): "The long-time properties of diffusion in a medium with static traps", J. Chem. Phys. **77**(12), 6281–6284.

P. Grassberger, I. Procaccia (1982): "Diffusion and drift in a medium with randomly distributed traps", Phys. Rev. A **26**(6), 3686–3688.

M. Grmela, H.C. Öttinger (1997): "Dynamics and thermodynamics of complex fluids I + II", Phys. Rev. E **56**, 6620–6632; 6633–6655.

D.H.E. Gross (2001): "Microcanonical Thermodynamics", World Scientific, Singapore.

P. Gruner-Bauer, F.G. Mertens (1988): "Excitation spectrum of the Toda lattice for finite temperatures", Z. Phys. B **70**, 435–445.

S. Grossmann, S. Thomae (1977): "Invariant distributions and stationary correlation functions", Z. Naturf. **32a**, 1353–63.

R. Haase (1963): "Thermodynamik der irreversiblen Prozesse", Verlag Steinkopf, Darmstadt.

H. Haken (1973): "Approximate solution of Fokker–Planck equations", Z. Phys. **263**, 267–277.

B. Havsteen (1991): "A stochastic attractor participates in chymotrypsin catalysis. A new facet of enzyme catalysis", J. Theor. Biol. **151**, 557–571.

H. Haken, ed. (1973): "Synergetics" (Proc. Int. Symp., Elmau 1972), B.G. Teubner, Stuttgart.

H. Haken, ed. (1977): "Synergetics. A Workshop", Springer, Berlin, Heidelberg, New York.

H. Haken (1983): "Synergetics. An Introduction", 3rd. Edn., Springer Berlin, Heidelberg, New York.

H. Haken (1983): "Advanced Synergetics", Springer, Berlin-Heidelberg.

H. Haken (1988): "Information and Self-organization", Springer, Berlin, Heidelberg, New York.

P. Hänggi, P. Talkner, M. Borkovec (1990): "Reaction-rate theory: fifty years after Kramers", Rev. Mod. Phys. **62**, 251–341.

P. Hänggi, P. Talkner, eds. (1995): "New trends in Kramer's reaction rate theory.", Kluwer, Boston.

P. Hänggi, G. Schmid, I. Goychuk (2003): "Statistical physics of biocomplexity", in F. Beck et al., l.c.

P. Hanggi, F. Marchesoni, F. Nori (2005): "Brownian Motors", Annlen der Physik **14**, 51–70.

P. Hanggi, F. Marchesoni (2005): "Introduction: 100 years of Brownian motion", Chaos **15**, XXX.

J.W. Haus, K.W. Kehr (1987): "Diffusion in regular and disordered lattices", *Phys. Rep.* **150**(5–6), 263–406.

B. Havsteen (1989): "A new principle of enzyme catalysis: coupled vibrations facilitate conformational changes", *J. Theor. Biol.* **140**, 101–109.

R.G. Helleman (1983): "Onset of large scale chaos", in: "Long-Time Prediction in Dynamics", ed. by C.H. Horton, L.E. Reichl, V.G. Szebehely, Wiley, New York.

H.v. Helmholtz (1895): "Wissenschaftliche Abhandlungen" 1–3, Teubner Leipzig.

M. Henon (1969): "Numerical study of quadratic area-preserving mappings", *Q. Appl. Math.* **27**, 291–312.

J.Hofbauer, K. Sigmund (1984): "Evolutionstheorie und dynamische Systeme", Parey, Hamburg.

A. Holden, ed. (1986): "Chaos", Manchester Univ. Press, Manchester.

J. Honerkamp (1998): "Statistical physics", Springer-Verlag, Berlin.

W. Horsthemke, R. Lefever (1984): "Noise induced transitions", Springer, Berlin.

D. Helbing (1997): "Verkehrsdynamik", Springer, Berlin.

D. Helbing (2001): "Traffic and related self-driven many-particle systems", *Rev. Mod. Phys.* **73**, 1067–1141.

R. Hilfer, (2000): "Applications of Fractional Calculus in Physics", World Scientific. River Edge, N.J.

A.V. Holden, ed. (1986): "Chaos", Manchester University Press, Manchester.

W.G. Hoover (1988): "Reversible mechanics and time's arrow", *Phys. Rev. A* **37**, 52–65.

W.G. Hoover (1998): "Time-reversibility in nonequilibrium thermomechanics", *Physica D* **112**, 225–240.

W.G. Hoover (2001): "Time reversibility, computer simulation, and chaos", World Scientific, Singapore.

W.G. Hoover, H.A. Posch, B.L. Holian, M.J. Gillan, M. Mareschal, C. Massobrio (1987): "Dissipative irreversibility from Nose's reversible mechanics", *Molecular Simulation* **1**, 79–86.

W.G. Hoover, H.A. Posch (1996): "Numerical heat conductivity in smooth particle applied mechanics", *Phys. Rev. E* **54**, 5142–5145.

K. Huang (1987): "Statistical Mechanics", Wiley, New York.

R.S. Ingarden, ed. (1986): "Polish men in science. Marian Smoluchowski, his life and scientific work", PWN, Warzawa.

M.B. Isichenko (1992): "Percolation, statistical topography, and transport in random media", *Rev. Mod. Phys.* **64**(4), 961–1043.

C. Jarzynski (1997): "Nonequilibrium equality for free energy differences", *Phys. Rev. Lett.* **78**, 2690–2693; "Equilibrium free-energy differences from nonequilibrium measurements: A master equation approach", *Phys. Rev. E* **56**, 5018–5035.

E.T. Jaynes (1957): "Maximum entropy principle", *Phys. Rev.* **106**, 620–630; **108**, 171–180.

E.T. Jaynes (1985): "Macroscopic prediction", in "Complex systems — operational approaches" (ed. H. Haken), Springer, Berlin.

M. Jenssen (1991): "Statistical thermodynamics of Toda systems", *Phys. Lett. A* **159**, 6–16.

M. Jenssen, W. Ebeling (1991): in "Far-from-Equilibrium Dynamics of Chemical Systems", eds. J. Popielawski and J. Gorecki, World Scientific, Singapore.

M. Jenssen, W. Ebeling (2000): "Distribution functions and excitation spectra of Toda systems at intermediate temperatures", *Physica D* **141**, 117–132.

D. Jou, J. Casas-Vazques, G. Lebon (1993): "Extended irreversible thermodynamics", Springer, Berlin.

N.G. van Kampen (1992): "Stochastic processes in physics and chemistry", North Holland, Amsterdam.

K. Kang, S. Redner (1984): "Scaling approach for the kinetics of recombination processes", *Phys. Rev. Lett.* **52**(12), 955–958.

M. Karplus (1982): "Dynamics of proteins", *Ber. Bunsenges. Phys. Chem.* **86**, 386–240.

J.U. Keller (1977): "Thermodynamik der irreversiblen Prozesse", De Gruyter, Berlin, New York.

V.M. Kenkre, R.S. Knox (1974): "Generalized Master-equation theory of excitation transfer", *Phys. Rev. B* **9**, 5279–5290.

D.A. Kessler, H. Levine (1998): "Fluctuation-induced diffusive instabilities", *Nature* **394** (6693) 556–558.

J. Klafter, R. Silbey (1980): "Derivation of the continuous-time random-walk equation", *Phys. Rev. Lett.* **44**, 55–58.

J. Klafter, G. Zumofen (1993) "Lévy statistics in a Hamiltonian system", *Phys. Rev. E* **49**, 4873–4877.

R. Klages, K. Rateitschak, G. Nicolis (2000): "Thermostating by deterministic scattering: Construction of nonequilibrium steady states", *Phys. Rev. Lett.* **84** 4268–4271.

O. Klein (1921): "Zur statistischen Theorie der Suspensionen und Lösungen", Arkiv für Matematik, *Astronomi och Fysik* **16**, No. 5, 1.

H. Kleinert (2003): "Path integrals in quantum mechanics, statistics and polymer science", 3rd ed., World Scientific, Singapore.

Yu. L. Klimontovich (1982, 1986): "Statistical physics", Nauka, Moscow (in Russian), Harwood Academic Publ., New York.

Yu. L. Klimontovich, M. Bonitz (1988): "Evolution of the entropy of stationary states in selforganization processes in the control parameter space", *Z. Phys. B — Condensed Matter* **70**, 241.

Yu. L. Klimontovich (1990, 1991): "Turbulent motion. Structure of chaos. New approach to the statistical theory of open systems", Nauka, Moscow; Kluwer Academic Publ., Dordrecht.

Yu. L. Klimontovich (1995): "Statistical theory of open systems", (in Russian), Janus, Moscow; Engl. transl. Kluwer Academic Publ., Amsterdam (1997).

A. Kolmogorov, I. Petrovskii, N. Piscounov (1937): "Étude de l équation de la diffusion avec croissance de la quantité de la matière et son application a un problèm biologique", *Moscow Univ. Bull. Math.* **1**, 1–25.

A.N. Kolmogorov (1958): "A new metric invariant of transitive dynamical systems", *Dokl. Akad. Nauk* **119**, 100.

R. Kopelman (1988): "Fractal reaction kinetics", *Science* **241**, 1620.

E. Kotomin, V. Kuzovkov (1996): "Modern Aspects of Diffusion-Controlled Reactions", Elsevier, Amsterdam.

H.A. Kramers (1940): "Brownian motion in a field of force and the diffusion model of chemical reactions", *Physica* **7**, 284–304.

Kubo R. (1957): "Statistical-mechanical theory of irreversible processes. 1. General theory and simple applications to magnetic and condunction problems", *J. Phys. Soc. Japan* **12**(6): 570–586.

R. Kupferman (2004): "Fractional kinetics in Kac–Zwanzig heat bath models", *J. Stat. Phys.* **114**, 291.

Y. Kuramoto (1984): "Chemical oscillations, waves, and turbulence", Springer, Berlin.

V. Kuzovkov, E. Kotomin (1988): "Kinetics of bimolecular reactions in condensed media — critical phenomena and microscopic self-organization", *Rep. Prog. Phys.* **51**(12): 1479–1523.

J.A. Krumhansl, J.R. Schrieffer (1975): "Dynamics and statistical-mechanics of a one-dimensional model Hamiltonian for structural phase-transitions", *Phys. Rev. B* **11**, 3535–3545.

L. Laloux, P. Le Doussal (1998): "Aging and diffusion in low dimensional environments", *Phys. Rev. E* **57**, 6296–6326.

L.D. Landau, E.M. Lifshits (1990): "Statistical Physics" (3rd ed.), Pergamon Press, New York.

P. Langevin (1908): "Sur la theéorie du mouvement brownien", *CR Hebd. Acad. Sci.* **146**, p. 530.

A. Lasota, M. Mackey (1985): "Probabilistic properties of deterministic systems", Cambridge University Press, Cambridge.

F. Ledrappier, L.-S. Young (1985): "The metric entropy of diffeomorphisms", *Annals of Mathematics* **122**, 509–550.

A. Lichtenberg, M.A. Liberman (1983): "Regular and stochastic motion", Springer, New York.

A.D. Linde (1984): "Elementary particles and cosmology", *Rep. Prog. Phys.* **47**(8).

H.P. Lu, L. Xun, X.S. Xie (1998): "Single-molecule enzymatic dynamics", *Science* **282**, 1877–1887.

A. Lyapunov (1892): "General problem of stability of motion", Charkov.

S. Machlup, L. Onsager (1953): "Fluctuations and irreversible process. 2. Systems with kinetic energy", *Phys. Rev.* **91**, 1512–1515.

M.C. Mackey (1989): "The dynamic origin of increasing entropy", *Rev. Mod. Phys.* **61**, 981–1015.

H. Malchow, L. Schimansky-Geier (1986): "Noise and diffusion in bistable systems", Teubner-Verlag, Leipzig.

M. Mareschal, E. Kestemont (1987): "Order and fluctuations in nonequilibrium molecular-dynamics simulations of two-dimensional fluids", *J. Stat. Phys.* **48**, 1187.

J.A. McCammon, S.C. Harvey (1987): "Dynamics of proteins and protein acids", Cambridge University Press.

M.C. Mackey (1989): "The dynamic origin of increasing entropy", *Rev. Mod. Phys.* **61**, 981–1015.

J. Mai, I.M. Sokolov, A. Blumen (1996): "Front propagation and local ordering in one-dimensional irreversible autocatalytic reaction", *Phys. Rev. Lett.* **77**, 4462.

J. Mai, I.M. Sokolov, A. Blumen (1998): "Discreteness effects on the front propagation in the $A + B \to 2A$ reactions in 3 dimensions", *Europhysics Letters* **44**, 7.

J. Mai, I.M. Sokolov, A. Blumen (2000): "Front propagation in one-dimensional autocatalytic reactions: The breakdown of the classical picture at small particle concentrations", *Phys. Rev. E* **62**, 141.

V. Makarov, W. Ebeling, M. Velarde (2000): "Soliton-like waves on dissipative Toda lattices", *Int. J. Bifurc. & Chaos* **10**, 1075–1085.

V. Makarov, E. del Rio, W. Ebeling, M. Velarde (2001): "Dissipative Toda-Rayleigh lattices and its oscillatory modes", *Phys. Rev. E* **64**, 136601.

D.C. Mattis, M.L. Glasser (1998): "The uses of quantum field theory in diffusion-limited reactions", *Rev. Mod. Phys.* **70**(3), 979–1001.

V.I. Melnikov and S.V. Meshkov (1986): "Theory of activated rate processes: exact solution of the Kramers problem", *J. Chem. Phys.* **85**, 1018.

V.I. Melnikov (1991): "Review of the Kramers problem", *Rev. Mod. Phys.* **62**, 251.

F.G. Mertens, H. Büttner (1986): "Solitons on the Toda lattice", in Trullinger et al., eds., l.c.

R. Metzler, J. Klafter (2000): The random walk's guide to anomalous diffusion: a fractional dynamics approach, *Phys. Rep.* **339**, 1–7.

A.S. Mikhailov, L. Schimansky-Geier, W. Ebeling (1983): "Stochastic motion of the propagating front in bistable media", *Phys. Lett. A* **96**, 153–455.

A.S. Mikhailov (1989) "Selected topics in fluctuational kinetics of reactions", *Phys. Rep.* **184**, 307.

A.S. Mikhailov (1990): "Foundations of Synergetics I", Springer, Berlin, Heidelberg, New York.

A.S. Mikhailov, A.Yu. Loskutov (1990): "Foundations of Synergetics II", Springer, Berlin, Heidelberg, New York.

A. Mikhailov, D. Meinköhn (1997): "Self-motion in physico-chemical systems far from thermal equilibrium", in "Stochastic dynamics", ed. by L. Schimansky-Geier, T. Pöschel, Springer, Berlin.

A.S. Mikhailov, D.H. Zanette (1999): "Noise-induced breakdown of coherent collective motion in swarms", *Phys. Rev. E* **60**, 4571–4575.

A.S. Mikhailov, V. Cahlenbuhr (2002): "From Cells to Societies", Springer, Berlin.

K.S. Miller, B. Ross (1993): "An Introduction to the Fractional Calculus and Fractional Differential Equations", Wiley, N.Y.

L. Molgedey, W. Ebeling (2000): "Local order, entropy and predictability of financial time series", *Eur. Phys. J. B* **15**, 733–737; Physica A **287** (2000) 420–427.

E. Monson, R. Kopelman (2000): "Observation of laser speckle effects and nonclassical kinetics in an elementary chemical reaction", *Phys. Rev. Lett.* **85**(3), 666–669.

E. Monson, R. Kopelman (2004): "Nonclassical kinetics of an elementary $A + B \to C$ reaction-diffusion system showing effects of a speckled initial reactant distribution and eventual self-segregation: Experiments", *Phys. Rev. E* **69**, 021103.

E.W. Montroll, G.H. Weiss (1965): "Random Walks on Lattices. 2", *J. Math. Phys.* **6**(2), 167.

E.W. Montroll (1969): "Random walks on lattices. 3. Calculation of first-passage time with application to exciton trapping on photosynthetic units", *J. Math. Phys.* **10**, 753.

C. Montus, J.P. Bouchaud (1996): "Models of traps and glass phenomenology", *J. Phys. A* **29**, 384–386.

I.V. Morosov, G.E. Norman, A.A. Valuev (2001): "Stochastic properties of strongly coupled plasmas", *Phys. Rev. E* **63**, 036405-1-9.

P. Morse (1929): "Diatomic molecules according to the wave mechanics", *Phys. Rev.* **34**, 57–64.

E. Mosekilde, L. Mosekilde, eds. (1991): "Complexity, Chaos and Biological Evolution", Plenum Press, New York and London.

F. Moss, P.V.E. McClintock, eds. (1989): "Noise in Nonlinear Dynamical Systems", Cambridge University Press, Cambridge.

J.D. Murray (1989): "Mathematical biology", Springer, Berlin.

W. Muschik (1988): "Aspects of non-equilibrium thermodynamics", Lectures on Fundamental Methods, McGill University.

W. Muschik (1990): "Aspects of Non-Equilibrium Thermodynamics", World Scientific, Singapore 1990.

V. Muto, A.C. Scott, P.L. Christiansen (1989): "Thermally generated solitons in a Toda lattice model of DNA", *Phys. Lett. A* **136**, 33–43.

F.J. Muzzio, J.M. Ottino (1989): "Evolution of a lamellar system with diffusion and reaction — a scaling approach", *Phys. Rev. Lett.* **63**(1), 47–50.

F.J. Muzzio, J.M. Ottino (1989): "Dynamics of a lamellar system with diffusion and reaction — scaling analysis and global kinetics", *Phys. Rev. A* **40**(12), 7182–7192.

F.J. Muzzio, J.M. Ottino (1990): "Diffusion and reaction in a lamellar system — self-similarity with finite rates of reaction", *Phys. Rev. A* **42**(10), 5873–5884.

Yu.I. Neimark (1972): "The method of point maps in the theory of nonlinear oscillations", (in Russian), Nauka, Moscow.

Yu.I. Neimark, P.S. Landa (1987): "Stochastic and chaotic oscillations (in Russian)", Nauka, Moscow.

G. Neugebauer (1980): "Relativistische Thermodynamik", Akademie-Verlag, Berlin.

G. Nicolis, I. Prigogine (1977): "Self-Organization in Non-Equilibrium Systems", Wiley-Interscience Publ., New York.

G. Nicolis, I. Prigogine (1987): "Die Erforschung des Komplexen", Piper-Verlag, München/Zürich.

G. Nicolis, C. Nicolis, J.S. Nicolis (1989): "Chaotic dynamics, Markov partitions and Zipf's law", *J. Stat. Phys.* **54**, 915–925.

J.S. Nicolis (1991): "Chaos and Information Processing", World Scientific, Singapore, 1991.

G. Nicolis, S. Martinez, E. Tirapegui (1991): "Finite coarse-graining and Chapman-Kolmogorov equation in conservative dynamical systems", *Chaos, Solitons and Fractals* **1**, 25.

G. Nicolis, J. Piasecki, D. McKernan (1992): "Towards a probabilistic description of deterministic chaos", in G. Györgi, I. Kondor, L. Sasvari, T. Tel (eds.): From phase transitions to chaos, World Scientific, Singapore.

C. Nicolis, W. Ebeling, C. Baraldi (1997): "Markov processes, dynamical entropies and the statistical prediction of mesoscale wheather regimes", *Tellus* **49 A**, 108–118.

G.E. Norman, V.V. Stegailov (2002): "The stochastic properties of a molecular-dynamical Lennard–Jones system in equilibrium and in non-equilibrium states, *J. Exp. Theor. Phys.* **92**, 879–886.

G.E. Norman, V.V. Stegailov (2002): "Stochastic and dynamic properties of molecular-dynamical systems, *Comp. Physics Comm.* **177**, 678–683.

I.I. Novikov (1958): "The Efficiency of nuclear power stations", *Journal of Nuclear Energy (USSR)* **7**, 125–128.

A. Okubo, S.A. Levin (2001): "Diffusion and ecological problems: Modern perspectives", Springer, Berlin.

K.B. Oldham, J. Spanier (1974): "The Fractional Calculus: Integrations and Differentiations of Arbitrary Order", Academic Press, N.Y.

L. Onsager (1931): "Reciprocal relations in irreversible processes. I", Phys. Rev. **37**, 405; "Reciprocal Relations in irreversible processes. II", **38**, 2265.

L. Onsager, A. Machlup (1953): "Fluctuations and irreversible processes", *Phys. Rev.* **91**, 1505–1512.

A. Ordemann, F. Moss, G. Balaszi (2003): "Motions of daphnia in a light field: random walks with a zooplankton", in F. Beck, M.T. Hütt, U. Lüttge (eds.), l.c.

J. Ortner, F. Schautz, W. Ebeling (1997): "Quasiclassical molecular dynamics simulations of the electron gas: dynamic properties", *Phys. Rev. E* **56**, 4665–4670.

E. Ott (1993): "Chaos in dynamical systems", Cambridge University Press, Cambridge.

H.C. Öttinger (2005): "Beyond Equilibrium Thermodynamics", Wiley Interscience, London.

J.M. Ottino (1989): "The Kinematics of Mixing, Stretching, Chaos, and Transport", Cambridge University Press, Cambridge.

A.A. Ovchinnikov, S.F. Timashev, A.A. Belyi (1986): "Kinetics of Diffusion-Controlled Chemical Processes", Khimiya, Moscow; English translation: Nova Science, NY, 1989.

A.A. Ovchinnikov, Ya.B. Zeldovich (1978): "Role of density fluctuations in bimolecular reaction kinetics", *Chem. Phys.* **28**(1–2), 215–218.

T. Palszegi, I.M. Sokolov, H.F. Kauffmann (1998): "Excitation trapping in dynamically disordered polymers", *Macromolecules* **31**, 2521.

D. Panja (2004): "Effects of fluctuations on propagating fronts", *Phys. Repts.* **393**(2), 87–174.

P.J. Park, W. Sung (1998): "Polymer release out of a spherical vesicle through a pore", *Phys. Rev. E* **57**, 730–734.

J.M.R. Parrondo, P. Espanol (1996): "Criticism of Feynman's analysis of the ratchet as an engine", *Am J. Phys.* **64**(9), 1125–1130.

J.M.R. Parrondo, B.J. De Cisneros (2002): "Energetics of Brownian motors: a review", *Appl. Phys. A — Mater.* **75** (2), 179–191.

K. Pearson (1905): "The problem of the random walk", *Nature* **72** (1865) 294, *Nature* **72** (1867) 342.

J.B. Pesin (1977): "Characteristic Lyapunov exponents and smooth ergodic theory", *Russ. Math. Surveys* **32**, 355–385.

T. Petrosky, I. Prigogine (1988): "Poincare's theorem and unitary transformation in classical and in quantum theory" *Physica A* **147**, 439.

M. Planck (1916): "Über einen Satz der Statistischen Dynamik und seine Erweiterung in der Quantentheorie", S.-B. Preuß. Akad. Wiss., 324–341.

T. Pohl, U. Feudel, W. Ebeling (2002): "Bifurcations of a semiclassical atom in a periodic field", *Phys. Rev. E* **65**, 046228.

H. Poincare (1892): "Methods nouvelles de la mechanique celeste", Paris; reprinted by Dover, New York (1957).

H. Poincare (1928): "Oevres", Gauthier-Villars, Paris.

E. Pollak, H. Grabert, P. Hänggi (1989): "Theory of activated rate processes for arbitrary frequency", *J. Chem. Phys.* **91**, 4073.

L.S. Pontryagin, A.A. Andronov, A.A. Vitt (1933): "On the statistical treatment of dynamical systems", *J. Eksp. Teor. Fiz.* **3**, 165 [English translation in "Noise in Nonlinear Dynamical Systems", ed. by F. Moss, P.V.E. McClintock, Cambridge Univ. Press, Cambridge, vol. 1, p. 329].

T. Pöschel, W. Ebeling, H. Rosé (1995): "Guessing probability distributions from small samples", *J. Stat. Phys.* **80**, 1443–1452.

I. Prigogine, R. Defay (1962): "Chemische Thermodynamik", Leipzig.

I. Prigogine (1967): "Introduction to Thermodynamics of irreversible processes", 3rd ed., Wiley, New York.

I. Prigogine (1969): "Structure, dissipation and life", in "Theoretical Physics and Biology", (Ed.: M. Marois), North Holland Publ., Amsterdam.

I. Prigogine, G. Nicolis, A. Babloyantz (1971): "Thermodynamics and Evolution", *Physics Today* **25**, 23, 38.

I. Prigogine (1980): "From being to becoming", Freeman, San Francisco.

I. Prigogine, I. Stengers (1984): "Order out of Chaos", Heinemann, London.

I. Prigogine (1989): "The microscopic meaning of irreversible processes", *Z. Physik. Chem. (Leipzig)* **270**, 477–490.

I. Prigogine et al. (1991): "Integrability and chaos in classical and quantum mechanics", *Chaos, Solitons and Fractals* **1**, 3.

A.P. Prudnikov, O.I. Marichev, and Yu.A. Brychkov (1990): "Integrals and Series, Vol. 3: More Special Functions", Newark, NJ: Gordon and Breach.

W.-J. Rappel, A. Nicol, A. Sarkissian, H. Levine (1999): "Self-organized vortex state in two-dimensional dictostelium dynamics", *Phys. Rev. Lett.* **83**, 1247–1255.

K. Rateitschak, R. Klages, W.G. Hoover (2000): "The Nose-Hoover thermostated Lorentz gas", *J. Stat. Phys.* **101**, 61–77.

J.W.S. Rayleigh (1880): "On the resultant of a large number of vibrations of the same pitch and of arbitrary phases", *Phil. Mag.* **10**(5), 73–78.

Lord Rayleigh (1883): "On maintained vibrations", *Phil. Mag.* **15**, 229.

J.W. Rayleigh (1894): "Theory of sound", New edition: Dover, New York, 1945.

J.W.S. Rayleigh (1905): "The problem of the random walk", *Nature* **72**(1866), 318.

S. Redner (2001): "A Guide to First-Passage Processes", Cambridge University Press, Cambridge.

F. Reif (1965): "Fundamentals of statistical and thermal physics", McGraw Hill, New York.

R. Reigada, F. Sagués, I.M. Sokolov, J.M. Sancho, A. Blumen (1996): "Kinetics of the $A + B \to 0$ Reaction under Steady and Turbulent Flows", *J. Chem. Phys.* **105**, 10925.

R. Reigada, F. Sagués, I.M. Sokolov, J.M. Sancho, A. Blumen (1997): "Fluctuation-Dominated Kinetics under Stirring", *Phys. Rev. Lett.* **78**, 741.

R. Reigada, A. Romero, A. Sarmiento, K. Lindenberg (1999): "One-dimensional arrays of oscillators: Energy localization in thermal equilibrium", *J. Chem. Phys.* **111**, 1373–1383.

R. Reigada, A. Sarmiento, A. Romero, J.M. Sancho, K. Lindenberg (2000): "Harvesting thermal fluctuations: Activation process induced by a nonlinear chain in thermal equilibrium", *J. Chem. Phys.* **112**, 10615–10624.

L.E. Reichl (1980): "A modern couse in statistical physics", Univ. Texas Press, Austin.

L.E. Reichl (1992): "The transition to chaos", Springer, New York.

P. Reimann (2002): "Brownian motors: noisy transport far from equilibrium", *Phys. Rep.* **361** (2-4): 57–265.

P. Reimann, P. Hanggi (2002): "Introduction to the physics of Brownian motors", *Appl. Phys. A — Mater.* **75**(2), 169–178.

J. Renn (2005): "Einstein's invention of Brownian motion", *Annalwen der Physik* **14**, Supplement, 23–27.

P.A. Rey, J. Cardy (1999): "Asymptotic form of the approach to equilibrium in reversible recombination reactions", *J. Phys. A — Math. Gen.* **32**, 1585–1603.

S.A. Rice (1985): "Diffusion limited reactions", Elsevier, Amsterdam.

B. Rinn, P. Maass, J.-P. Bouchaud (2000): "Multiple scaling regimes in simple aging models", *Phys. Rev. Lett.* **84**, 5403–5406.

B. Rinn, P. Maass, J.-P. Bouchaud (2001): "Hopping in the glass configuration space: Subaging and generalized scaling laws", *Phys. Rev. B* **64**, 104417.

H. Risken (1988, 1994): "The Fokker–Planck equation", Springer, Berlin.

O.E. Rössler (1976): "An equation for continous chaos", *Phys. Lett.* **57**, 397–398.

Yu.M. Romanovsky (1997): "Some problems of cluster dynamics of biological macromolecules", in "Stochastic Dynamics", L. Schimansky-Geier, T. Poeschel (eds.), *Ser. Lecture Notes on Physics*, Springer Verlag, Berlin, pp. 1–13.

Yu.M. Romanovsky, W. Ebeling (2002): "Some problems of cluster dynamics of biological macromolecules", MGU, Moscow.

J.M. Sancho, A.H. Romero, K. Lindenberg, F. Sagues, R. Reigada, A.M. Lacasta (1996): "$A + B \to 0$ reaction with different initial patterns", *J. Phys. Chem.* **100**(49), 19066–19074.

W. van Saarloos (2003): "Front propagation into unstable states", *Phys. Rep.* **386**, 29–222.

Scher H., Montroll E.W. (1975): "Anomalous transit-time dispersion in amorphous solids", *Phys. Rev. B* **12**(6), 2455–2477.

M. Schienbein, H. Gruler (1993): "Langevin equation, Fokker–Planck equation and cell migration", *Bull. Math. Biol.* **55**, 585–608.

M. Schienbein, K. Franke, H. Gruler (1994): "Random walk and directed movement: comparison between inert particles and self-organized molecule machines", *Phys. Rev. E* **49**, 5462–5471.

L. Schimansky-Geier, T. Pöschel, T., eds. (1997): "Stochastic Dynamics", Springer, Berlin.

L. Schimansky-Geier, P. Talkner (2002): "Tools of stochastic dynamics", in: W. Ebeling et al., eds. "Stochastic dynamics of reacting biomolecules", l.c.

L. Schimansky-Geier, W. Ebeling, U. Erdmann (2005): "Stationary distribution densities of active Brownian particles", *Acta Physica Polonica B* **36**(5), 1757–1769.

T. Schneider (1986): "Classical statistical mechanics of lattice dynamic model systems", in Trullinger et al., eds., l.c.

W.R. Schneider, W. Wyss (1987): "Fractional diffusion equation", *Helv. Phys. Acta* **60**(2), 358–358.

W.R. Schneider, W. Wyss (1989): "Fractional diffusion and wave-equations", *J. Math. Phys.* **30**(1), 134–144.

H.G. Schöpf (1978): "Von Kirchhoff bis Planck", Akademie-Verlag, Berlin.

H.G. Schöpf (1984): "Rudolf Clausius. Ein Versuch, ihn zu verstehen", *Ann. Physik* **41**, 185.

H.G. Schuster (1984): "Deterministic Chaos", Physik-Verlag, Weinheim, 3rd augmented edition, Weinheim, 1995.

F. Schwabl (2000): "Statistische Mechanik", Springer, Heidelberg.

F. Schweitzer, L. Schimansky-Geier (1994): "Clustering of active walkers in a two-component system", *Physica A* **206**, 359–379.

F. Schweitzer, K. Lao, F. Family (1997): "Active random walkers simulate trunk trail formation by ants", *BioSystems* **41**, 153–166.

F. Schweitzer, W. Ebeling, B. Tilch (1998): "Complex motion of Brownian particles with energy depots", *Phys. Rev. Lett.* **80**, 5044–5048.

F. Schweitzer, W. Ebeling, B. Tilch (2001): "Statistical mechanics of canonical-dissipative systems and applications to swarm dynamics", *Phys. Rev. E* **64**, 02110-1-12.

F. Schweitzer (2003): "Brownian agents and active particles", Springer, Berlin.

S.K. Scott (1994): "Oscillations, waves, and chaos in chemical kinetics", Oxford Univ. Press, Oxford.

L.P. Shilnikov (1984): "Bifurcation theory and turbulence", in R.E. Sagdeev, ed., "Nonlinear and turbulent processes in physics", Vol. 3, Harwood Academic Publishers, 1627–1635.

L.P. Shilnikov (1997): "Mathematical problems of nonlinear dynamics: A tutorial", *Int. J. Bifurcation & Chaos* **7**, 1953–2001.

M. Shlesinger, B.J. West, J. Klafter (1987): "Levy dynamics of enhanced diffusion — application to turbulence", *Phys. Rev. Lett.* **58**, 1100.

M. F. Shlesinger, U. Frisch, G. M. Zaslavskii (1996) (eds.): "Levy Flights and Related Topics in Physics: Proceedings of the International Workshop Held at Nice, France, 27–30 June 1994", (Lecture Notes in Physics).

M.F. Shlesinger, G.M. Zaslavsky, J. Klafter (1993): "Strange Kinetics", *Nature* **363**(6424), 31–37.

Y. Sinai (1970,1972): "Ergodic theory", *Uspheki Math. Nauk* **25**, 141–160; **27**, 137–150.

Y. Sinai (1977): "Introduction to ergodic theory", Princeton University Press, Princeton.

M. von Smoluchowski (1906): "Zur kinetischen Theorie der Brownschen Molekularbewegung und der Suspensionen", *Ann. der Physik* **21**, 756–780.

M. von Smoluchowski (1913) "Einige Beispiele Brownscher Molekularbewegung unter Einflußäußerer Kr̆fte", *Bull. Int. Acad. Sci. Cracovie*, Mat.-Naturw. Klasse A, pp. 418–434, reprinted in "Ostwalds Klassiker der exacten Wissenschaften", Band 199, 3. Auflage, Harri Deutsch, Frankfurt am Main, 1997.

M. von Smoluchowski (1915): "Über Brownsche Molekularbewegung unter Einwirkung äusserer Kräfte und deren Zusammenhang mit der verallgemeinerten Diffusionsgleichung", *Ann. der Physik*, pp. 1103–1112.

M. von Smoluchowski (1917): "Versuch einer mathematischen Theorie der Kogulationskinetik kolloidaler Lösungen", *Z. Phys. Chem.* **92**, 129–135

M. von Smoluchowski (1924–28): "Oeuvres", (Ed. by W. Natanson), PAU, Krakow.

I.M. Sokolov (1986): "Dimensions and other geometrical critical exponents in the percolation theory", *Usp. Fiz. Nauk* **150**, 221; English translation: *Sov. Phys. Uspekhi* **29**, 924.

I.M. Sokolov (1986): "Spatial and temporal asymptotic behavior of annihilation reactions", *JETP Lett.+* **44**(1), 67–70.

I.M. Sokolov, A. Blumen (1990): "Diffusion-controlled reactions in lamellar systems", *Phys. Rev. A* **43**, 6, 2714.

I.M. Sokolov, A. Blumen (1991) a: "Reactions in systems with mixing", *J. Phys. A: Math. Gen.* **24**, 3687.

I.M. Sokolov, A. Blumen (1991) b: "Mixing in the reaction-diffusion scheme", *Phys. Rev. Lett.* **66**, 1942.

I.M. Sokolov, A. Blumen (1991) c: "Mixing in reaction-diffusion problems", *Int. J. Mod. Phys. B* **5**, 3127.

I.M. Sokolov, A. Blumen (1992) in: S.M. Aharony, ed., "Synthesis, Characterization, and Theory of Polymeric Networks and Gels", Plenum Press, NY, pp. 53–65.

I.M. Sokolov, A. Blumen (1994): "Memory effects in diffusion-controlled reactions", *Europhys. Lett.* **27**, 495.

I.M. Sokolov (1998): "On the energetics of a nonlinear system rectifying thermal fluctuations", *Europhysics Letters* **44**, 278.

I.M. Sokolov (1999): "Reversible fluctuation rectifier", *Phys. Rev. E* **60**, 4946.

I.M. Sokolov, A. Blumen, J. Klafter (2001): "Dynamics of annealed systems under external fields: CTRW and the fractional Fokker–Planck equations", *Europhys. Lett.* **56**, 175.

I.M. Sokolov, A. Blumen, J. Klafter (2001): "Linear response in complex systems: CTRW and the fractional Fokker–Planck equations", *Physica A* **302**, 268.

I.M. Sokolov, J. Klafter, A. Blumen (2002): "Fractional kinetics", *Phys. Today* **55**(11) 48–54.

I.M. Sokolov, R. Metzler (2003): "Towards deterministic equations for Lévy walks: the fractional material derivative", *Phys. Rev. E* **67**, 010101(R).

I.M. Sokolov (2003): "Cyclization of a polymer: A first passage problem for a non-Markovian process", *Phys. Rev. Lett.* **90**, 080601.

F. Spitzer (1976): "Principles of random walks", Springer, N.Y., 2nd ed. 2001.

H.E. Stanley (2003): "Statistical physics and economic fluctuations: do outliers exist?", *Physica A* **318**(1–2): 279–292.

W.-H. Steeb (1991): "A Handbook of Terms Used in Chaos and Quantum Chaos," BI Wissenschaftsverlag, Mannheim-Wien-Zürich.

R. Steuer. L. Molgedey, W. Ebeling, M.A. Jimenez-Montano (2001): "Entropy and optimal partition for data analysis", *Eur. Phys. J. B* **19**, 265–269.

R. Steuer, W. Ebeling, D. Russel, S. Bahar, A. Neiman, F. Moss (2001): "Entropy and local uncertainty of data from sensory neurons", *Phys. Rev. E* **64**, 265–269.

O. Steuernagel, W. Ebeling, V. Cahlenbuhr (1994): "An elementary model for directed active motion", *Chaos, Soliton & Fractals*, **4**, 1917–1930.

R. L. Stratonovich (1961, 1963, 1967): "Selected Topics in the theory of random noise", Sov. Radio, Moscow (in Russ.); Gordon & Breach, New York (Vol. 1, 1963, Vol. 2, 1967).

R. L. Stratonovich (1994): "Nonlinear nonequilibrium thermodynamics", Springer-Verlag, Berlin.

D.N. Subarev (1976): "Statistische Thermodynamik des Nichgleichgewichts", Akademie-Verlag, Berlin.

W. Sung, P.J. Park (1996): "Polymer translocation through a pore in a membrane", *Phys. Rev. Lett.* **77**, 783–786.

A. Szabo, R. Zwanzig, N. Agmon (1988): "Diffusion-controlled reactions with mobile traps", *Phys. Rev. Lett.* **61**, 2496.

M.S. Taqqu (2001): "Bachelier and his times: A Conversation with Bernard Bru", in "Mathematical Finance — Bachelier Congress 2000", H. Geman et al. editors, Springer, Berlin.

G.I. Taylor, "Diffusion by continuous movements", *Proc. Lond. Math Soc.* (2) 20, 196–212 (1921).

N. Theodorakopoulos (1984): "Finite-temperature excitations of the classical Toda chain", *Phys. Rev. Lett.* **53**, 871–881.

H. Thomas, G.E. Morfill, V. Demmel, J. Goree, B. Feuerbacher, D. Möhlmann (1994): "Plasma crystal — Coulomb crystallization in a dusty plasma", *Phys. Rev. Lett.* **73**, 652.

H.M. Thomas, G.E. Morfill (1996): "Melting dynamics of a plasma crystal", *Nature* **379**, 806.

B. Tilch, F. Schweitzer, W. Ebeling (1999): "Directed motion of Brownian particles with internal ehergy depots", *Physica A* **273**, 294–314.

M. Toda (1981): "Theory of Nonlinear lattices", Springer Berlin.

M. Toda (1983): "Nonlinear Waves and Solitons", Kluwer, Dordrecht.

M. Toda, N. Saitoh (1983): "The classical specific-heat of the exponential lattice", *J. Phys. Soc. Japan* **52**, 3703–3713.

D. Toussaint, F. Wilczek (1983): "Particle-antiparticle annihilation in diffusive motion", *J. Chem. Phys.* **78**(5), 2642–2647.

M. Tribus, R.D. Levine, eds. (1978): "The maximum entropy formalism", MIT Press, Cambridge.

S.A. Trigger, A.G. Zagorodny (2002, 2003): "Negative friction in dusty plasmas", *Contr. Plasma Physics* **43**, 381–383.

S.A. Trigger (2003): "Fokker–Planck equation for Boltzmann-type and active particles: transfer probability approach", *Phys. Rev. E* **67**, 046403-1-10.

S.E. Trullinger, V.E. Zakharov, V.L. Pokrovsky, eds. (1986): "Solitons", North Holland, Amsterdam.

J.K.E. Tunaley (1974): "Theory of a.c. conductivity based on random-walks", Phys. Rev. Lett. **33**, 1037–1039.

V.V. Uchaikin (2003): "Self-similar anomalous diffusion and Levy-stable laws", *Uspekhi Fiz. Nauk* **173** 848–875, English translation: *Phys. Usp.+* **46**(8), 821–849 (2003).

G. E. Uhlenbeck, L. S. Ornstein (1930): "On the Theory of the Brownian Motion", *Phys. Rev.* **36**, 823–841.

B. Van der Pol (1920): "A theory of the amplitude of free and forced triode vibration", *Radio Review* **1**, 701.

N. Van Kampen (1981, 1992): "Stochastic processes in physics and chemistry", North Holland, Amsterdam.

M. Velarde, W. Ebeling, A. Chetverikov (2005): "Dissipative solitons and currents in active lattices", *Int. J. Bifurc. Chaos* **15**, 245–251.

T. Vicsek, A. Czirok, E. Ben-Jacob, I. Cohen, O. Shochet (1995): "Novel type of phase transition in a system of self-driven particles", *Phys. Rev. Lett.* **75**, 1226–1229.

T. Vicsek (2001): "Fluctuations and scaling in biology", University Press, Oxford.

C.P. Warren, G. Mikus, E. Somfai, L.M. Sander (2001): "Fluctuation effects in an epidemic model", *Phys. Rev. E* **63**(5), 056103.

N. Wax, ed. (1954): "Selected papers on noise", Dover, New York.

B.H. Weber, D.J. Depew, J.D. Smith, eds. (1988): "Enropy, Infomation and Evolution", MIT Press, Cambridge (MA).

A. Wehrl (1978): "The Entropy Concept", *Rev. Mod. Phys.* **50**, 221.

G.H. Weiss (1994): "Aspects and Applications of the Random Walk", North Holland, Amsterdam.

U. Weiss (1993): "Quantum Dissipative Systems", World Scientific, Singapore.

G.R. Welch, B. Somogyi, S. Damjanovich (1982): "The role of protein fluctuation in enzyme action", *Prog. Biophys. Molec. Biol.* **39**, 109–146.

P.C. Werner, F.-W. Gerstengarbe, W. Ebeling (1999): "Changes in the probability of sequences, exit time distribution and dynamical entropy in the Potsdam temperature record", *Theor. Appl. Climatol.* **62**, 125–132.

G. Wilemski, M. Fixman (1974): "Diffusion-controlled interchain reactions of polymers. 1. Theory", *J. Chem. Phys.* **60** (3), 866–877; "Diffusion-controlled interchain reactions of polymers. 2. Results for a pair of terminal reactive groups", *ibid.*, 878–890.

S. Winkle (2000): "Kulturgeschichte der Seuchen", Komet, Frechen.

G.M. Zaslavsky (1985): "Chaos in Dynamic Systems", Harwood, New York.

W.Q. Zhu (2003): "Nonlinear stochstic dynamics and control — A Hamiltonian theoretical framework", Science Presss, Beijing.

D.N. Zubarev, G. Röpke, A.A. Morozov (1996):, "Statistical Mechanics of Nonequilibrium Processes, Vol. I", VCH Wiley, Weinheim and Berlin.

D.N. Zubarev, G. Röpke, A.A. Morozov (1997): "Statistical Mechanics of Nonequilibrium Processes, Vol. II", VCH Wiley, Weinheim and Berlin.

G. Zumofen, J. Klafter (1993): "Power spectra and random-walks in intermittent chaotic systems", *Physica D* **69**, 436–446.

W.H. Zurek (1991): "Decoherence and the transition from quantum to classical", *Physics Today* **44**, 36.

R.W. Zwanzig (1959): "Contribution to the theory of Brownian motion", *Phys. Fluids* **2**(1), 12–19.

R. Zwanzig (1973): "Generalized Langevin equation in equilibrium", *J. Stat. Phys.* **9**, 215.

R. Zwanzig (1980): in "Problems in nonlinear transport theory", in L. Garrido, ed., Springer, N.Y.

R. Zwanzig (2001): "Nonequilibrium statistical mechanics", University Press, Oxford.

Index

Active Brownian particle, 261
Active cluster, 285
Adiabatic system, 19
Aging, 241
Annihilation reaction, 208, 215
Anomalous diffusion, 144, 251
Arrhenius law, 191, 238
Attractor, 43
Autocatalytic reaction, 208, 221
Avogadro number, 20

Backward Kolmogorov equation, 164, 195
Bolltzmann H-theorem, 60
Boltzmann constant, 35
Boltzmann entropy, 85
Boltzmann equation, 60
Boltzmann–Planck entropy, 62
Boltzmann–Planck principle, 83, 97
Brownian motion, 2, 8, 12, 139, 261

Canonical distribution, 90
Canonical ensemble, 94
Canonical-dissipative system, 115
Caputo derivative, 246
Carnot efficiency, 180
Central limit theorem, 228
Chapman–Kolmogorov equation, 53, 162, 253
Characteristic function, 226
Closed system, 19, 89
Continuous time random walk, 232, 235, 239, 240, 245
Correlation function, 105

Detailed balance, 175

Diffusion equation, 140
Diffusion-controlled reaction, 204
Dissipative system, 43
Distribution function, 59, 84, 111, 268
Double-well potential, 192
Driven Brownian motion, 111
Dynamical system, 39, 41, 68

Einstein relation, 112
Energy, 3, 23, 32
Energy depot, 120, 262
Energy diffusion, 173, 196
Entropy, 4, 25, 32, 85
Entropy production, 28
Ergodic hypothesis, 5
Ergodic theorem, 64
Ersatzprozess, 33, 99
Extensive variable, 20
External variable, 20

First Law, 24
First passage time, 183, 194
Flow-over-population approach, 189
Fluctuation-dissipation theorem, 142
Fluctuation-dominated kinetics, 215
Fluctuations, 97, 100, 207, 215
Fokker–Planck equation, 50, 112, 161, 245, 268
Fractional Fokker–Planck equation, 225, 245, 253
Free energy, 8, 30, 38
Free enthalpy, 31, 200, 201
Frobenius–Perron equation, 72
Front propagation, 221

Generating function, 229
Gibbs distribution, 89
Gibbs fundamental relation, 26
Gradient system, 44
Grand canonical ensemble, 94

Hamiltonian, 41
Hamiltonian system, 46, 64, 73
Heat, 3, 24, 25
Heat bath, 89, 159, 176

Intensive variable, 20
Internal energy, 8
Internal variable, 20
Irreversible process, 22, 26, 28
Irreversible thermodynamics, 9
Isolated system, 19, 87

Jarzynski identity, 38

Kinetic coefficient, 103
Klein–Kramers equation, 167, 196
Kolmogorov entropy, 78, 79
Kramers–Moyal expansion, 140, 163

Lévy flight, 251, 253, 255
Lévy walk, 251, 256, 257
Langevin equation, 15, 112, 141, 155
Limit cycle, 274
Linear irreversible process, 97, 102
Liouville equation, 73, 76, 169
Lyapunov exponent, 46, 71
Lyapunov function, 23, 28

Map, 40, 69
Markov model, 79
Markov partition, 81, 82
Markov process, 52
Mass action law, 201
Master equation, 53, 172, 235, 236
Maximum entropy principle, 92, 132
Maxwell–Boltzmann distribution, 59
Microcanonical distribution, 89
Microcanonical ensemble, 87, 93, 118
Mixing procedure, 217

Noise, 141, 144, 254
Nonmarkovian Fokker–Planck equation, 247
Nonmarkovian Langevin equation, 155

Nose–Hoover thermostat, 126
Number of different sites visited, 231
Nyquist theorem, 157

Onsager theory, 102
Open system, 19, 89
Operational time, 233
Ornstein–Uhlenbeck process, 150, 165, 250
Oseen interaction, 300
Overdamped motion, 148

Path integral, 153
Period doubling, 71
Pesin entropy, 79
Phase space, 57
Power law, 236

Quasistatic process, 22
Quenching reaction, 208

Random walk, 16, 225, 228, 232
Rayleigh friction, 268
Reaction kinetics, 199, 202
Reaction rate, 202
Reaction-diffusion equation, 203, 217, 221
Relaxation coefficient, 104
Renewal approach, 185
Resonance, 108
Return probability, 231
Reversible process, 22
Riemann–Liouville derivative, 246
Riesz derivative, 246, 253

Scavenging reaction, 207
Second Law, 8, 26
Slowly changing function, 230
Smoluchowski equation, 162, 184
Smoluchowski rate, 205
Stationary process, 105
Subordinated Markov process, 233
Swarm, 285
Symmetry relation, 109

Tauberian theorems, 230
Taylor–Kubo formula, 146, 158
Temperature, 21
Thermodynamic equilibrium, 22
Thermodynamic flux, 103
Thermodynamic force, 103
Third Law, 8, 28

Toda system, 65, 116
Transition rate, 172, 193
Trapping reaction, 207, 208, 212

Underdamped motion, 196

Vaks–Balagurov kinetics, 208
Valoric interpretation, 33

Waiting time distribution, 233, 236
Weyl derivative, 247, 253
Wiener–Khintchin theorem, 106
Work, 3, 24, 32, 38

Zeroth Law, 22

Printed in the United States
By Bookmasters